Landolt-Börnstein / New Series

Landolt-Börnstein

Numerical Data and Functional Relationships
in Science and Technology

New Series
Editor in Chief: O. Madelung

Units and Fundamental Constants in Physics and Chemistry

Nuclear and Particle Physics (Group I)

Atomic and Molecular Physics (Group II)

Crystal and Solid State Physics (Group III)

Macroscopic and Technical Properties of Matter (Group IV)

Geophysics and Space Research (Group V)

Astronomy, Astrophysics, and Space Research (Group VI)

Biophysics (Group VII)

Landolt-Börnstein
Numerical Data and Functional Relationships in Science and Technology
New Series / Editor in Chief: O. Madelung

Group I: Nuclear and Particle Physics
Volume 13

Production of Radionuclides at Intermediate Energies

Subvolume b
Interactions of Protons with Targets from Kr to Te

Editor: H. Schopper

Contributors:
A.S. Iljinov, V.G. Semenov, M.P. Semenova, N.M Sobolevsky, L.V. Udovenko

Springer-Verlag
Berlin Heidelberg New York
London Paris Tokyo
Hong Kong Barcelona Budapest

ISSN 0942-797 (Nuclear and Particle Physics)

ISBN 3-540-55143-3 Springer-Verlag Berlin Heidelberg New York
ISBN 0-387-55143-3 Springer-Verlag New York Berlin Heidelberg

Library of Congress Cataloging in Publication Data
Zahlenwerte und Funktionen aus Naturwissenschaften und Technik, Neue Serie. Gesamtherausgabe: K.H. Hellwege.
Vol. I/13b Gesamtherausgabe, K.H. Hellwege, O. Madelung, or K.H. Hellwege and A.M. Hellwege.
At head of title: Landolt-Börnstein. Added t.p.: Numerical data and functional relationships in science and technology.
Tables chiefly in English.
Intended to supersede the Physikalisch-chemische Tabellen by H.H. Landolt and R. Börnstein of which the 6th ed.
began publication in 1950 under title: Zahlenwerte und Funktionen aus Physik, Chemie, Astronomie,
Geophysik und Technik.
Vols. published after v. 1 of group I have imprint: Berlin, NewYork, Springer-Verlag
Errata slip laid in Gruppe III, Bd. 17a; Bd. 18
Includes bibliographies.
1. Physics--Tables. 2. Chemistry--Tables. 3. Engineering Tables.
I. Hellwege, K.H., ed. II. Börnstein, R.(Richard), 1852-1913. III. Landolt, H.(Hans), 1831-1910.
Physikalisch-chemische Tabellen. IV. Title: Numerical data and functional relationships in science and
technology
QC61.23 502`.12 62-53136

Typesetting: Camera ready by authors and editor
Printing: Mercedes-Druck, Berlin
Binding: Schöneberger Buchbinderei, Berlin

63/3040 - 5 4 3 2 1 0 - Printed on acid-free paper

Editor

H. Schopper
CERN, CH-1211 Geneva 23, Switzerland

Contributors

A.S. Iljinov, N.M. Sobolevsky, L.V. Udovenko
Institute of Nuclear Research of the Academy of Sciences of Russia, Moscow, Russia

M.P. Semenova, V.G. Semenov
Research Centre of Spacecraft Radiation Safety of the Health Ministry of Russia, Moscow, Russia

Preface

The precise knowledge of production cross sections for specific radionuclides for medium energies (10 MeV to 100 GeV) of the incident particle gains ever more importance, not only for nuclear physics, but also for many applications, e. g. astrophysics, radiation protection at accelerators but also on space flights, radioisotope techniques in medicine and for industrial applications and perhaps even for removal of nuclear waste by accelerators.

Since the last systematic review of these production cross sections in 1973 (LB I/5a, b, c) an impressive amount of new experimental data has been accumulated. In the first subvolume almost 10 000 data have been compiled for the production of radioisotopes by medium energy protons on targets from He to Br. In this second subvolume more than 7 000 data for havier targets from Kr to Te are compiled. The precise values of the cross sections can be taken from the tables, whereas the figures show the general trend of the cross sections and can be more readily used for interpolations.

Two further subvolumes containing data for heavier targets and other projectiles will follow.

The origin of the authors made it possible to include also data from East European countries which sometimes are not easy to access. The tables and figures have been compiled with extraordinary care and should present the most complete compilation for this type of nuclear reactions.

My thanks are due to the authors for their carefully collecting and evaluating of the data. I would also like to express my gratitude to the members of the Landolt-Börnstein editorial staff and the production department of the publisher who have helped to bring the edition of this volume to a successful end.

Geneva, October 1992 **The Editor**

Contents

1 Introduction 1
1.1 General remarks 1
1.2 Selection criteria 1
1.3 Explanation of the tables 2
1.4 Explanation of the figures 3
1.5 Data evaluation 3
1.6 List of information sources 3
1.7 References 4

2 Excitation functions: Survey for incident protons 5

3 Excitation functions 37
3.1 Tables 37
3.2 Figures 247
3.3 References 307

1 Introduction

1.1 General remarks

The present subvolume is a continuation of the subvolume LB I/13a [Ilj91]. It includes the experimental data on the cross section for radionuclide production in proton interactions at the kinetic energies up to 100 GeV with target-nuclei from Kr to Te (7200 experimental values of the cross section). The data published as papers in 33 scientific periodicals and as preprints of a few national scientific centres and laboratories by the beginning of 1992 have been included. Section 1.6 presents the list of the information sources used in this compilation.

The preponderance of the data is presented in the form of excitation functions of the reactions which produce a given nuclide in its final state. (The excitation function gives the dependence of the cross section for the formation of a reaction product on projectile energy.) The excitation functions are presented in tables and in figures. The tabulated data are more informative because they show not only the energy and cross section values, but also the errors involved. The errors are not shown in the figures not to overload the latter with details. On the other hand, the plots are more pictorial and make it readily possible to demonstrate the behaviour of an excitation function, to interpolate the data, and to estimate the data uncertainty by comparing the results obtained by different experimental groups.

The proton data for heavier targets from I to Am will be presented in the third subvolume LB I/13c, and the data for other projectiles, namely pions and antiprotons, in the fourth subvolume LB I/13d. Besides, the subvolume LB I/13d will include brief methodological comments on each experimental paper mentioned in the reference list. The comments will give information about accelerators, measurement techniques, monitoring and other experimental features.

The data on the cross section for radionuclide production by neutrons are not included in this handbook. The matter is that the main part of the measured cross sections belongs to the low-energy region $E < 20$ MeV, and there are the ample elaborate compilations [Byc82, Gar76] for this energy domain. There are few data for the energy region 20 MeV $< E <$ 40 MeV. However, these data correspond to the high-energy "tails" of those excitation functions which are included in the low-energy compilations. So, it is not reasonable to include these data in the present handbook. As to the isolated scarce measurements performed at intermediate energies $E > 100$ MeV, they were averaged over a very wide spectrum of the primary energy. For this reason these data were not included in the handbook.

1.2 Selection criteria

The data on the cross section for the production of radionuclides are included in the present volume if some selection criteria are satisfied. These criteria are the same which were used in the previous subvolume LB I/13a. Nevertheless, we repeat these criteria to provide a certain convenience of using this subvolume.

1. Target. The volume includes the experimental data obtained with targets of natural isotopic composition and those enriched by a studied isotope. As a rule, the enrichment of a target by the isotope under study was rather high (> 90···95%) thereby permitting the authors of a cited paper to refer the data to a monoisotopic target with proton number Z_t and nucleon number A_t. In this case the nucleon number A_t together with the chemical symbol and the number Z_t of a chemical element are shown simultaneously in a table and in a figure (for example, ${}^{82}{}_{36}$Kr). A considerable fraction of the excitation functions have been measured with targets consisting of a natural isotopic mixture. In this case the nucleon number A_t is omitted in the notation of the target in a table and in a figure (for example, $_{36}$Kr). The same notation is used for a chemical element which has a single stable isotope.

2. Projectile. The projectile is a proton with $A_i = 1$ and $Z_i = 1$.

3. Product. The product is a radionuclide with nucleon number A_j and proton number Z_j . If a product nuclide has a metastable state, then the cross section for the formation of the state may be measured (in this case a notation of the type ^{111m}In is used), or the sum of the cross sections for formation of the ground and metastable states may be measured (in such a case the notation is $^{111m+g}$In). Besides, the cumulative cross section for nuclide formation may be measured (i.e. when a product nucleus is formed together with a short-lived isobaric nucleus which decays subsequently to a product nucleus). In this case we set off such a data into its own table to form the excitation function for the cumulative cross section and introduce a special notation in a subtitle of this table. The volume contains only those data on the cumulative yield of radionuclides which correspond to the case when the decay of the daughter nucleus (product nucleus) can be disregarded within the period of the (practically total) decay of the isobaric mother nucleus.

4. Initial energy. We have not restricted the minimal energy of projectile protons, E, but, instead, included all the available data. So, most of the excitation functions are presented from the threshold of a respective reaction.

5. Shape of excitation function. Each excitation function is described by a smooth curve in the intermediate energy range. If the experimental results of an original paper are presented in graphical form only, we assigned numerical values to the data and, then, introduced them in a figure and in a table. We carried out the given procedure for experimental points only, rather than for a curve drawn sometimes by the authors through the points.

1.3 Explanation of the tables

The table of chapter 2 presents a joint survey of excitation functions for subvolumes LB I/13a and LB I/13b, ordered according to increasing Z_t and A_t values of a target nucleus starting from the natural isotopic mixture. The product nuclei are ordered according to increasing Z_j values in the case of a target with given Z_t and A_t values, and with increasing A_j values in the case of a fixed Z_j value. Since we do not discriminate the particular production channel of a given radionuclide and, besides, different production channels of a final nucleus are normally not discriminated experimentally, we designate the excitation functions using the notation $Z_tA_t \rightarrow Z_jA_j$ even if a particular channel can be well identified. The second column of the table contains the notation of an excitation function. The third column of the table shows the number of experimental points measured for a given excitation function. The next two columns present the page numbers where the reader can find a table with experimental data and a plot for a given excitation function. There is also a letter a or b preceding these numbers. This letter indicates the subvolume LB I/13a or LB I/13b which contains a table and a plot for a given excitation function. Such a presentation could provide a certain convenience of using both subvolumes.

The table of section 3.1 contains the experimental data for the excitation functions ordered in the same manner as in the table of chapter 2. The proton energy E in the laboratory system is given in MeV in the first column. The energy error, ΔE, (r.m.s.) if any is presented in MeV in the second column. The next column gives the cross section σ in mb, and the 4th column the cross section error Δσ in mb if it is contained in the original paper. The last column shows the abbreviations of the references to the respective experimental paper. The first three letters of an abbreviation correspond to the initial letters of the first author's surname, the numbers give the year of publication. If all five symbols prove to be the same for different papers, then the sixth symbol is introduced, namely, a letter a, b, c, etc. (In this case it is possible that one of the papers with identical abbreviation is not provided with the additional sixth letter, for example, Lag 86, Lag 86a).

The list of references can be found in section 3.3. An asterisk in the tables means that the numerical values of cross sections had been extracted not from the numerical tables of the original paper but from its figures using the graphic digitizer or, simply, compass and ruler.

1.4 Explanation of the figures

The figures showing the excitation functions in section 3.2 are ordered in the same manner as the tables. The cross section σ in mb (1 mb = 10^{-27} cm) is plotted as the ordinate. A logarithmic scale is used. The proton kinetic energy E_p in the laboratory system in MeV is plotted as the abscissa. As a rule, we use a linear scale up to E_p = 100 MeV and a logarithmic scale at higher energies.

The data points in the figures show the experimental cross section values. Each of the cross sections taken from individual papers is labelled with its own mark indicating the reference abbreviation used in the figure. If a lot of experimental groups have carried out numerous measurements for a given excitation function, the function may be presented in two figures. Some data presented in the tables of section 3.1 may be absent in a figure if they are located very densely. The single data points on an excitation function located far away in energy from the main group of data points are not, as a rule, shown in a figure. An excitation function is plotted usually if the number of measured points exceeds 10. However, figures with small numbers of data points are sometimes presented if the data points describe properly the threshold behaviour or cover a representative energy range of the excitation function.

1.5 Data evaluation

The volume contains all the data collected by the authors. An expert evaluation of the material which would result in the elimination of obsolete or explicitly conflicting data was not made. The point is that such an expert evaluation is subjective, especially in the case where the data are scanty and the measurements were taken by a single experimental group. Since all the well-studied excitation functions are plotted, the reader himself can easily exclude the explicitly conflicting data, i.e. he can make his own evaluation. The methodological comments on experimental papers which will be included to the subvolume LB I/13d may help the reader to evaluate the data.

1.6 List of the information sources

Title	Abbreviation
Analyst (London)	Analyst
Annals of Physics (New York)	Ann. Phys. (N.Y.)
Applied Radiation and Isotopes	Appl. Radiat. Isot.
Atomnaya Energiya	At. Energ.
Canadian Journal of Chemistry	Can. J. Chem.
Canadian Journal of Physics	Can. J. Phys.
Comptes Rendus	Compt. Rend.
Czechoslovak Journal of Physics	Czech. J. Phys.

Title	Abbreviation
Geokhimiya	Geokhimiya
Geophysical Journal (Oxford)	Geophys. J.
Helvetica Physica Acta	Helv. Phys. Acta
IEEE Transaction on Nuclear Science	IEEE Trans. Nucl. Sci.
International Journal of Applied Radiation and Isotopes	Int. J. Appl. Radiat. Isot.
Izvestiya Akademii Nauk SSSR, Seriya Fizicheskaya	Izv. Akad. Nauk SSSR, Ser. Fiz.
Journal of Inorganic and Nuclear Chemistry	J. Inorg. Nucl. Chem.
Journal of Nuclear Medicine	J. Nucl. Med.
Journal de Physique (Paris)	J. Phys. (Paris)
Journal de Physique et le Radium	J. Phys. Radium
Journal of the Physical Society of Japan	J. Phys. Soc. Jpn.
Journal of Radioanalytical Chemistry	J. Radioanal. Chem.
Journal of Radioanalytical and Nuclear Chemistry Letters	J. Radioanal. Nucl. Chem. Lett.
Nuclear Instruments and Methods in Physics Research	Nucl. Instrum. Methods Phys. Res.
Nuclear Physics	Nucl. Phys.
Nuovo Cimento	Nuovo Cimento
Physical Review	Phys. Rev.
Physics Letters	Phys. Lett.
Radiation Effects	Radiat. Eff.
Radiochemical and Radioanalytical Letters	Radiochem. Radioanal. Lett.
Radiochimica Acta	Radiochim. Acta
Radiokhimiya	Radiokhimiya
Yadernaya Fizika	Yad. Fiz.
Zeitschrift für Physik	Z. Phys.
Zhurnal Eksperimental'noi i Teoreticheskoi Fiziki	JETP

Preprints, Joint Institute for Nuclear Research (JINR) Dubna
Preprints, Leningrad Institute of Nuclear Physics (LINP)
Preprints, University of California Radiation Laboratory (UCRL)
Private Communications
Workshop Abstracts

1.7 References for chapter 1

Byc82 Bychkov, V.M., Manokhin, V.N., Paschenko, A.B., Pliaskin, V.I.: Cross Sections of Neutron-Induced Threshold Reactions. Handbook. Energoatomizdat, Moscow, 1982.

Gar76 Garber, D.I., Kinsley R.R.: Neutron Cross Sections, Vol. 2, Curves. BNL 325, Third Edition, Brookhaven, 1976.

Ilj91 Iljinov, A.S., Semenov, V.G., Semenova, M.P., Sobolevsky, N.M., Udovenko, L.V.: Production of Radionuclides at Intermediate Energies. Landolt-Börnstein, New Series, Group I, Vol. 13a, Berlin-Heidelberg-New York: Springer-Verlag, 1991.

No.	Excitation function	No. of points	Table on page	Figure on page

2 Excitation functions: Survey for incident protons

(for reactions listed in subvolumes 13a and 13b)

No.	Excitation function	No. of points	Table on page	Figure on page
1	${}_2He \rightarrow {}^3_1H$	2	a19	
2	${}_3Li \rightarrow {}^7_4Be$	19	a19	a271
3	${}^7_3Li \rightarrow {}^7_4Be$	76	a19	a271
4	${}_4Be \rightarrow {}^7_4Be$	31	a21	a271
5	${}_5B \rightarrow {}^7_4Be$	19	a22	a272
6	${}_5B \rightarrow {}^{11}_6C$	33	a22	a272
7	${}^{10}_5B \rightarrow {}^7_4Be$	3	a23	
8	${}^{11}_5B \rightarrow {}^7_4Be$	3	a23	
9	${}^{11}_5B \rightarrow {}^{11}_6C$	106	a23	a272
10	${}_6C \rightarrow {}^3_1H$	9	a25	a272
11	${}_6C \rightarrow {}^7_4Be$	116	a26	a273
12	${}_6C \rightarrow {}^{10}_6C$	8	a28	a273
13	${}_6C \rightarrow {}^{11}_6C$	152	a28	a274
14	${}^{13}_6C \rightarrow {}^{13}_7N$	13	a32	a274
15	${}_7N \rightarrow {}^3_1H$	4	a32	
16	${}_7N \rightarrow {}^7_4Be$	60	a32	a275
17	${}_7N \rightarrow {}^{10}_6C$	1	a34	
18	${}_7N \rightarrow {}^{11}_6C$	68	a34	a275
19	${}_7N \rightarrow {}^{13}_7N$	25	a35	a275
20	${}_7N \rightarrow {}^{14}_8O$	53	a36	a275
21	${}^{15}_7N \rightarrow {}^7_4Be$	2	a37	
22	${}^{15}_7N \rightarrow {}^{15}_8O$	20	a37	a276
23	${}_8O \rightarrow {}^3_1H$	4	a38	
24	${}_8O \rightarrow {}^7_4Be$	53	a38	a276
25	${}_8O \rightarrow {}^{10}_4Be$	3	a39	
26	${}_8O \rightarrow {}^{10}_6C$	3	a39	
27	${}_8O \rightarrow {}^{11}_6C$	29	a39	a276
28	${}_8O \rightarrow {}^{14}_6C$	6	a40	
29	${}_8O \rightarrow {}^{13}_7N$	43	a40	a276
30	${}_8O \rightarrow {}^{14}_8O$	2	a41	
31	${}_8O \rightarrow {}^{15}_8O$	49	a41	a277

No.	Excitation function	No. of points	Table on page	Figure on page
32	$^{18}_{8}O \rightarrow {}^{7}_{4}Be$	2	a42	
33	$^{18}_{8}O \rightarrow {}^{18}_{9}F$	19	a42	a277
34	$_{9}F \rightarrow {}^{7}_{4}Be$	21	a43	a277
35	$_{9}F \rightarrow {}^{11}_{6}C$	8	a43	a277
36	$_{9}F \rightarrow {}^{13}_{7}N$	2	a44	
37	$_{9}F \rightarrow {}^{15}_{8}O$	6	a44	
38	$_{9}F \rightarrow {}^{18}_{9}F$ (cumulative)	57	a44	a278
39	$_{9}F \rightarrow {}^{19}_{10}Ne$	11	a45	a278
40	$_{11}Na \rightarrow {}^{7}_{4}Be$	16	a45	a278
41	$_{11}Na \rightarrow {}^{18}_{9}F$	15	a46	a279
42	$_{11}Na \rightarrow {}^{22}_{11}Na$	37	a46	a279
43	$_{12}Mg \rightarrow {}^{3}_{1}H$	6	a47	
44	$_{12}Mg \rightarrow {}^{7}_{4}Be$	32	a47	a279
45	$_{12}Mg \rightarrow {}^{10}_{4}Be$	3	a48	
46	$_{12}Mg \rightarrow {}^{18}_{9}F$	72	a48	a280
47	$_{12}Mg \rightarrow {}^{21}_{11}Na$	3	a50	
48	$_{12}Mg \rightarrow {}^{22}_{11}Na$	89	a50	a280
49	$_{12}Mg \rightarrow {}^{24}_{11}Na$	82	a52	a281
50	$_{12}Mg \rightarrow {}^{23}_{12}Mg$	3	a54	
51	$_{12}Mg \rightarrow {}^{24}_{13}Al$	3	a54	
52	$_{12}Mg \rightarrow {}^{25}_{13}Al$	2	a54	
53	$_{12}Mg \rightarrow {}^{26}_{13}Al$	12	a54	a281
54	$^{24}_{12}Mg \rightarrow {}^{18}_{9}F$	2	a54	
55	$^{24}_{12}Mg \rightarrow {}^{22}_{11}Na$	6	a54	
56	$^{24}_{12}Mg \rightarrow {}^{24}_{11}Na$	2	a55	
57	$^{25}_{12}Mg \rightarrow {}^{18}_{9}F$	24	a55	a281
58	$^{25}_{12}Mg \rightarrow {}^{22}_{11}Na$	38	a55	a281
59	$^{25}_{12}Mg \rightarrow {}^{24}_{11}Na$	28	a56	a282
60	$^{26}_{12}Mg \rightarrow {}^{18}_{9}F$	10	a57	a282
61	$^{26}_{12}Mg \rightarrow {}^{22}_{11}Na$	22	a57	a282
62	$^{26}_{12}Mg \rightarrow {}^{24}_{11}Na$	11	a58	a282
63	$_{13}Al \rightarrow {}^{3}_{1}H$	6	a58	
64	$_{13}Al \rightarrow {}^{7}_{4}Be$	92	a58	a283
65	$_{13}Al \rightarrow {}^{10}_{4}Be$	3	a60	
66	$_{13}Al \rightarrow {}^{11}_{6}C$	4	a60	
67	$_{13}Al \rightarrow {}^{13}_{7}N$	4	a60	
68	$_{13}Al \rightarrow {}^{15}_{8}O$	2	a61	
69	$_{13}Al \rightarrow {}^{19}_{8}O$	1	a61	
70	$_{13}Al \rightarrow {}^{18}_{9}F$	65	a61	a283

No.	Excitation function	No. of points	Table on page	Figure on page
71	${}_{13}Al \rightarrow {}^{20}{}_{9}F$	1	a62	
72	${}_{13}Al \rightarrow {}^{23}{}_{10}Ne$	1	a62	
73	${}_{13}Al \rightarrow {}^{24}{}_{10}Ne$	3	a62	
74	${}_{13}Al \rightarrow {}^{22}{}_{11}Na$	217	a63	a284
75	${}_{13}Al \rightarrow {}^{24}{}_{11}Na$	205	a67	a285
76	${}_{13}Al \rightarrow {}^{25}{}_{11}Na$	1	a72	
77	${}_{13}Al \rightarrow {}^{23}{}_{12}Mg$	4	a72	
78	${}_{13}Al \rightarrow {}^{27}{}_{12}Mg$	2	a72	
79	${}_{13}Al \rightarrow {}^{26}{}_{13}Al$	16	a72	a286
80	${}_{13}Al \rightarrow {}^{26}{}_{14}Si$	1	a72	
81	${}_{13}Al \rightarrow {}^{27}{}_{14}Si$	4	a72	
82	${}_{14}Si \rightarrow {}^{3}{}_{1}H$	5	a73	
83	${}_{14}Si \rightarrow {}^{7}{}_{4}Be$	47	a73	a286
84	${}_{14}Si \rightarrow {}^{10}{}_{4}Be$	3	a74	
85	${}_{14}Si \rightarrow {}^{18}{}_{9}F$	14	a74	a286
86	${}_{14}Si \rightarrow {}^{22}{}_{11}Na$	102	a74	a287
87	${}_{14}Si \rightarrow {}^{24}{}_{11}Na$	62	a77	a288
88	${}_{14}Si \rightarrow {}^{27}{}_{12}Mg$	14	a78	a287
89	${}_{14}Si \rightarrow {}^{28}{}_{12}Mg$	8	a78	
90	${}_{14}Si \rightarrow {}^{25}{}_{13}Al$	5	a79	
91	${}_{14}Si \rightarrow {}^{26}{}_{13}Al$	10	a79	a287
92	${}_{14}Si \rightarrow {}^{28}{}_{13}Al$	20	a79	a288
93	${}_{14}Si \rightarrow {}^{29}{}_{13}Al$	20	a79	a288
94	${}_{14}Si \rightarrow {}^{27}{}_{14}Si$	4	a80	
95	${}_{14}Si \rightarrow {}^{28}{}_{15}P$	3	a80	
96	${}_{14}Si \rightarrow {}^{29}{}_{15}P$	4	a80	
97	${}^{28}{}_{14}Si \rightarrow {}^{18}{}_{9}F$	2	a80	
98	${}^{29}{}_{14}Si \rightarrow {}^{18}{}_{9}F$	2	a80	
99	${}^{29}{}_{14}Si \rightarrow {}^{22}{}_{11}Na$	2	a81	
100	${}^{29}{}_{14}Si \rightarrow {}^{24}{}_{11}Na$	2	a81	
101	${}^{30}{}_{14}Si \rightarrow {}^{28}{}_{12}Mg$	5	a81	
102	${}^{30}{}_{14}Si \rightarrow {}^{29}{}_{13}Al$	5	a81	
103	${}_{15}P \rightarrow {}^{22}{}_{11}Na$	4	a81	
104	${}_{15}P \rightarrow {}^{24}{}_{11}Na$	8	a81	a289
105	${}_{15}P \rightarrow {}^{28}{}_{12}Mg$	14	a81	a289
106	${}_{15}P \rightarrow {}^{30}{}_{15}P$	11	a82	a289
107	${}_{16}S \rightarrow {}^{22}{}_{11}Na$	4	a82	
108	${}_{16}S \rightarrow {}^{24}{}_{11}Na$	9	a82	a289
109	${}_{16}S \rightarrow {}^{28}{}_{12}Mg$	10	a83	a289
110	${}^{34}{}_{16}S \rightarrow {}^{34}{}_{17}Cl$	19	a83	a290

No.	Excitation function	No. of points	Table on page	Figure on page
111	$_{17}Cl \rightarrow {}^{22}_{11}Na$	4	a83	
112	$_{17}Cl \rightarrow {}^{24}_{11}Na$	12	a83	a290
113	$_{17}Cl \rightarrow {}^{28}_{12}Mg$	8	a84	a290
114	$_{18}Ar \rightarrow {}^{24}_{11}Na$	8	a84	a290
115	$_{18}Ar \rightarrow {}^{28}_{12}Mg$	8	a84	a290
116	$_{19}K \rightarrow {}^{22}_{11}Na$	4	a84	
117	$_{19}K \rightarrow {}^{24}_{11}Na$	11	a85	a291
118	$_{19}K \rightarrow {}^{28}_{12}Mg$	5	a85	
119	$_{20}Ca \rightarrow {}^{22}_{11}Na$	6	a85	a291
120	$_{20}Ca \rightarrow {}^{24}_{11}Na$	4	a85	
121	$_{20}Ca \rightarrow {}^{37}_{18}Ar$	11	a85	a291
122	$_{20}Ca \rightarrow {}^{39}_{18}Ar$	11	a86	a291
123	$_{20}Ca \rightarrow {}^{42}_{18}Ar$	5	a86	
124	$_{21}Sc \rightarrow {}^{22}_{11}Na$	4	a86	
125	$_{21}Sc \rightarrow {}^{24}_{11}Na$	4	a86	
126	$_{21}Sc \rightarrow {}^{43}_{19}K$	9	a86	a291
127	$_{21}Sc \rightarrow {}^{44}_{21}Sc$	12	a87	a292
128	$_{21}Sc \rightarrow {}^{44m}_{21}Sc$	11	a87	a292
129	$_{21}Sc \rightarrow {}^{44m+g}_{21}Sc$	11	a87	a292
130	$_{21}Sc \rightarrow {}^{44}_{22}Ti$	10	a88	a292
131	$_{21}Sc \rightarrow {}^{45}_{22}Ti$	39	a88	a293
132	$_{22}Ti \rightarrow {}^{7}_{4}Be$	3	a89	
133	$_{22}Ti \rightarrow {}^{22}_{11}Na$	7	a89	a293
134	$_{22}Ti \rightarrow {}^{24}_{11}Na$	4	a89	
135	$_{22}Ti \rightarrow {}^{42}_{19}K$	4	a89	
136	$_{22}Ti \rightarrow {}^{43}_{19}K$	13	a89	a293
137	$_{22}Ti \rightarrow {}^{43}_{21}Sc$	22	a90	a293
138	$_{22}Ti \rightarrow {}^{44}_{21}Sc$	22	a90	a293
139	$_{22}Ti \rightarrow {}^{44m}_{21}Sc$	32	a91	a294
140	$_{22}Ti \rightarrow {}^{46}_{21}Sc$	6	a91	a294
141	$_{22}Ti \rightarrow {}^{46m+g}_{21}Sc$	27	a92	a294
142	$_{22}Ti \rightarrow {}^{47}_{21}Sc$	33	a92	a294
143	$_{22}Ti \rightarrow {}^{48}_{21}Sc$	25	a93	a295
144	$_{22}Ti \rightarrow {}^{48}_{23}V$	74	a94	a295
145	${}^{47}_{22}Ti \rightarrow {}^{44}_{21}Sc$	1	a95	
146	${}^{47}_{22}Ti \rightarrow {}^{46}_{21}Sc$	1	a95	
147	${}^{47}_{22}Ti \rightarrow {}^{47}_{23}V$	39	a95	a295

No.	Excitation function	No. of points	Table on page	Figure on page
148	$^{48}_{22}Ti \rightarrow {}^{39}_{17}Cl$	5	a96	
149	$^{48}_{22}Ti \rightarrow {}^{42}_{19}K$	10	a97	a296
150	$^{48}_{22}Ti \rightarrow {}^{43}_{19}K$	11	a97	a296
151	$^{48}_{22}Ti \rightarrow {}^{43}_{21}Sc$	12	a97	a296
152	$^{48}_{22}Ti \rightarrow {}^{44}_{21}Sc$	13	a97	a296
153	$^{48}_{22}Ti \rightarrow {}^{46}_{21}Sc$	14	a98	a296
154	$^{48}_{22}Ti \rightarrow {}^{47}_{21}Sc$	23	a98	a297
155	$^{48}_{22}Ti \rightarrow {}^{47}_{23}V$	11	a99	a297
156	$^{48}_{22}Ti \rightarrow {}^{48}_{23}V$	28	a99	a297
157	$^{49}_{22}Ti \rightarrow {}^{44}_{21}Sc$	1	a100	
158	$^{49}_{22}Ti \rightarrow {}^{46}_{21}Sc$	1	a100	
159	$^{49}_{22}Ti \rightarrow {}^{47}_{21}Sc$	1	a100	
160	$^{49}_{22}Ti \rightarrow {}^{48}_{21}Sc$	1	a100	
161	$^{49}_{22}Ti \rightarrow {}^{49}_{23}V$	61	a100	a297
162	$^{50}_{22}Ti \rightarrow {}^{38}_{17}Cl$	5	a101	
163	$^{50}_{22}Ti \rightarrow {}^{39}_{17}Cl$	8	a102	a298
164	$^{50}_{22}Ti \rightarrow {}^{42}_{19}K$	10	a102	a298
165	$^{50}_{22}Ti \rightarrow {}^{43}_{19}K$	12	a102	a298
166	$^{50}_{22}Ti \rightarrow {}^{43}_{21}Sc$	6	a102	
167	$^{50}_{22}Ti \rightarrow {}^{44}_{21}Sc$	10	a103	a298
168	$^{50}_{22}Ti \rightarrow {}^{46}_{21}Sc$	14	a103	a298
169	$^{50}_{22}Ti \rightarrow {}^{47}_{21}Sc$	15	a103	a299
170	$^{50}_{22}Ti \rightarrow {}^{48}_{21}Sc$	11	a104	a299
171	$^{50}_{22}Ti \rightarrow {}^{50}_{23}V$	47	a104	a299
172	$_{23}V \rightarrow {}^{7}_{10}Be$	3	a105	
173	$_{23}V \rightarrow {}^{22}_{11}Na$	8	a105	a299
174	$_{23}V \rightarrow {}^{24}_{11}Na$	9	a105	a299
175	$_{23}V \rightarrow {}^{27}_{12}Mg$	1	a105	
176	$_{23}V \rightarrow {}^{31}_{14}Si$	1	a106	
177	$_{23}V \rightarrow {}^{32}_{15}P$	5	a106	a300
178	$_{23}V \rightarrow {}^{33}_{15}P$	1	a106	
179	$_{23}V \rightarrow {}^{35}_{16}S$	1	a106	
180	$_{23}V \rightarrow {}^{34}_{17}Cl$	1	a106	
181	$_{23}V \rightarrow {}^{38}_{17}Cl$	11	a106	a300
182	$_{23}V \rightarrow {}^{39}_{17}Cl$	11	a106	a300
183	$_{23}V \rightarrow {}^{42}_{19}K$	15	a107	a300
184	$_{23}V \rightarrow {}^{43}_{19}K$	42	a107	a301
185	$_{23}V \rightarrow {}^{45}_{20}Ca$	5	a108	
186	$_{23}V \rightarrow {}^{47}_{20}Ca$	28	a108	a301
187	$_{23}V \rightarrow {}^{43}_{21}Sc$	9	a109	a301
188	$_{23}V \rightarrow {}^{44}_{21}Sc$	32	a109	a301
189	$_{23}V \rightarrow {}^{44m}_{21}Sc$	16	a110	a302

No.	Excitation function	No. of points	Table on page	Figure on page
190	$_{23}V \rightarrow {}^{46}{}_{21}Sc$	29	a110	a302
191	$_{23}V \rightarrow {}^{46m+g}{}_{21}Sc$	43	a111	a304
192	$_{23}V \rightarrow {}^{47}{}_{21}Sc$	66	a112	a303
193	$_{23}V \rightarrow {}^{48}{}_{21}Sc$	40	a113	a303
194	$_{23}V \rightarrow {}^{45}{}_{22}Ti$	14	a114	a304
195	$_{23}V \rightarrow {}^{47}{}_{23}V$	10	a115	a304
196	$_{23}V \rightarrow {}^{48}{}_{23}V$ (cumulative)	59	a115	a305
197	$_{23}V \rightarrow {}^{48}{}_{24}Cr$	25	a116	a306
198	$_{23}V \rightarrow {}^{49}{}_{24}Cr$	31	a117	a306
199	$_{23}V \rightarrow {}^{51}{}_{24}Cr$	156	a118	a307
200	$_{24}Cr \rightarrow {}^{51}{}_{24}Cr$	16	a121	a307
201	$_{24}Cr \rightarrow {}^{52}{}_{25}Mn$	41	a121	a308
202	$_{24}Cr \rightarrow {}^{52m}{}_{25}Mn$	91	a122	a308
203	$_{24}Cr \rightarrow {}^{52m+g}{}_{25}Mn$	11	a124	a308
204	${}^{52}{}_{24}Cr \rightarrow {}^{52}{}_{25}Mn$	36	a125	a308
205	${}^{52}{}_{24}Cr \rightarrow {}^{52m}{}_{25}Mn$	39	a125	a309
206	${}^{52}{}_{24}Cr \rightarrow {}^{52m+g}{}_{25}Mn$	11	a126	a309
207	$_{25}Mn \rightarrow {}^{7}{}_{4}Be$	3	a127	
208	$_{25}Mn \rightarrow {}^{22}{}_{11}Na$	5	a127	
209	$_{25}Mn \rightarrow {}^{24}{}_{11}Na$	2	a127	
210	$_{25}Mn \rightarrow {}^{44m}{}_{21}Sc$	19	a127	a309
211	$_{25}Mn \rightarrow {}^{46m+g}{}_{21}Sc$	27	a127	a309
212	$_{25}Mn \rightarrow {}^{47}{}_{21}Sc$	20	a128	a309
213	$_{25}Mn \rightarrow {}^{48}{}_{21}Sc$	20	a129	a310
214	$_{25}Mn \rightarrow {}^{48}{}_{23}V$	27	a129	a310
215	$_{25}Mn \rightarrow {}^{48}{}_{24}Cr$	23	a130	a310
216	$_{25}Mn \rightarrow {}^{51}{}_{24}Cr$	40	a130	a310
217	$_{25}Mn \rightarrow {}^{52}{}_{25}Mn$	37	a131	a310
218	$_{25}Mn \rightarrow {}^{54}{}_{25}Mn$	49	a132	a311
219	$_{25}Mn \rightarrow {}^{52}{}_{26}Fe$	53	a133	a311
220	$_{25}Mn \rightarrow {}^{55}{}_{26}Fe$	118	a134	a311
221	$_{26}Fe \rightarrow {}^{7}{}_{4}Be$	14	a137	a312
222	$_{26}Fe \rightarrow {}^{10}{}_{4}Be$	3	a137	
223	$_{26}Fe \rightarrow {}^{22}{}_{11}Na$	7	a137	a312
224	$_{26}Fe \rightarrow {}^{24}{}_{11}Na$	13	a138	a312
225	$_{26}Fe \rightarrow {}^{32}{}_{15}P$	10	a138	a312
226	$_{26}Fe \rightarrow {}^{42}{}_{19}K$	10	a138	a313
227	$_{26}Fe \rightarrow {}^{43}{}_{19}K$	10	a138	a313
228	$_{26}Fe \rightarrow {}^{44m}{}_{21}Sc$	12	a139	a313
229	$_{26}Fe \rightarrow {}^{47}{}_{21}Sc$	13	a139	a313

No.	Excitation function	No. of points	Table on page	Figure on page
230	${}_{26}Fe \rightarrow {}^{48}_{23}V$ (cumulative)	25	a139	a313
231	${}_{26}Fe \rightarrow {}^{51}_{24}Cr$	52	a140	a314
232	${}_{26}Fe \rightarrow {}^{52m+g}_{25}Mn$	45	a141	a314
233	${}_{26}Fe \rightarrow {}^{54}_{25}Mn$	48	a142	a315
234	${}_{26}Fe \rightarrow {}^{52}_{26}Fe$	9	a143	a315
235	${}_{26}Fe \rightarrow {}^{55}_{27}Co$	64	a143	a315
236	${}_{26}Fe \rightarrow {}^{56}_{27}Co$	154	a145	a316
237	${}_{26}Fe \rightarrow {}^{57}_{27}Co$	44	a148	a316
238	${}_{26}Fe \rightarrow {}^{58m+g}_{27}Co$	21	a149	a317
239	${}^{54}_{26}Fe \rightarrow {}^{51}_{25}Mn$	19	a150	a317
240	${}^{56}_{26}Fe \rightarrow {}^{52}_{25}Mn$	13	a150	a317
241	${}^{56}_{26}Fe \rightarrow {}^{52m+g}_{25}Mn$	9	a151	a317
242	${}^{56}_{26}Fe \rightarrow {}^{54}_{25}Mn$	9	a151	a317
243	${}^{56}_{26}Fe \rightarrow {}^{55}_{26}Fe$	9	a151	a317
244	${}^{56}_{26}Fe \rightarrow {}^{55}_{27}Co$	11	a151	a318
245	${}^{56}_{26}Fe \rightarrow {}^{56}_{27}Co$	27	a152	a318
246	${}^{57}_{26}Fe \rightarrow {}^{54}_{25}Mn$	18	a152	a318
247	${}^{57}_{26}Fe \rightarrow {}^{57}_{27}Co$	78	a153	a318
248	${}^{58}_{26}Fe \rightarrow {}^{54}_{25}Mn$	10	a154	a319
249	${}_{27}Co \rightarrow {}^{7}_{4}Be$	3	a155	
250	${}_{27}Co \rightarrow {}^{22}_{11}Na$	7	a155	a319
251	${}_{27}Co \rightarrow {}^{24}_{11}Na$	5	a155	
252	${}_{27}Co \rightarrow {}^{42}_{19}K$	7	a155	a319
253	${}_{27}Co \rightarrow {}^{43}_{19}K$	6	a155	a319
254	${}_{27}Co \rightarrow {}^{44m}_{21}Sc$	20	a156	a319
255	${}_{27}Co \rightarrow {}^{46}_{21}Sc$	4	a156	
256	${}_{27}Co \rightarrow {}^{46m+g}_{21}Sc$	21	a156	a320
257	${}_{27}Co \rightarrow {}^{47}_{21}Sc$	20	a157	a320
258	${}_{27}Co \rightarrow {}^{48}_{23}V$	23	a157	a320
259	${}_{27}Co \rightarrow {}^{51}_{24}Cr$	27	a158	a320
260	${}_{27}Co \rightarrow {}^{52}_{25}Mn$	29	a158	a320
261	${}_{27}Co \rightarrow {}^{54}_{25}Mn$	58	a159	a321
262	${}_{27}Co \rightarrow {}^{56}_{25}Mn$	13	a160	a321
263	${}_{27}Co \rightarrow {}^{55}_{27}Co$ (cumulative)	19	a161	a321
264	${}_{27}Co \rightarrow {}^{56}_{27}Co$	58	a161	a321
265	${}_{27}Co \rightarrow {}^{57}_{27}Co$	85	a162	a322
266	${}_{27}Co \rightarrow {}^{58}_{27}Co$	45	a164	a322
267	${}_{27}Co \rightarrow {}^{58m+g}_{27}Co$	76	a165	a322
268	${}_{27}Co \rightarrow {}^{56}_{28}Ni$	7	a167	a322
269	${}_{27}Co \rightarrow {}^{57}_{28}Ni$	57	a167	a323
270	${}_{27}Co \rightarrow {}^{59}_{28}Ni$	92	a168	a323

No.	Excitation function	No. of points	Table on page	Figure on page
271	$_{28}Ni \rightarrow {}^{7}_{4}Be$	14	a171	a323
272	$_{28}Ni \rightarrow {}^{22}_{11}Na$	7	a171	a324
273	$_{28}Ni \rightarrow {}^{24}_{11}Na$	10	a171	a324
274	$_{28}Ni \rightarrow {}^{32}_{15}P$	9	a171	a324
275	$_{28}Ni \rightarrow {}^{48}_{23}V$	11	a172	a324
276	$_{28}Ni \rightarrow {}^{51}_{24}Cr$	17	a172	a325
277	$_{28}Ni \rightarrow {}^{52}_{25}Mn$	18	a172	a325
278	$_{28}Ni \rightarrow {}^{54}_{25}Mn$	16	a173	a325
279	$_{28}Ni \rightarrow {}^{52}_{26}Fe$	45	a173	a326
280	$_{28}Ni \rightarrow {}^{55}_{27}Co$ (cumulative)	82	a174	a327
281	$_{28}Ni \rightarrow {}^{56}_{27}Co$	34	a176	a326
282	$_{28}Ni \rightarrow {}^{57}_{27}Co$ (cumulative)	36	a177	a327
283	$_{28}Ni \rightarrow {}^{58m+g}_{27}Co$	36	a178	a327
284	$_{28}Ni \rightarrow {}^{60m+g}_{27}Co$	34	a178	a327
285	$_{28}Ni \rightarrow {}^{56}_{28}Ni$	25	a179	a328
286	$_{28}Ni \rightarrow {}^{57}_{28}Ni$	51	a180	a329
287	$_{28}Ni \rightarrow {}^{60}_{29}Cu$	31	a181	a328
288	$_{28}Ni \rightarrow {}^{61}_{29}Cu$	65	a182	a328
289	${}^{58}_{28}Ni \rightarrow {}^{51}_{24}Cr$	4	a183	
290	${}^{58}_{28}Ni \rightarrow {}^{52}_{25}Mn$	8	a183	a329
291	${}^{58}_{28}Ni \rightarrow {}^{54}_{25}Mn$	4	a184	
292	${}^{58}_{28}Ni \rightarrow {}^{52}_{26}Fe$	6	a184	
293	${}^{58}_{28}Ni \rightarrow {}^{53}_{26}Fe$	16	a184	a329
294	${}^{58}_{28}Ni \rightarrow {}^{55}_{26}Fe$	9	a184	
295	${}^{58}_{28}Ni \rightarrow {}^{55}_{27}Co$	77	a185	a330
296	${}^{58}_{28}Ni \rightarrow {}^{56}_{27}Co$	23	a186	a330
297	${}^{58}_{28}Ni \rightarrow {}^{57}_{27}Co$ (cumulative)	53	a187	a331
298	${}^{58}_{28}Ni \rightarrow {}^{56}_{28}Ni$	24	a188	a330
299	${}^{58}_{28}Ni \rightarrow {}^{57}_{28}Ni$ (cumulative)	61	a189	a331
300	${}^{60}_{28}Ni \rightarrow {}^{51}_{24}Cr$	2	a190	
301	${}^{60}_{28}Ni \rightarrow {}^{54}_{25}Mn$	2	a190	
302	${}^{60}_{28}Ni \rightarrow {}^{53}_{26}Fe$	2	a190	
303	${}^{60}_{28}Ni \rightarrow {}^{55}_{27}Co$ (cumulative)	5	a190	
304	${}^{60}_{28}Ni \rightarrow {}^{56}_{27}Co$	18	a191	a332
305	${}^{60}_{28}Ni \rightarrow {}^{57}_{27}Co$ (cumulative)	43	a191	a332
306	${}^{60}_{28}Ni \rightarrow {}^{58}_{27}Co$	13	a192	a332
307	${}^{60}_{28}Ni \rightarrow {}^{57}_{28}Ni$	4	a192	
308	${}^{60}_{28}Ni \rightarrow {}^{60}_{29}Cu$	10	a192	a332
309	${}^{61}_{28}Ni \rightarrow {}^{58}_{27}Co$	9	a193	a332
310	${}^{61}_{28}Ni \rightarrow {}^{61}_{29}Cu$	66	a193	a333
311	${}^{62}_{28}Ni \rightarrow {}^{51}_{24}Cr$	4	a194	
312	${}^{62}_{28}Ni \rightarrow {}^{54}_{25}Mn$	4	a195	
313	${}^{62}_{28}Ni \rightarrow {}^{57}_{27}Co$	21	a195	a333

No.	Excitation function	No. of points	Table on page	Figure on page
314	$^{62}_{28}Ni \rightarrow ^{58}_{27}Co$	39	a195	a333
315	$^{62}_{28}Ni \rightarrow ^{61}_{27}Co$	15	a196	a333
316	$^{62}_{28}Ni \rightarrow ^{61}_{29}Cu$	28	a196	a334
317	$^{62}_{28}Ni \rightarrow ^{62}_{29}Cu$	25	a197	a334
318	$^{64}_{28}Ni \rightarrow ^{51}_{24}Cr$	2	a198	
319	$^{64}_{28}Ni \rightarrow ^{54}_{25}Mn$	2	a198	
320	$^{64}_{28}Ni \rightarrow ^{57}_{27}Co$	8	a198	
321	$^{64}_{28}Ni \rightarrow ^{60}_{27}Co$	10	a198	a334
322	$^{64}_{28}Ni \rightarrow ^{61}_{27}Co$	19	a198	a334
323	$^{64}_{28}Ni \rightarrow ^{64}_{29}Cu$	46	a199	a334
324	$_{29}Cu \rightarrow ^{7}_{4}Be$	17	a200	a335
325	$_{29}Cu \rightarrow ^{18}_{9}F$	10	a200	a335
326	$_{29}Cu \rightarrow ^{22}_{11}Na$	15	a201	a335
327	$_{29}Cu \rightarrow ^{24}_{11}Na$	33	a201	a336
328	$_{29}Cu \rightarrow ^{38}_{17}Cl$	11	a202	a336
329	$_{29}Cu \rightarrow ^{39}_{18}Ar$	10	a202	a337
330	$_{29}Cu \rightarrow ^{42}_{18}Ar$	9	a202	a337
331	$_{29}Cu \rightarrow ^{42}_{19}K$	14	a203	a337
332	$_{29}Cu \rightarrow ^{43}_{19}K$	13	a203	a337
333	$_{29}Cu \rightarrow ^{44}_{21}Sc$	4	a203	
334	$_{29}Cu \rightarrow ^{44m}_{21}Sc$	9	a203	a338
335	$_{29}Cu \rightarrow ^{46}_{21}Sc$	8	a204	a338
336	$_{29}Cu \rightarrow ^{47}_{21}Sc$	9	a204	a338
337	$_{29}Cu \rightarrow ^{48}_{21}Sc$	7	a204	
338	$_{29}Cu \rightarrow ^{48}_{23}V$	8	a204	a338
339	$_{29}Cu \rightarrow ^{51}_{24}Cr$	9	a205	a338
340	$_{29}Cu \rightarrow ^{52}_{25}Mn$	17	a205	a339
341	$_{29}Cu \rightarrow ^{54}_{25}Mn$	21	a205	a339
342	$_{29}Cu \rightarrow ^{56}_{25}Mn$	7	a206	
343	$_{29}Cu \rightarrow ^{52}_{26}Fe$	13	a206	a339
344	$_{29}Cu \rightarrow ^{55}_{26}Fe$	6	a206	
345	$_{29}Cu \rightarrow ^{59}_{26}Fe$	29	a207	a339
346	$_{29}Cu \rightarrow ^{55}_{27}Co$ (cumulative)	11	a207	a340
347	$_{29}Cu \rightarrow ^{56}_{27}Co$	25	a208	a340
348	$_{29}Cu \rightarrow ^{57}_{27}Co$ (cumulative)	31	a208	a340
349	$_{29}Cu \rightarrow ^{58}_{27}Co$	28	a209	a340
350	$_{29}Cu \rightarrow ^{58m+g}_{27}Co$	20	a210	a340
351	$_{29}Cu \rightarrow ^{60}_{27}Co$	27	a210	a340
352	$_{29}Cu \rightarrow ^{61}_{27}Co$	9	a211	
353	$_{29}Cu \rightarrow ^{56}_{28}Ni$	13	a211	a341
354	$_{29}Cu \rightarrow ^{57}_{28}Ni$	29	a211	a341
355	$_{29}Cu \rightarrow ^{65}_{28}Ni$	10	a212	a341

No.	Excitation function	No. of points	Table on page	Figure on page
356	$_{29}Cu \rightarrow {}^{60}_{29}Cu$ (cumulative)	27	a212	a341
357	$_{29}Cu \rightarrow {}^{61}_{29}Cu$ (cumulative)	42	a213	a341
358	$_{29}Cu \rightarrow {}^{62}_{29}Cu$	7	a214	
359	$_{29}Cu \rightarrow {}^{64}_{29}Cu$	38	a214	a342
360	$_{29}Cu \rightarrow {}^{61}_{30}Zn$	7	a215	
361	$_{29}Cu \rightarrow {}^{62}_{30}Zn$	69	a215	a342
362	$_{29}Cu \rightarrow {}^{63}_{30}Zn$	84	a217	a343
363	$_{29}Cu \rightarrow {}^{65}_{30}Zn$	104	a219	a343
364	${}^{63}_{29}Cu \rightarrow {}^{60}_{29}Cu$	6	a221	
365	${}^{63}_{29}Cu \rightarrow {}^{61}_{29}Cu$ (cumulative)	19	a221	a344
366	${}^{63}_{29}Cu \rightarrow {}^{62}_{29}Cu$	24	a221	a344
367	${}^{63}_{29}Cu \rightarrow {}^{61}_{30}Zn$	9	a222	a345
368	${}^{63}_{29}Cu \rightarrow {}^{62}_{30}Zn$	41	a222	a344
369	${}^{63}_{29}Cu \rightarrow {}^{63}_{30}Zn$	165	a223	a345,346
370	${}^{65}_{29}Cu \rightarrow {}^{61}_{29}Cu$ (cumulative)	9	a227	a347
371	${}^{65}_{29}Cu \rightarrow {}^{62}_{29}Cu$	14	a227	a347
372	${}^{65}_{29}Cu \rightarrow {}^{64}_{29}Cu$	101	a227	a347
373	${}^{65}_{29}Cu \rightarrow {}^{62}_{30}Zn$	10	a230	a347
374	${}^{65}_{29}Cu \rightarrow {}^{63}_{30}Zn$	13	a230	a348
375	${}^{65}_{29}Cu \rightarrow {}^{65}_{30}Zn$	246	a230	a348
376	$_{30}Zn \rightarrow {}^{22}_{11}Na$	4	a236	
377	$_{30}Zn \rightarrow {}^{24}_{11}Na$	4	a236	
378	$_{30}Zn \rightarrow {}^{66}_{31}Ga$	33	a236	a349
379	$_{30}Zn \rightarrow {}^{67}_{31}Ga$	32	a237	a349
380	${}^{66}_{30}Zn \rightarrow {}^{65}_{30}Zn$	8	a237	a349
381	${}^{66}_{30}Zn \rightarrow {}^{66}_{31}Ga$	30	a238	a349
382	${}^{67}_{30}Zn \rightarrow {}^{66}_{31}Ga$	14	a238	a350
383	${}^{67}_{30}Zn \rightarrow {}^{67}_{31}Ga$	25	a239	a350
384	${}^{68}_{30}Zn \rightarrow {}^{67}_{29}Cu$	16	a239	a350
385	${}^{68}_{30}Zn \rightarrow {}^{65}_{30}Zn$ (cumulative)	6	a240	a350
386	${}^{68}_{30}Zn \rightarrow {}^{66}_{31}Ga$	9	a240	a350
387	${}^{68}_{30}Zn \rightarrow {}^{67}_{31}Ga$	37	a240	a351
388	${}^{68}_{30}Zn \rightarrow {}^{68}_{31}Ga$	29	a241	a351
389	$_{31}Ga \rightarrow {}^{22}_{11}Na$	4	a242	
390	$_{31}Ga \rightarrow {}^{24}_{11}Na$	4	a242	
391	${}^{69}_{31}Ga \rightarrow {}^{64}_{29}Cu$	14	a242	a351
392	${}^{69}_{31}Ga \rightarrow {}^{67}_{29}Cu$	5	a242	a352
393	${}^{69}_{31}Ga \rightarrow {}^{63}_{30}Zn$	7	a242	a352
394	${}^{69}_{31}Ga \rightarrow {}^{65}_{30}Zn$ (cumulative)	19	a243	a352
395	${}^{69}_{31}Ga \rightarrow {}^{66}_{31}Ga$	13	a243	a352
396	${}^{69}_{31}Ga \rightarrow {}^{67}_{31}Ga$	16	a243	a353

No.	Excitation function	No. of points	Table on page	Figure on page
397	${}^{69}_{31}Ga \rightarrow {}^{68}_{31}Ga$	15	a244	a353
398	${}^{69}_{31}Ga \rightarrow {}^{66}_{32}Ge$	9	a244	a353
399	${}^{69}_{31}Ga \rightarrow {}^{67}_{32}Ge$	7	a244	
400	${}^{69}_{31}Ga \rightarrow {}^{68}_{32}Ge$	12	a245	a353
401	${}^{69}_{31}Ga \rightarrow {}^{69}_{32}Ge$	13	a245	a353
402	${}^{71}_{31}Ga \rightarrow {}^{67}_{29}Cu$ (cumulative)	13	a245	a354
403	${}^{71}_{31}Ga \rightarrow {}^{65}_{30}Zn$	5	a246	
404	${}^{71}_{31}Ga \rightarrow {}^{69}_{30}Zn$ (cumulative)	7	a246	
405	${}^{71}_{31}Ga \rightarrow {}^{69m}_{30}Zn$	15	a246	a354
406	${}^{71}_{31}Ga \rightarrow {}^{68}_{31}Ga$	4	a246	
407	${}^{71}_{31}Ga \rightarrow {}^{70}_{31}Ga$	14	a246	a354
408	${}^{71}_{31}Ga \rightarrow {}^{68}_{32}Ge$	8	a247	a354
409	${}^{71}_{31}Ga \rightarrow {}^{69}_{32}Ge$	14	a247	a354
410	${}_{32}Ge \rightarrow {}^{22}_{11}Na$	4	a247	
411	${}_{32}Ge \rightarrow {}^{24}_{11}Na$	4	a248	
412	${}^{74}_{32}Ge \rightarrow {}^{73}_{31}Ga$	6	a248	
413	${}_{33}As \rightarrow {}^{22}_{11}Na$	4	a248	
414	${}_{33}As \rightarrow {}^{24}_{11}Na$	4	a248	
415	${}_{33}As \rightarrow {}^{77}_{31}Ga$	7	a248	
416	${}_{33}As \rightarrow {}^{74}_{33}As$	7	a248	
417	${}_{34}Se \rightarrow {}^{22}_{11}Na$	4	a249	
418	${}_{34}Se \rightarrow {}^{24}_{11}Na$	4	a249	
419	${}^{76}_{34}Se \rightarrow {}^{74}_{35}Br$	4	a249	
420	${}^{76}_{34}Se \rightarrow {}^{75}_{35}Br$	10	a249	a354
421	${}^{76}_{34}Se \rightarrow {}^{76}_{35}Br$	10	a249	a355
422	${}_{35}Br \rightarrow {}^{75}_{35}Br$	14	a250	a355
423	${}_{35}Br \rightarrow {}^{80}_{35}Br$	5	a250	
424	${}_{35}Br \rightarrow {}^{80m}_{35}Br$	5	a250	
425	${}_{35}Br \rightarrow {}^{76}_{36}Kr$	15	a250	a355
426	${}_{35}Br \rightarrow {}^{77}_{36}Kr$	30	a251	a355
427	${}_{35}Br \rightarrow {}^{79}_{36}Kr$	23	a251	a355
428	${}^{79}_{35}Br \rightarrow {}^{74}_{35}Br$	4	a252	
429	${}^{79}_{35}Br \rightarrow {}^{75}_{35}Br$	7	a252	a356
430	${}^{79}_{35}Br \rightarrow {}^{76}_{35}Br$	8	a252	a356
431	${}^{79}_{35}Br \rightarrow {}^{77}_{35}Br$	11	a253	a356
432	${}^{79}_{35}Br \rightarrow {}^{78}_{35}Br$	11	a253	a356
433	${}^{79}_{35}Br \rightarrow {}^{76}_{36}Kr$	9	a253	a356
434	${}^{79}_{35}Br \rightarrow {}^{77}_{36}Kr$	11	a253	a357
435	${}^{79}_{35}Br \rightarrow {}^{79}_{36}Kr$	12	a254	a357

No.	Excitation function	No. of points	Table on page	Figure on page
436	${}^{81}_{35}Br \rightarrow {}^{75}_{35}Br$	4	a254	
437	${}^{81}_{35}Br \rightarrow {}^{76}_{35}Br$	4	a254	
438	${}^{81}_{35}Br \rightarrow {}^{77}_{35}Br$ (cumulative)	6	a254	a357
439	${}^{81}_{35}Br \rightarrow {}^{78}_{35}Br$	7	a254	a358
440	${}^{81}_{35}Br \rightarrow {}^{80}_{35}Br$	16	a255	a358
441	${}^{81}_{35}Br \rightarrow {}^{80m}_{35}Br$	11	a255	a358
442	${}^{81}_{35}Br \rightarrow {}^{76}_{36}Kr$	3	a255	
443	${}^{81}_{35}Br \rightarrow {}^{77}_{36}Kr$	6	a256	a358
444	${}^{81}_{35}Br \rightarrow {}^{79}_{36}Kr$	10	a256	a358
445	${}_{36}Kr \rightarrow {}^{77}_{35}Br$ (cumulative)	29	b37	b247
446	${}_{36}Kr \rightarrow {}^{82}_{35}Br$	26	b37	b247
447	${}_{36}Kr \rightarrow {}^{77}_{36}Kr$	27	b38	b247
448	${}_{36}Kr \rightarrow {}^{79}_{36}Kr$ (cumulative)	28	b39	b247
449	${}_{36}Kr \rightarrow {}^{85m}_{36}Kr$	28	b39	b248
450	${}_{36}Kr \rightarrow {}^{79}_{37}Rb$	49	b40	b248
451	${}_{36}Kr \rightarrow {}^{81}_{37}Rb$ (cumulative)	69	b41	b248
452	${}_{36}Kr \rightarrow {}^{81m}_{37}Rb$	30	b43	b248
453	${}_{36}Kr \rightarrow {}^{82m}_{37}Rb$	77	b43	b248
454	${}_{36}Kr \rightarrow {}^{83}_{37}Rb$	78	b45	b249
455	${}_{36}Kr \rightarrow {}^{84}_{37}Rb$ (cumulative)	78	b47	b249
456	${}_{36}Kr \rightarrow {}^{84m}_{37}Rb$	63	b49	b249
457	${}_{36}Kr \rightarrow {}^{86}_{37}Rb$	10	b50	b249
458	${}^{78}_{36}Kr \rightarrow {}^{75}_{35}Br$	9	b50	b249
459	${}^{78}_{36}Kr \rightarrow {}^{77}_{36}Kr$	8	b51	b250
460	${}^{78}_{36}Kr \rightarrow {}^{77}_{37}Rb$	6	b51	b250
461	${}^{80}_{36}Kr \rightarrow {}^{76}_{35}Br$	4	b51	
462	${}^{80}_{36}Kr \rightarrow {}^{77}_{35}Br$	7	b51	b250
463	${}^{80}_{36}Kr \rightarrow {}^{79}_{37}Rb$	5	b51	
464	${}^{82}_{36}Kr \rightarrow {}^{81}_{37}Rb$	10	b52	b250
465	${}^{82}_{36}Kr \rightarrow {}^{82m}_{37}Rb$	15	b52	b250
466	${}^{82}_{36}Kr \rightarrow {}^{83}_{37}Rb$	16	b52	b251
467	${}^{83}_{36}Kr \rightarrow {}^{81}_{37}Rb$	10	b53	b251
468	${}^{83}_{36}Kr \rightarrow {}^{82m}_{37}Rb$	15	b53	b251
469	${}^{83}_{36}Kr \rightarrow {}^{83}_{37}Rb$	14	b53	b251
470	${}^{83}_{36}Kr \rightarrow {}^{84}_{37}Rb$	14	b54	b251
471	${}_{37}Rb \rightarrow {}^{74}_{33}As$	1	b54	
472	${}_{37}Rb \rightarrow {}^{77}_{35}Br$ (cumulative)	1	b54	
473	${}_{37}Rb \rightarrow {}^{76}_{36}Kr$ (cumulative)	1	b54	
474	${}_{37}Rb \rightarrow {}^{77}_{36}Kr$ (cumulative)	1	b54	
475	${}_{37}Rb \rightarrow {}^{79}_{36}Kr$ (cumulative)	1	b54	
476	${}_{37}Rb \rightarrow {}^{84}_{37}Rb$	10	b55	b252
477	${}_{37}Rb \rightarrow {}^{86}_{37}Rb$	8	b55	b252

No.	Excitation function	No. of points	Table on page	Figure on page
478	$_{37}Rb \rightarrow {}^{82}{}_{38}Sr$	29	b55	b252
479	$_{37}Rb \rightarrow {}^{83m+g}{}_{38}Sr$	8	b56	b252
480	$_{37}Rb \rightarrow {}^{85m}{}_{38}Sr$	5	b56	
481	$_{37}Rb \rightarrow {}^{85m+g}{}_{38}Sr$	8	b56	b252
482	$_{37}Rb \rightarrow {}^{87m}{}_{38}Sr$	10	b56	b253
483	$^{85}{}_{37}Rb \rightarrow {}^{79}{}_{36}Kr$	17	b57	b253
484	$^{85}{}_{37}Rb \rightarrow {}^{81}{}_{37}Rb$	17	b57	b253
485	$^{85}{}_{37}Rb \rightarrow {}^{81m}{}_{37}Rb$	15	b58	b253
486	$^{85}{}_{37}Rb \rightarrow {}^{82m}{}_{37}Rb$	17	b58	b253
487	$^{85}{}_{37}Rb \rightarrow {}^{83}{}_{37}Rb$	26	b58	b254
488	$^{85}{}_{37}Rb \rightarrow {}^{84}{}_{37}Rb$	20	b59	b254
489	$^{85}{}_{37}Rb \rightarrow {}^{84m}{}_{37}Rb$	38	b59	b254
490	$^{85}{}_{37}Rb \rightarrow {}^{84m+g}{}_{37}Rb$	19	b60	b254
491	$^{85}{}_{37}Rb \rightarrow {}^{81}{}_{38}Sr$	13	b61	b255
492	$^{85}{}_{37}Rb \rightarrow {}^{82}{}_{38}Sr$	19	b61	b255
493	$^{85}{}_{37}Rb \rightarrow {}^{83}{}_{38}Sr$	29	b62	b255
494	$^{85}{}_{37}Rb \rightarrow {}^{85m}{}_{38}Sr$	25	b62	b255
495	$^{85}{}_{37}Rb \rightarrow {}^{85m+g}{}_{38}Sr$	13	b63	b256
496	$^{85}{}_{37}Rb \rightarrow {}^{87}{}_{38}Sr$	9	b63	
497	$_{38}Sr \rightarrow {}^{48}{}_{23}V$	1	b64	
498	$_{38}Sr \rightarrow {}^{51}{}_{24}Cr$ (cumulative)	1	b64	
499	$_{38}Sr \rightarrow {}^{52m+g}{}_{25}Mn$	1	b64	
500	$_{38}Sr \rightarrow {}^{54}{}_{25}Mn$	1	b64	
501	$_{38}Sr \rightarrow {}^{59}{}_{26}Fe$	1	b64	
502	$_{38}Sr \rightarrow {}^{56}{}_{27}Co$	1	b64	
503	$_{38}Sr \rightarrow {}^{57}{}_{27}Co$	1	b64	
504	$_{38}Sr \rightarrow {}^{58m+g}{}_{27}Co$	1	b64	
505	$_{38}Sr \rightarrow {}^{60m+g}{}_{27}Co$	1	b64	
506	$_{38}Sr \rightarrow {}^{65}{}_{30}Zn$ (cumulative)	1	b64	
507	$_{38}Sr \rightarrow {}^{71m+g}{}_{30}Zn$	1	b64	
508	$_{38}Sr \rightarrow {}^{67}{}_{31}Ga$ (cumulative)	1	b64	
509	$_{38}Sr \rightarrow {}^{68}{}_{32}Ge$ (cumulative)	1	b65	
510	$_{38}Sr \rightarrow {}^{69}{}_{32}Ge$ (cumulative)	1	b65	
511	$_{38}Sr \rightarrow {}^{71}{}_{33}As$ (cumulative)	1	b65	
512	$_{38}Sr \rightarrow {}^{74}{}_{33}As$	3	b65	
513	$_{38}Sr \rightarrow {}^{72}{}_{34}Se$ (cumulative)	1	b65	
514	$_{38}Sr \rightarrow {}^{73m+g}{}_{34}Se$	1	b65	
515	$_{38}Sr \rightarrow {}^{75}{}_{34}Se$	1	b65	
516	$_{38}Sr \rightarrow {}^{75}{}_{35}Br$	1	b65	
517	$_{38}Sr \rightarrow {}^{77m+g}{}_{35}Br$ (cumulative)	1	b65	
518	$_{38}Sr \rightarrow {}^{82m+g}{}_{35}Br$	1	b65	
519	$_{38}Sr \rightarrow {}^{76}{}_{36}Kr$ (cumulative)	1	b65	

No.	Excitation function	No. of points	Table on page	Figure on page
520	$_{38}Sr \rightarrow {}^{77}_{36}Kr$ (cumulative)	2	b66	
521	$_{38}Sr \rightarrow {}^{79m+g}_{36}Kr$ (cumulative)	2	b66	
522	$_{38}Sr \rightarrow {}^{85m}_{36}Kr$ (cumulative)	1	b66	
523	$_{38}Sr \rightarrow {}^{81}_{37}Rb$ (cumulative)	1	b66	
524	$_{38}Sr \rightarrow {}^{82m}_{37}Rb$	2	b66	
525	$_{38}Sr \rightarrow {}^{83}_{37}Rb$	1	b66	
526	$_{38}Sr \rightarrow {}^{84}_{37}Rb$	3	b66	
527	$_{38}Sr \rightarrow {}^{84m}_{37}Rb$	1	b66	
528	$_{38}Sr \rightarrow {}^{86m+g}_{37}Rb$	1	b66	
529	$_{38}Sr \rightarrow {}^{81}_{38}Sr$	1	b66	
530	$_{38}Sr \rightarrow {}^{85m}_{38}Sr$	1	b66	
531	$_{38}Sr \rightarrow {}^{87m}_{38}Sr$	2	b67	
532	$_{38}Sr \rightarrow {}^{87}_{39}Y$ (cumulative)	2	b67	
533	$_{38}Sr \rightarrow {}^{87m}_{39}Y$ (cumulative)	1	b67	
534	$_{38}Sr \rightarrow {}^{88}_{39}Y$	2	b67	
535	${}^{86}_{38}Sr \rightarrow {}^{82m}_{37}Rb$	8	b67	
536	${}^{86}_{38}Sr \rightarrow {}^{83}_{37}Rb$	16	b67	b256
537	${}^{86}_{38}Sr \rightarrow {}^{85m+g}_{38}Sr$	14	b68	b256
538	${}^{86}_{38}Sr \rightarrow {}^{85}_{39}Y$	15	b68	b256
539	${}^{86}_{38}Sr \rightarrow {}^{85m}_{39}Y$	16	b68	b256
540	${}^{86}_{38}Sr \rightarrow {}^{86}_{39}Y$	26	b69	b256
541	${}^{86}_{38}Sr \rightarrow {}^{86m}_{39}Y$	25	b69	b257
542	${}^{87}_{38}Sr \rightarrow {}^{83}_{37}Rb$	10	b70	
543	${}^{87}_{38}Sr \rightarrow {}^{84}_{37}Rb$	1	b70	
544	${}^{87}_{38}Sr \rightarrow {}^{84m}_{37}Rb$	14	b70	b257
545	${}^{87}_{38}Sr \rightarrow {}^{84m+g}_{37}Rb$	20	b71	b257
546	${}^{87}_{38}Sr \rightarrow {}^{86}_{39}Y$	17	b71	b257
547	${}^{87}_{38}Sr \rightarrow {}^{86m}_{39}Y$	17	b72	b257
548	${}^{87}_{38}Sr \rightarrow {}^{87}_{39}Y$	26	b72	b258
549	${}^{87}_{38}Sr \rightarrow {}^{87m}_{39}Y$	25	b73	b258
550	${}^{88}_{38}Sr \rightarrow {}^{83}_{37}Rb$	11	b73	b258
551	${}^{88}_{38}Sr \rightarrow {}^{84}_{37}Rb$	13	b73	b258
552	${}^{88}_{38}Sr \rightarrow {}^{84m}_{37}Rb$	8	b74	
553	${}^{88}_{38}Sr \rightarrow {}^{84m+g}_{37}Rb$	9	b74	
554	${}^{88}_{38}Sr \rightarrow {}^{86}_{37}Rb$	11	b74	b258
555	${}^{88}_{38}Sr \rightarrow {}^{85}_{38}Sr$	8	b75	b259
556	${}^{88}_{38}Sr \rightarrow {}^{85m}_{38}Sr$	8	b75	b259
557	${}^{88}_{38}Sr \rightarrow {}^{85m+g}_{38}Sr$	8	b75	b259
558	${}^{88}_{38}Sr \rightarrow {}^{84}_{39}Y$	6	b75	b259
559	${}^{88}_{38}Sr \rightarrow {}^{85}_{39}Y$	9	b75	b259
560	${}^{88}_{38}Sr \rightarrow {}^{85m}_{39}Y$	9	b76	b260
561	${}^{88}_{38}Sr \rightarrow {}^{85m+g}_{39}Y$	9	b76	b260
562	${}^{88}_{38}Sr \rightarrow {}^{86}_{39}Y$	15	b76	b260

No.	Excitation function	No. of points	Table on page	Figure on page
563	$^{88}_{38}Sr \rightarrow {}^{86m}_{39}Y$	14	b77	b260
564	$^{88}_{38}Sr \rightarrow {}^{86m+g}_{39}Y$	12	b77	b260
565	$^{88}_{38}Sr \rightarrow {}^{87}_{39}Y$	26	b77	b260
566	$^{88}_{38}Sr \rightarrow {}^{87m}_{39}Y$	26	b78	b261
567	$^{88}_{38}Sr \rightarrow {}^{87m+g}_{39}Y$	8	b78	b261
568	$^{88}_{38}Sr \rightarrow {}^{88}_{39}Y$	54	b79	b261
569	$_{39}Y \rightarrow {}^{7}_{4}Be$	3	b80	
570	$_{39}Y \rightarrow {}^{22}_{11}Na$	3	b80	
571	$_{39}Y \rightarrow {}^{39}_{18}Ar$	3	b80	
572	$_{39}Y \rightarrow {}^{46}_{21}Sc$	2	b80	
573	$_{39}Y \rightarrow {}^{48}_{23}V$	2	b80	
574	$_{39}Y \rightarrow {}^{51}_{24}Cr$	2	b81	
575	$_{39}Y \rightarrow {}^{52}_{25}Mn$	2	b81	
576	$_{39}Y \rightarrow {}^{54}_{25}Mn$	3	b81	
577	$_{39}Y \rightarrow {}^{59}_{26}Fe$	2	b81	
578	$_{39}Y \rightarrow {}^{56}_{27}Co$	3	b81	
579	$_{39}Y \rightarrow {}^{58}_{27}Co$	3	b81	
580	$_{39}Y \rightarrow {}^{60}_{27}Co$	1	b81	
581	$_{39}Y \rightarrow {}^{65}_{30}Zn$	4	b81	
582	$_{39}Y \rightarrow {}^{67}_{31}Ga$	1	b81	
583	$_{39}Y \rightarrow {}^{69}_{32}Ge$	1	b82	
584	$_{39}Y \rightarrow {}^{74}_{33}As$ (cumulative)	9	b82	b261
585	$_{39}Y \rightarrow {}^{72}_{34}Se$	5	b82	
586	$_{39}Y \rightarrow {}^{73m}_{34}Se$	5	b82	
587	$_{39}Y \rightarrow {}^{75}_{34}Se$ (cumulative)	7	b82	
588	$_{39}Y \rightarrow {}^{77}_{35}Br$ (cumulative)	1	b82	
589	$_{39}Y \rightarrow {}^{76}_{36}Kr$ (cumulative)	1	b82	
590	$_{39}Y \rightarrow {}^{77}_{36}Kr$ (cumulative)	1	b83	
591	$_{39}Y \rightarrow {}^{79}_{36}Kr$ (cumulative)	1	b83	
592	$_{39}Y \rightarrow {}^{81}_{36}Kr$	9	b83	b262
593	$_{39}Y \rightarrow {}^{85}_{36}Kr$	8	b83	b262
594	$_{39}Y \rightarrow {}^{83}_{37}Rb$ (cumulative)	9	b83	b262
595	$_{39}Y \rightarrow {}^{84}_{37}Rb$	10	b83	b262
596	$_{39}Y \rightarrow {}^{84m+g}_{37}Rb$	5	b84	
597	$_{39}Y \rightarrow {}^{86}_{37}Rb$	8	b84	b262
598	$_{39}Y \rightarrow {}^{82}_{38}Sr$	5	b84	
599	$_{39}Y \rightarrow {}^{83}_{38}Sr$	6	b84	b263
600	$_{39}Y \rightarrow {}^{85}_{38}Sr$ (cumulative)	9	b85	b263
601	$_{39}Y \rightarrow {}^{85m+g}_{38}Sr$	13	b85	b263
602	$_{39}Y \rightarrow {}^{87m}_{38}Sr$	5	b85	
603	$_{39}Y \rightarrow {}^{84}_{39}Y$	3	b85	
604	$_{39}Y \rightarrow {}^{85}_{39}Y$	5	b85	

No.	Excitation function	No. of points	Table on page	Figure on page
605	$_{39}Y \rightarrow {}^{85m+g}{}_{39}Y$	5	b86	
606	$_{39}Y \rightarrow {}^{86m+g}{}_{39}Y$	11	b86	b263
607	$_{39}Y \rightarrow {}^{87}{}_{39}Y$	7	b86	b263
608	$_{39}Y \rightarrow {}^{87m+g}{}_{39}Y$	15	b86	b264
609	$_{39}Y \rightarrow {}^{88}{}_{39}Y$	106	b87	b264,265
610	$_{39}Y \rightarrow {}^{86}{}_{40}Zr$	16	b89	b265
611	$_{39}Y \rightarrow {}^{87}{}_{40}Zr$	20	b89	b265
612	$_{39}Y \rightarrow {}^{88}{}_{40}Zr$	93	b90	b266
613	$_{39}Y \rightarrow {}^{89}{}_{40}Zr$	175	b92	b267,268
614	$_{40}Zr \rightarrow {}^{7}{}_{4}Be$	4	b96	
615	$_{40}Zr \rightarrow {}^{22}{}_{11}Na$	3	b96	
616	$_{40}Zr \rightarrow {}^{39}{}_{18}Ar$	3	b96	
617	$_{40}Zr \rightarrow {}^{42}{}_{18}Ar$	2	b96	
618	$_{40}Zr \rightarrow {}^{46}{}_{21}Sc$	2	b96	
619	$_{40}Zr \rightarrow {}^{48}{}_{23}V$	2	b96	
620	$_{40}Zr \rightarrow {}^{51}{}_{24}Cr$	2	b96	
621	$_{40}Zr \rightarrow {}^{54}{}_{25}Mn$	3	b97	
622	$_{40}Zr \rightarrow {}^{56}{}_{27}Co$	3	b97	
623	$_{40}Zr \rightarrow {}^{58}{}_{27}Co$	3	b97	
624	$_{40}Zr \rightarrow {}^{65}{}_{30}Zn$	4	b97	
625	$_{40}Zr \rightarrow {}^{67}{}_{31}Ga$	1	b97	
626	$_{40}Zr \rightarrow {}^{74}{}_{33}As$	4	b97	
627	$_{40}Zr \rightarrow {}^{75}{}_{34}Se$ (cumulative)	6	b97	
628	$_{40}Zr \rightarrow {}^{77}{}_{35}Br$ (cumulative)	1	b97	
629	$_{40}Zr \rightarrow {}^{76}{}_{36}Kr$ (cumulative)	1	b98	
630	$_{40}Zr \rightarrow {}^{77}{}_{36}Kr$ (cumulative)	1	b98	
631	$_{40}Zr \rightarrow {}^{79}{}_{36}Kr$ (cumulative)	1	b98	
632	$_{40}Zr \rightarrow {}^{81}{}_{36}Kr$	7	b98	b267
633	$_{40}Zr \rightarrow {}^{85}{}_{36}Kr$	7	b98	b268
634	$_{40}Zr \rightarrow {}^{83}{}_{37}Rb$	24	b98	b268
635	$_{40}Zr \rightarrow {}^{83}{}_{37}Rb$ (cumulative)	8	b99	b268
636	$_{40}Zr \rightarrow {}^{84}{}_{37}Rb$	21	b99	b269
637	$_{40}Zr \rightarrow {}^{86}{}_{37}Rb$	2	b100	
638	$_{40}Zr \rightarrow {}^{85}{}_{38}Sr$ (cumulative)	7	b100	
639	$_{40}Zr \rightarrow {}^{86}{}_{39}Y$	24	b100	b269
640	$_{40}Zr \rightarrow {}^{87}{}_{39}Y$	24	b100	b269
641	$_{40}Zr \rightarrow {}^{88}{}_{39}Y$	5	b101	
642	$_{40}Zr \rightarrow {}^{91}{}_{39}Y$	1	b101	
643	$_{40}Zr \rightarrow {}^{88}{}_{40}Zr$ (cumulative)	8	b101	b269
644	$_{40}Zr \rightarrow {}^{89}{}_{40}Zr$	1	b101	
645	$_{40}Zr \rightarrow {}^{95}{}_{40}Zr$ (cumulative)	8	b102	b270
646	$_{40}Zr \rightarrow {}^{90}{}_{41}Nb$	39	b102	b270

No.	Excitation function	No. of points	Table on page	Figure on page
647	${}_{40}Zr \rightarrow {}^{92m}_{41}Nb$ (cumulative)	3	b103	
648	${}^{90}_{40}Zr \rightarrow {}^{71}_{33}As$ (cumulative)	1	b103	
649	${}^{90}_{40}Zr \rightarrow {}^{72}_{33}As$	1	b103	
650	${}^{90}_{40}Zr \rightarrow {}^{74}_{33}As$	1	b103	
651	${}^{90}_{40}Zr \rightarrow {}^{72}_{34}Se$ (cumulative)	1	b103	
652	${}^{90}_{40}Zr \rightarrow {}^{73}_{34}Se$ (cumulative)	1	b103	
653	${}^{90}_{40}Zr \rightarrow {}^{75}_{34}Se$ (cumulative)	1	b103	
654	${}^{90}_{40}Zr \rightarrow {}^{77}_{35}Br$ (cumulative)	1	b103	
655	${}^{90}_{40}Zr \rightarrow {}^{74}_{37}Rb$	1	b103	
656	${}^{90}_{40}Zr \rightarrow {}^{75}_{37}Rb$	1	b103	
657	${}^{90}_{40}Zr \rightarrow {}^{76}_{37}Rb$	1	b104	
658	${}^{90}_{40}Zr \rightarrow {}^{77}_{37}Rb$	1	b104	
659	${}^{90}_{40}Zr \rightarrow {}^{78}_{37}Rb$	1	b104	
660	${}^{90}_{40}Zr \rightarrow {}^{79}_{37}Rb$	1	b104	
661	${}^{90}_{40}Zr \rightarrow {}^{80}_{37}Rb$	1	b104	
662	${}^{90}_{40}Zr \rightarrow {}^{81}_{37}Rb$ (cumulative)	2	b104	
663	${}^{90}_{40}Zr \rightarrow {}^{82}_{37}Rb$	2	b104	
664	${}^{90}_{40}Zr \rightarrow {}^{82m}_{37}Rb$	8	b104	b270
665	${}^{90}_{40}Zr \rightarrow {}^{83}_{37}Rb$	11	b104	b270
666	${}^{90}_{40}Zr \rightarrow {}^{84}_{37}Rb$	9	b105	b270
667	${}^{90}_{40}Zr \rightarrow {}^{86}_{37}Rb$	1	b105	
668	${}^{90}_{40}Zr \rightarrow {}^{82}_{38}Sr$ (cumulative)	1	b105	
669	${}^{90}_{40}Zr \rightarrow {}^{83}_{38}Sr$ (cumulative)	10	b105	b271
670	${}^{90}_{40}Zr \rightarrow {}^{85}_{38}Sr$	9	b105	b271
671	${}^{90}_{40}Zr \rightarrow {}^{83}_{39}Y$	4	b106	
672	${}^{90}_{40}Zr \rightarrow {}^{84}_{39}Y$	6	b106	b271
673	${}^{90}_{40}Zr \rightarrow {}^{85}_{39}Y$	9	b106	b271
674	${}^{90}_{40}Zr \rightarrow {}^{86}_{39}Y$	12	b106	b271
675	${}^{90}_{40}Zr \rightarrow {}^{86m}_{39}Y$	7	b107	
676	${}^{90}_{40}Zr \rightarrow {}^{86m+g}_{39}Y$	9	b107	
677	${}^{90}_{40}Zr \rightarrow {}^{87}_{39}Y$	25	b107	b272
678	${}^{90}_{40}Zr \rightarrow {}^{87m}_{39}Y$	16	b108	b272
679	${}^{90}_{40}Zr \rightarrow {}^{87}_{39}Y$ (cumulative)	1	b108	
680	${}^{90}_{40}Zr \rightarrow {}^{88}_{39}Y$	12	b108	b272
681	${}^{90}_{40}Zr \rightarrow {}^{86}_{40}Zr$ (cumulative)	1	b109	
682	${}^{90}_{40}Zr \rightarrow {}^{88}_{40}Zr$ (cumulative)	1	b109	
683	${}^{90}_{40}Zr \rightarrow {}^{89}_{40}Zr$	21	b109	b272
684	${}^{90}_{40}Zr \rightarrow {}^{89}_{40}Zr$ (cumulative)	1	b109	
685	${}^{90}_{40}Zr \rightarrow {}^{89m}_{41}Nb$	16	b109	b272
686	${}^{90}_{40}Zr \rightarrow {}^{90}_{41}Nb$	33	b110	b273
687	${}^{90}_{40}Zr \rightarrow {}^{90m}_{41}Nb$	4	b111	
688	${}^{91}_{40}Zr \rightarrow {}^{71}_{33}As$ (cumulative)	1	b111	
689	${}^{91}_{40}Zr \rightarrow {}^{72}_{33}As$	1	b111	

No.	Excitation function	No. of points	Table on page	Figure on page
690	${}^{91}_{40}Zr \rightarrow {}^{74}_{33}As$	1	b111	
691	${}^{91}_{40}Zr \rightarrow {}^{72}_{34}Se$ (cumulative)	1	b111	
692	${}^{91}_{40}Zr \rightarrow {}^{73}_{34}Se$ (cumulative)	1	b111	
693	${}^{91}_{40}Zr \rightarrow {}^{75}_{34}Se$ (cumulative)	1	b111	
694	${}^{91}_{40}Zr \rightarrow {}^{71}_{35}Br$	1	b111	
695	${}^{91}_{40}Zr \rightarrow {}^{72}_{35}Br$	1	b111	
696	${}^{91}_{40}Zr \rightarrow {}^{73}_{35}Br$	1	b111	
697	${}^{91}_{40}Zr \rightarrow {}^{74}_{35}Br$	1	b111	
698	${}^{91}_{40}Zr \rightarrow {}^{75}_{35}Br$	1	b112	
699	${}^{91}_{40}Zr \rightarrow {}^{76}_{35}Br$	1	b112	
700	${}^{91}_{40}Zr \rightarrow {}^{77}_{35}Br$ (cumulative)	2	b112	
701	${}^{91}_{40}Zr \rightarrow {}^{78}_{35}Br$	1	b112	
702	${}^{91}_{40}Zr \rightarrow {}^{80}_{35}Br$	1	b112	
703	${}^{91}_{40}Zr \rightarrow {}^{82}_{35}Br$	1	b112	
704	${}^{91}_{40}Zr \rightarrow {}^{83}_{35}Br$	1	b112	
705	${}^{91}_{40}Zr \rightarrow {}^{84}_{35}Br$	1	b112	
706	${}^{91}_{40}Zr \rightarrow {}^{76}_{37}Rb$	1	b112	
707	${}^{91}_{40}Zr \rightarrow {}^{77}_{37}Rb$	1	b112	
708	${}^{91}_{40}Zr \rightarrow {}^{78}_{37}Rb$	1	b112	
709	${}^{91}_{40}Zr \rightarrow {}^{79}_{37}Rb$	1	b112	
710	${}^{91}_{40}Zr \rightarrow {}^{80}_{37}Rb$	1	b113	
711	${}^{91}_{40}Zr \rightarrow {}^{81}_{37}Rb$ (cumulative)	2	b113	
712	${}^{91}_{40}Zr \rightarrow {}^{82}_{37}Rb$	2	b113	
713	${}^{91}_{40}Zr \rightarrow {}^{83}_{37}Rb$	2	b113	
714	${}^{91}_{40}Zr \rightarrow {}^{84}_{37}Rb$	2	b113	
715	${}^{91}_{40}Zr \rightarrow {}^{86}_{37}Rb$	1	b113	
716	${}^{91}_{40}Zr \rightarrow {}^{88}_{37}Rb$	1	b113	
717	${}^{91}_{40}Zr \rightarrow {}^{82}_{38}Sr$ (cumulative)	1	b113	
718	${}^{91}_{40}Zr \rightarrow {}^{83}_{38}Sr$	1	b113	
719	${}^{91}_{40}Zr \rightarrow {}^{86}_{39}Y$	1	b113	
720	${}^{91}_{40}Zr \rightarrow {}^{87}_{39}Y$	9	b113	
721	${}^{91}_{40}Zr \rightarrow {}^{87}_{39}Y$ (cumulative)	1	b114	
722	${}^{91}_{40}Zr \rightarrow {}^{87m}_{39}Y$	10	b114	
723	${}^{91}_{40}Zr \rightarrow {}^{88}_{39}Y$	23	b114	b273
724	${}^{91}_{40}Zr \rightarrow {}^{86}_{40}Zr$ (cumulative)	1	b115	
725	${}^{91}_{40}Zr \rightarrow {}^{88}_{40}Zr$ (cumulative)	1	b115	
726	${}^{91}_{40}Zr \rightarrow {}^{89}_{40}Zr$	8	b115	
727	${}^{91}_{40}Zr \rightarrow {}^{89}_{40}Zr$ (cumulative)	1	b115	
728	${}^{91}_{40}Zr \rightarrow {}^{90}_{41}Nb$	17	b115	b273
729	${}^{91}_{40}Zr \rightarrow {}^{91}_{41}Nb$	12	b116	b273
730	${}^{92}_{40}Zr \rightarrow {}^{88}_{39}Y$	13	b116	b273
731	${}^{92}_{40}Zr \rightarrow {}^{90}_{41}Nb$	8	b116	
732	${}^{92}_{40}Zr \rightarrow {}^{92}_{41}Nb$	11	b116	b274

No.	Excitation function	No. of points	Table on page	Figure on page
733	${}^{92}_{40}Zr \rightarrow {}^{92m}_{41}Nb$	10	b117	b274
734	${}^{94}_{40}Zr \rightarrow {}^{71}_{33}As$ (cumulative)	1	b117	
735	${}^{94}_{40}Zr \rightarrow {}^{72}_{33}As$	1	b117	
736	${}^{94}_{40}Zr \rightarrow {}^{74}_{33}As$	1	b117	
737	${}^{94}_{40}Zr \rightarrow {}^{72}_{34}Se$ (cumulative)	1	b117	
738	${}^{94}_{40}Zr \rightarrow {}^{73}_{34}Se$ (cumulative)	1	b117	
739	${}^{94}_{40}Zr \rightarrow {}^{75}_{34}Se$ (cumulative)	1	b117	
740	${}^{94}_{40}Zr \rightarrow {}^{77}_{35}Br$ (cumulative)	1	b118	
741	${}^{94}_{40}Zr \rightarrow {}^{76}_{37}Rb$	1	b118	
742	${}^{94}_{40}Zr \rightarrow {}^{77}_{37}Rb$	1	b118	
743	${}^{94}_{40}Zr \rightarrow {}^{78}_{37}Rb$	1	b118	
744	${}^{94}_{40}Zr \rightarrow {}^{79}_{37}Rb$	1	b118	
745	${}^{94}_{40}Zr \rightarrow {}^{80}_{37}Rb$	1	b118	
746	${}^{94}_{40}Zr \rightarrow {}^{81}_{37}Rb$ (cumulative)	2	b118	
747	${}^{94}_{40}Zr \rightarrow {}^{82}_{37}Rb$	2	b118	
748	${}^{94}_{40}Zr \rightarrow {}^{83}_{37}Rb$	2	b118	
749	${}^{94}_{40}Zr \rightarrow {}^{84}_{37}Rb$	2	b118	
750	${}^{94}_{40}Zr \rightarrow {}^{86}_{37}Rb$	1	b118	
751	${}^{94}_{40}Zr \rightarrow {}^{88}_{37}Rb$	1	b119	
752	${}^{94}_{40}Zr \rightarrow {}^{89}_{37}Rb$	1	b119	
753	${}^{94}_{40}Zr \rightarrow {}^{90}_{37}Rb$	1	b119	
754	${}^{94}_{40}Zr \rightarrow {}^{82}_{38}Sr$ (cumulative)	1	b119	
755	${}^{94}_{40}Zr \rightarrow {}^{83}_{38}Sr$ (cumulative)	1	b119	
756	${}^{94}_{40}Zr \rightarrow {}^{86}_{39}Y$	1	b119	
757	${}^{94}_{40}Zr \rightarrow {}^{87}_{39}Y$	1	b119	
758	${}^{94}_{40}Zr \rightarrow {}^{87}_{39}Y$ (cumulative)	1	b119	
759	${}^{94}_{40}Zr \rightarrow {}^{88}_{39}Y$	1	b119	
760	${}^{94}_{40}Zr \rightarrow {}^{86}_{40}Zr$ (cumulative)	1	b119	
761	${}^{94}_{40}Zr \rightarrow {}^{88}_{40}Zr$	1	b119	
762	${}^{94}_{40}Zr \rightarrow {}^{89}_{40}Zr$	1	b119	
763	${}^{94}_{40}Zr \rightarrow {}^{89}_{40}Zr$ (cumulative)	1	b120	
764	${}^{96}_{40}Zr \rightarrow {}^{72}_{30}Zn$ (cumulative)	1	b120	
765	${}^{96}_{40}Zr \rightarrow {}^{66}_{31}Ga$	1	b120	
766	${}^{96}_{40}Zr \rightarrow {}^{67}_{31}Ga$ (cumulative)	1	b120	
767	${}^{96}_{40}Zr \rightarrow {}^{72}_{31}Ga$	1	b120	
768	${}^{96}_{40}Zr \rightarrow {}^{73}_{31}Ga$ (cumulative)	1	b120	
769	${}^{96}_{40}Zr \rightarrow {}^{71}_{33}As$ (cumulative)	1	b120	
770	${}^{96}_{40}Zr \rightarrow {}^{72}_{33}As$	1	b120	
771	${}^{96}_{40}Zr \rightarrow {}^{74}_{33}As$	1	b120	
772	${}^{96}_{40}Zr \rightarrow {}^{72}_{34}Se$ (cumulative)	1	b120	
773	${}^{96}_{40}Zr \rightarrow {}^{95}_{39}Y$	5	b120	
774	${}^{96}_{40}Zr \rightarrow {}^{95}_{40}Zr$	25	b121	b274
775	${}^{96}_{40}Zr \rightarrow {}^{95}_{41}Nb$	32	b121	b274

No.	Excitation function	No. of points	Table on page	Figure on page
776	$^{96}_{40}Zr \rightarrow ^{95m}_{41}Nb$	26	b122	b274
777	$^{96}_{40}Zr \rightarrow ^{96}_{41}Nb$	39	b123	b275
778	$_{41}Nb \rightarrow ^{3}_{1}H$	2	b123	
779	$_{41}Nb \rightarrow ^{7}_{4}Be$	4	b124	
780	$_{41}Nb \rightarrow ^{22}_{11}Na$	1	b124	
781	$_{41}Nb \rightarrow ^{24}_{11}Na$ (cumulative)	1	b124	
782	$_{41}Nb \rightarrow ^{47}_{20}Ca$ (cumulative)	1	b124	
783	$_{41}Nb \rightarrow ^{44m}_{21}Sc$	1	b124	
784	$_{41}Nb \rightarrow ^{46}_{21}Sc$	1	b124	
785	$_{41}Nb \rightarrow ^{47}_{21}Sc$	1	b124	
786	$_{41}Nb \rightarrow ^{48}_{21}Sc$	1	b124	
787	$_{41}Nb \rightarrow ^{48}_{23}V$	1	b124	
788	$_{41}Nb \rightarrow ^{52}_{25}Mn$ (cumulative)	1	b124	
789	$_{41}Nb \rightarrow ^{54}_{25}Mn$	1	b124	
790	$_{41}Nb \rightarrow ^{56}_{25}Mn$ (cumulative)	1	b125	
791	$_{41}Nb \rightarrow ^{56}_{27}Co$	1	b125	
792	$_{41}Nb \rightarrow ^{58}_{27}Co$	1	b125	
793	$_{41}Nb \rightarrow ^{65}_{30}Zn$ (cumulative)	1	b125	
794	$_{41}Nb \rightarrow ^{66}_{31}Ga$	1	b125	
795	$_{41}Nb \rightarrow ^{67}_{31}Ga$ (cumulative)	1	b125	
796	$_{41}Nb \rightarrow ^{69}_{32}Ge$ (cumulative)	1	b125	
797	$_{41}Nb \rightarrow ^{71}_{33}As$ (cumulative)	1	b125	
798	$_{41}Nb \rightarrow ^{72}_{33}As$	1	b125	
799	$_{41}Nb \rightarrow ^{74}_{33}As$	2	b125	
800	$_{41}Nb \rightarrow ^{73}_{34}Se$ (cumulative)	1	b125	
801	$_{41}Nb \rightarrow ^{75}_{34}Se$	1	b125	
802	$_{41}Nb \rightarrow ^{75}_{35}Br$ (cumulative)	1	b126	
803	$_{41}Nb \rightarrow ^{76}_{35}Br$	1	b126	
804	$_{41}Nb \rightarrow ^{77}_{35}Br$ (cumulative)	2	b126	
805	$_{41}Nb \rightarrow ^{76}_{36}Kr$ (cumulative)	1	b126	
806	$_{41}Nb \rightarrow ^{77}_{36}Kr$ (cumulative)	1	b126	
807	$_{41}Nb \rightarrow ^{79}_{36}Kr$ (cumulative)	2	b126	
808	$_{41}Nb \rightarrow ^{81}_{37}Rb$ (cumulative)	1	b126	
809	$_{41}Nb \rightarrow ^{82m}_{37}Rb$	1	b126	
810	$_{41}Nb \rightarrow ^{83}_{37}Rb$	1	b126	
811	$_{41}Nb \rightarrow ^{84}_{37}Rb$	2	b126	
812	$_{41}Nb \rightarrow ^{82}_{38}Sr$ (cumulative)	1	b126	
813	$_{41}Nb \rightarrow ^{83}_{38}Sr$ (cumulative)	1	b127	
814	$_{41}Nb \rightarrow ^{84}_{39}Y$ (cumulative)	1	b127	
815	$_{41}Nb \rightarrow ^{86}_{39}Y$	1	b127	
816	$_{41}Nb \rightarrow ^{86m}_{39}Y$	1	b127	
817	$_{41}Nb \rightarrow ^{87}_{39}Y$	1	b127	

No.	Excitation function	No. of points	Table on page	Figure on page
818	$_{41}Nb \rightarrow {}^{87m}{}_{39}Y$ (cumulative)	1	b127	
819	$_{41}Nb \rightarrow {}^{88}{}_{39}Y$	1	b127	
820	$_{41}Nb \rightarrow {}^{90m}{}_{39}Y$	1	b127	
821	$_{41}Nb \rightarrow {}^{86}{}_{40}Zr$ (cumulative)	1	b127	
822	$_{41}Nb \rightarrow {}^{88}{}_{40}Zr$ (cumulative)	1	b127	
823	$_{41}Nb \rightarrow {}^{89}{}_{40}Zr$	15	b127	b275
824	$_{41}Nb \rightarrow {}^{89}{}_{40}Zr$ (cumulative)	1	b128	
825	$_{41}Nb \rightarrow {}^{89}{}_{41}Nb$	1	b128	
826	$_{41}Nb \rightarrow {}^{90}{}_{41}Nb$	1	b128	
827	$_{41}Nb \rightarrow {}^{92m}{}_{41}Nb$	20	b128	b275
828	$_{41}Nb \rightarrow {}^{93}{}_{42}Mo$	19	b129	b275
829	$_{41}Nb \rightarrow {}^{93m}{}_{42}Mo$	26	b129	b276
830	$_{42}Mo \rightarrow {}^{46}{}_{21}Sc$ (cumulative)	1	b130	
831	$_{42}Mo \rightarrow {}^{48}{}_{23}V$ (cumulative)	1	b130	
832	$_{42}Mo \rightarrow {}^{51}{}_{24}Cr$ (cumulative)	1	b130	
833	$_{42}Mo \rightarrow {}^{52}{}_{25}Mn$ (cumulative)	1	b130	
834	$_{42}Mo \rightarrow {}^{59}{}_{26}Fe$ (cumulative)	1	b130	
835	$_{42}Mo \rightarrow {}^{56}{}_{27}Co$ (cumulative)	1	b130	
836	$_{42}Mo \rightarrow {}^{57}{}_{27}Co$ (cumulative)	1	b130	
837	$_{42}Mo \rightarrow {}^{58}{}_{27}Co$ (cumulative)	1	b130	
838	$_{42}Mo \rightarrow {}^{65}{}_{30}Zn$ (cumulative)	1	b130	
839	$_{42}Mo \rightarrow {}^{71}{}_{33}As$ (cumulative)	1	b130	
840	$_{42}Mo \rightarrow {}^{74}{}_{33}As$	1	b131	
841	$_{42}Mo \rightarrow {}^{75}{}_{34}Se$ (cumulative)	1	b131	
842	$_{42}Mo \rightarrow {}^{77}{}_{35}Br$ (cumulative)	1	b131	
843	$_{42}Mo \rightarrow {}^{76}{}_{36}Kr$ (cumulative)	1	b131	
844	$_{42}Mo \rightarrow {}^{77}{}_{36}Kr$ (cumulative)	1	b131	
845	$_{42}Mo \rightarrow {}^{79}{}_{36}Kr$ (cumulative)	1	b131	
846	$_{42}Mo \rightarrow {}^{83}{}_{37}Rb$ (cumulative)	1	b131	
847	$_{42}Mo \rightarrow {}^{84}{}_{37}Rb$ (cumulative)	1	b131	
848	$_{42}Mo \rightarrow {}^{86}{}_{37}Rb$ (cumulative)	1	b131	
849	$_{42}Mo \rightarrow {}^{82}{}_{38}Sr$ (cumulative)	1	b131	
850	$_{42}Mo \rightarrow {}^{85}{}_{38}Sr$ (cumulative)	1	b131	
851	$_{42}Mo \rightarrow {}^{87}{}_{39}Y$ (cumulative)	1	b131	
852	$_{42}Mo \rightarrow {}^{88}{}_{40}Zr$ (cumulative)	1	b132	
853	$_{42}Mo \rightarrow {}^{89}{}_{40}Zr$ (cumulative)	1	b132	
854	$_{42}Mo \rightarrow {}^{91m}{}_{41}Nb$ (cumulative)	1	b132	
855	$_{42}Mo \rightarrow {}^{92m}{}_{41}Nb$	1	b132	
856	$_{42}Mo \rightarrow {}^{90}{}_{42}Mo$	39	b132	b276
857	$_{42}Mo \rightarrow {}^{93m}{}_{42}Mo$	42	b133	b276
858	$_{42}Mo \rightarrow {}^{99}{}_{42}Mo$	50	b134	b277
859	$_{42}Mo \rightarrow {}^{99}{}_{42}Mo$ (cumulative)	1	b135	
860	$_{42}Mo \rightarrow {}^{93}{}_{43}Tc$	46	b135	b277

No.	Excitation function	No. of points	Table on page	Figure on page
861	$_{42}Mo \rightarrow {}^{94}_{43}Tc$	50	b136	b277
862	$_{42}Mo \rightarrow {}^{95}_{43}Tc$	50	b137	b277
863	$_{42}Mo \rightarrow {}^{95m}_{43}Tc$	50	b138	b278
864	$_{42}Mo \rightarrow {}^{95m}_{43}Tc$ (cumulative)	1	b140	
865	$_{42}Mo \rightarrow {}^{96}_{43}Tc$	51	b140	b278
866	$_{42}Mo \rightarrow {}^{96}_{43}Tc$ (cumulative)	1	b141	
867	$_{42}Mo \rightarrow {}^{99m}_{43}Tc$	50	b141	b278
868	${}^{92}_{42}Mo \rightarrow {}^{88}_{40}Zr$	10	b142	
869	${}^{92}_{42}Mo \rightarrow {}^{88}_{41}Nb$	4	b142	
870	${}^{92}_{42}Mo \rightarrow {}^{88m}_{41}Nb$	4	b142	
871	${}^{92}_{42}Mo \rightarrow {}^{89}_{41}Nb$	17	b143	b278
872	${}^{92}_{42}Mo \rightarrow {}^{89m}_{41}Nb$	17	b143	b278
873	${}^{92}_{42}Mo \rightarrow {}^{89m+g}_{41}Nb$	17	b143	b279
874	${}^{92}_{42}Mo \rightarrow {}^{90}_{41}Nb$	7	b144	
875	${}^{92}_{42}Mo \rightarrow {}^{90}_{42}Mo$	4	b144	
876	${}^{94}_{42}Mo \rightarrow {}^{90}_{41}Nb$	22	b144	b279
877	${}^{94}_{42}Mo \rightarrow {}^{91}_{41}Nb$	1	b145	
878	${}^{94}_{42}Mo \rightarrow {}^{93}_{42}Mo$	1	b145	
879	${}^{94}_{42}Mo \rightarrow {}^{93m}_{42}Mo$	8	b145	
880	${}^{94}_{42}Mo \rightarrow {}^{93}_{43}Tc$	17	b145	b279
881	${}^{94}_{42}Mo \rightarrow {}^{93m}_{43}Tc$	18	b146	b279
882	${}^{94}_{42}Mo \rightarrow {}^{94}_{43}Tc$	21	b146	b279
883	${}^{94}_{42}Mo \rightarrow {}^{94m}_{43}Tc$	21	b147	b280
884	${}^{95}_{42}Mo \rightarrow {}^{92m}_{41}Nb$	17	b147	b280
885	${}^{95}_{42}Mo \rightarrow {}^{93}_{43}Tc$	9	b147	
886	${}^{95}_{42}Mo \rightarrow {}^{93m}_{43}Tc$	7	b148	
887	${}^{95}_{42}Mo \rightarrow {}^{94}_{43}Tc$	19	b148	b280
888	${}^{95}_{42}Mo \rightarrow {}^{94m}_{43}Tc$	19	b148	b280
889	${}^{95}_{42}Mo \rightarrow {}^{95}_{43}Tc$	46	b149	b281
890	${}^{95}_{42}Mo \rightarrow {}^{95m}_{43}Tc$	46	b150	b281
891	${}^{95}_{42}Mo \rightarrow {}^{95m+g}_{43}Tc$	22	b151	b282
892	${}^{96}_{42}Mo \rightarrow {}^{72}_{30}Zn$ (cumulative)	1	b152	
893	${}^{96}_{42}Mo \rightarrow {}^{66}_{31}Ga$	1	b152	
894	${}^{96}_{42}Mo \rightarrow {}^{67}_{31}Ga$	1	b152	
895	${}^{96}_{42}Mo \rightarrow {}^{67}_{31}Ga$ (cumulative)	1	b152	
896	${}^{96}_{42}Mo \rightarrow {}^{72}_{31}Ga$	2	b152	
897	${}^{96}_{42}Mo \rightarrow {}^{73}_{31}Ga$ (cumulative)	1	b152	
898	${}^{96}_{42}Mo \rightarrow {}^{71}_{33}As$	1	b152	
899	${}^{96}_{42}Mo \rightarrow {}^{71}_{33}As$ (cumulative)	1	b152	
900	${}^{96}_{42}Mo \rightarrow {}^{72}_{33}As$	1	b152	
901	${}^{96}_{42}Mo \rightarrow {}^{74}_{33}As$	1	b152	
902	${}^{96}_{42}Mo \rightarrow {}^{72}_{34}Se$	1	b152	
903	${}^{96}_{42}Mo \rightarrow {}^{72}_{34}Se$ (cumulative)	1	b152	

No.	Excitation function	No. of points	Table on page	Figure on page
904	$^{96}_{42}Mo \rightarrow {}^{92m}_{41}Nb$	13	b153	b282
905	$^{96}_{42}Mo \rightarrow {}^{95m}_{41}Nb$	5	b153	
906	$^{96}_{42}Mo \rightarrow {}^{95m+g}_{41}Nb$	5	b153	
907	$^{96}_{42}Mo \rightarrow {}^{93}_{43}Tc$	6	b153	b282
908	$^{96}_{42}Mo \rightarrow {}^{93m}_{43}Tc$	6	b153	b282
909	$^{96}_{42}Mo \rightarrow {}^{93m+g}_{43}Tc$	7	b154	b283
910	$^{96}_{42}Mo \rightarrow {}^{94}_{43}Tc$	15	b154	b283
911	$^{96}_{42}Mo \rightarrow {}^{94m}_{43}Tc$	15	b154	b283
912	$^{96}_{42}Mo \rightarrow {}^{94m+g}_{43}Tc$	9	b155	b283
913	$^{96}_{42}Mo \rightarrow {}^{95}_{43}Tc$	29	b155	b283
914	$^{96}_{42}Mo \rightarrow {}^{95m}_{43}Tc$	29	b155	b284
915	$^{96}_{42}Mo \rightarrow {}^{95m+g}_{43}Tc$	10	b156	b284
916	$^{96}_{42}Mo \rightarrow {}^{96}_{43}Tc$	34	b156	b284
917	$^{97}_{42}Mo \rightarrow {}^{92m}_{41}Nb$	7	b157	
918	$^{97}_{42}Mo \rightarrow {}^{96}_{41}Nb$	7	b157	
919	$^{97}_{42}Mo \rightarrow {}^{95}_{43}Tc$	11	b158	
920	$^{97}_{42}Mo \rightarrow {}^{95m}_{43}Tc$	10	b158	
921	$^{97}_{42}Mo \rightarrow {}^{96}_{43}Tc$	20	b158	b284
922	$^{98}_{42}Mo \rightarrow {}^{95}_{41}Nb$	15	b159	b284
923	$^{98}_{42}Mo \rightarrow {}^{95m}_{41}Nb$	14	b159	b285
924	$^{98}_{42}Mo \rightarrow {}^{95m+g}_{41}Nb$	25	b160	b285
925	$^{98}_{42}Mo \rightarrow {}^{96}_{41}Nb$	5	b160	
926	$^{98}_{42}Mo \rightarrow {}^{97}_{41}Nb$	6	b160	
927	$^{98}_{42}Mo \rightarrow {}^{96}_{43}Tc$	11	b161	
928	$^{98}_{42}Mo \rightarrow {}^{98}_{43}Tc$	15	b161	b285
929	$^{100}_{42}Mo \rightarrow {}^{95}_{41}Nb$	5	b161	
930	$^{100}_{42}Mo \rightarrow {}^{95m}_{41}Nb$	5	b161	
931	$^{100}_{42}Mo \rightarrow {}^{95m+g}_{41}Nb$	8	b162	
932	$^{100}_{42}Mo \rightarrow {}^{96}_{41}Nb$	30	b162	b285
933	$^{100}_{42}Mo \rightarrow {}^{97}_{41}Nb$	39	b162	b286
934	$^{100}_{42}Mo \rightarrow {}^{99}_{42}Mo$	33	b163	b286
935	$^{100}_{42}Mo \rightarrow {}^{99m}_{43}Tc$	24	b164	b286
936	$^{100}_{42}Mo \rightarrow {}^{100}_{43}Tc$	23	b165	b287
937	$^{96}_{44}Ru \rightarrow {}^{72}_{30}Zn$ (cumulative)	1	b165	
938	$^{96}_{44}Ru \rightarrow {}^{66}_{31}Ga$	1	b165	
939	$^{96}_{44}Ru \rightarrow {}^{67}_{31}Ga$ (cumulative)	1	b165	
940	$^{96}_{44}Ru \rightarrow {}^{72}_{31}Ga$	1	b166	
941	$^{96}_{44}Ru \rightarrow {}^{73}_{31}Ga$ (cumulative)	1	b166	
942	$^{96}_{44}Ru \rightarrow {}^{71}_{33}As$ (cumulative)	1	b166	
943	$^{96}_{44}Ru \rightarrow {}^{72}_{33}As$	1	b166	
944	$^{96}_{44}Ru \rightarrow {}^{74}_{33}As$	1	b166	
945	$^{96}_{44}Ru \rightarrow {}^{72}_{34}Se$ (cumulative)	1	b166	

No.	Excitation function	No. of points	Table on page	Figure on page
946	${}^{100}_{44}\mathrm{Ru} \rightarrow {}^{100}_{45}\mathrm{Rh}$	7	b166	
947	${}^{101}_{44}\mathrm{Ru} \rightarrow {}^{101}_{45}\mathrm{Rh}$	10	b166	b287
948	${}_{45}\mathrm{Rh} \rightarrow {}^{22}_{11}\mathrm{Na}$	1	b167	
949	${}_{45}\mathrm{Rh} \rightarrow {}^{69}_{32}\mathrm{Ge}$	1	b167	b287
950	${}_{45}\mathrm{Rh} \rightarrow {}^{74}_{33}\mathrm{As}$	1	b167	
951	${}_{45}\mathrm{Rh} \rightarrow {}^{75}_{34}\mathrm{Se}$	1	b167	
952	${}_{45}\mathrm{Rh} \rightarrow {}^{83}_{37}\mathrm{Rb}$	1	b167	
953	${}_{45}\mathrm{Rh} \rightarrow {}^{84}_{37}\mathrm{Rb}$	1	b167	
954	${}_{45}\mathrm{Rh} \rightarrow {}^{85}_{38}\mathrm{Sr}$	1	b167	
955	${}_{45}\mathrm{Rh} \rightarrow {}^{87}_{39}\mathrm{Y}$	1	b167	
956	${}_{45}\mathrm{Rh} \rightarrow {}^{91}_{39}\mathrm{Y}$	1	b167	
957	${}_{45}\mathrm{Rh} \rightarrow {}^{88}_{40}\mathrm{Zr}$	1	b167	
958	${}_{45}\mathrm{Rh} \rightarrow {}^{89}_{40}\mathrm{Zr}$	1	b167	
959	${}_{45}\mathrm{Rh} \rightarrow {}^{95}_{40}\mathrm{Zr}$	1	b167	
960	${}_{45}\mathrm{Rh} \rightarrow {}^{90}_{41}\mathrm{Nb}$	1	b168	
961	${}_{45}\mathrm{Rh} \rightarrow {}^{92m}_{41}\mathrm{Nb}$	1	b168	
962	${}_{45}\mathrm{Rh} \rightarrow {}^{95}_{41}\mathrm{Nb}$	1	b168	
963	${}_{45}\mathrm{Rh} \rightarrow {}^{94}_{43}\mathrm{Tc}$	7	b168	b287
964	${}_{45}\mathrm{Rh} \rightarrow {}^{95}_{43}\mathrm{Tc}$	10	b168	b287
965	${}_{45}\mathrm{Rh} \rightarrow {}^{95m}_{43}\mathrm{Tc}$	9	b168	b288
966	${}_{45}\mathrm{Rh} \rightarrow {}^{96}_{43}\mathrm{Tc}$	11	b169	b288
967	${}_{45}\mathrm{Rh} \rightarrow {}^{97}_{44}\mathrm{Ru}$ (cumulative)	43	b169	b288
968	${}_{45}\mathrm{Rh} \rightarrow {}^{99}_{45}\mathrm{Rh}$	13	b170	b288
969	${}_{45}\mathrm{Rh} \rightarrow {}^{99m}_{45}\mathrm{Rh}$	13	b170	b289
970	${}_{45}\mathrm{Rh} \rightarrow {}^{100}_{45}\mathrm{Rh}$	49	b171	b289
971	${}_{45}\mathrm{Rh} \rightarrow {}^{101}_{45}\mathrm{Rh}$	24	b172	b289
972	${}_{45}\mathrm{Rh} \rightarrow {}^{101m}_{45}\mathrm{Rh}$	39	b172	b290
973	${}_{45}\mathrm{Rh} \rightarrow {}^{102}_{45}\mathrm{Rh}$	1	b173	
974	${}_{45}\mathrm{Rh} \rightarrow {}^{102m}_{45}\mathrm{Rh}$	1	b173	
975	${}_{45}\mathrm{Rh} \rightarrow {}^{100}_{46}\mathrm{Pd}$	28	b173	b291
976	${}_{45}\mathrm{Rh} \rightarrow {}^{101}_{46}\mathrm{Pd}$	55	b174	b290
977	${}_{45}\mathrm{Rh} \rightarrow {}^{103}_{46}\mathrm{Pd}$	10	b175	b291
978	${}_{46}\mathrm{Pd} \rightarrow {}^{81}_{36}\mathrm{Kr}$	3	b176	
979	${}_{46}\mathrm{Pd} \rightarrow {}^{85}_{36}\mathrm{Kr}$	2	b176	
980	${}_{46}\mathrm{Pd} \rightarrow {}^{109}_{46}\mathrm{Pd}$	13	b176	b291
981	${}^{104}_{46}\mathrm{Pd} \rightarrow {}^{104}_{47}\mathrm{Ar}$	4	b176	
982	${}^{105}_{46}\mathrm{Pd} \rightarrow {}^{105}_{47}\mathrm{Ar}$	5	b176	
983	${}^{108}_{46}\mathrm{Pd} \rightarrow {}^{108}_{47}\mathrm{Ar}$	5	b176	
984	${}^{110}_{46}\mathrm{Pd} \rightarrow {}^{110}_{47}\mathrm{Ar}$	4	b177	
985	${}_{47}\mathrm{Ag} \rightarrow {}^{3}_{1}\mathrm{H}$	2	b177	

No.	Excitation function	No. of points	Table on page	Figure on page
986	$_{47}Ag \rightarrow {}^{7}_{4}Be$	13	b177	b291
987	$_{47}Ag \rightarrow {}^{7}_{4}Be$ (cumulative)	2	b177	
988	$_{47}Ag \rightarrow {}^{18}_{9}F$	6	b177	b292
989	$_{47}Ag \rightarrow {}^{24}_{10}Ne$ (cumulative)	3	b177	
990	$_{47}Ag \rightarrow {}^{22}_{11}Na$ (cumulative)	6	b178	
991	$_{47}Ag \rightarrow {}^{24}_{11}Na$	10	b178	b292
992	$_{47}Ag \rightarrow {}^{24}_{11}Na$ (cumulative)	7	b178	
993	$_{47}Ag \rightarrow {}^{27}_{12}Mg$ (cumulative)	1	b178	
994	$_{47}Ag \rightarrow {}^{28}_{12}Mg$	2	b178	
995	$_{47}Ag \rightarrow {}^{28}_{12}Mg$ (cumulative)	5	b178	
996	$_{47}Ag \rightarrow {}^{29}_{13}Al$ (cumulative)	1	b179	
997	$_{47}Ag \rightarrow {}^{32}_{15}P$	1	b179	
998	$_{47}Ag \rightarrow {}^{38}_{17}Cl$	2	b179	
999	$_{47}Ag \rightarrow {}^{38}_{17}Cl$ (cumulative)	1	b179	
1000	$_{47}Ag \rightarrow {}^{39}_{17}Cl$ (cumulative)	1	b179	
1001	$_{47}Ag \rightarrow {}^{41}_{18}Ar$ (cumulative)	3	b179	
1002	$_{47}Ag \rightarrow {}^{42}_{19}K$	4	b179	
1003	$_{47}Ag \rightarrow {}^{43}_{19}K$	2	b179	
1004	$_{47}Ag \rightarrow {}^{43}_{19}K$ (cumulative)	2	b179	
1005	$_{47}Ag \rightarrow {}^{45}_{20}Ca$	2	b179	
1006	$_{47}Ag \rightarrow {}^{47}_{20}Ca$	2	b180	
1007	$_{47}Ag \rightarrow {}^{47}_{20}Ca$ (cumulative)	1	b180	
1008	$_{47}Ag \rightarrow {}^{43}_{21}Sc$ (cumulative)	3	b180	
1009	$_{47}Ag \rightarrow {}^{44}_{21}Sc$	5	b180	
1010	$_{47}Ag \rightarrow {}^{44m}_{21}Sc$	7	b180	
1011	$_{47}Ag \rightarrow {}^{46}_{21}Sc$	6	b180	
1012	$_{47}Ag \rightarrow {}^{47}_{21}Sc$	5	b180	
1013	$_{47}Ag \rightarrow {}^{48}_{21}Sc$	5	b181	
1014	$_{47}Ag \rightarrow {}^{49}_{21}Sc$ (cumulative)	2	b181	
1015	$_{47}Ag \rightarrow {}^{45}_{22}Ti$ (cumulative)	2	b181	
1016	$_{47}Ag \rightarrow {}^{48}_{23}V$ (cumulative)	6	b181	
1017	$_{47}Ag \rightarrow {}^{49}_{23}V$ (cumulative)	2	b181	
1018	$_{47}Ag \rightarrow {}^{48}_{24}Cr$	2	b181	
1019	$_{47}Ag \rightarrow {}^{49}_{24}Cr$	1	b181	
1020	$_{47}Ag \rightarrow {}^{49}_{24}Cr$ (cumulative)	1	b181	
1021	$_{47}Ag \rightarrow {}^{51}_{24}Cr$ (cumulative)	2	b181	
1022	$_{47}Ag \rightarrow {}^{52}_{25}Mn$	4	b182	
1023	$_{47}Ag \rightarrow {}^{52}_{25}Mn$ (cumulative)	1	b182	
1024	$_{47}Ag \rightarrow {}^{52m}_{25}Mn$	2	b182	
1025	$_{47}Ag \rightarrow {}^{52m}_{25}Mn$ (cumulative)	1	b182	
1026	$_{47}Ag \rightarrow {}^{54}_{25}Mn$	7	b182	
1027	$_{47}Ag \rightarrow {}^{56}_{25}Mn$	1	b182	
1028	$_{47}Ag \rightarrow {}^{55}_{26}Fe$ (cumulative)	2	b182	

No.	Excitation function	No. of points	Table on page	Figure on page
1029	$_{47}Ag \rightarrow {}^{59}_{26}Fe$ (cumulative)	7	b182	
1030	$_{47}Ag \rightarrow {}^{55}_{27}Co$	2	b183	
1031	$_{47}Ag \rightarrow {}^{55}_{27}Co$ (cumulative)	2	b183	
1032	$_{47}Ag \rightarrow {}^{56}_{27}Co$	3	b183	
1033	$_{47}Ag \rightarrow {}^{56}_{27}Co$ (cumulative)	3	b183	
1034	$_{47}Ag \rightarrow {}^{57}_{27}Co$ (cumulative)	6	b183	
1035	$_{47}Ag \rightarrow {}^{58}_{27}Co$	6	b183	
1036	$_{47}Ag \rightarrow {}^{60}_{27}Co$	7	b183	
1037	$_{47}Ag \rightarrow {}^{61}_{27}Co$ (cumulative)	2	b184	
1038	$_{47}Ag \rightarrow {}^{60}_{29}Cu$ (cumulative)	3	b184	
1039	$_{47}Ag \rightarrow {}^{61}_{29}Cu$ (cumulative)	2	b184	
1040	$_{47}Ag \rightarrow {}^{64}_{29}Cu$	2	b184	
1041	$_{47}Ag \rightarrow {}^{67}_{29}Cu$ (cumulative)	2	b184	
1042	$_{47}Ag \rightarrow {}^{65}_{30}Zn$ (cumulative)	7	b184	
1043	$_{47}Ag \rightarrow {}^{69m}_{30}Zn$	1	b184	
1044	$_{47}Ag \rightarrow {}^{66}_{31}Ga$ (cumulative)	4	b184	
1045	$_{47}Ag \rightarrow {}^{67}_{31}Ga$ (cumulative)	5	b185	
1046	$_{47}Ag \rightarrow {}^{72}_{31}Ga$	2	b185	
1047	$_{47}Ag \rightarrow {}^{67}_{32}Ge$ (cumulative)	1	b185	
1048	$_{47}Ag \rightarrow {}^{69}_{32}Ge$ (cumulative)	1	b185	
1049	$_{47}Ag \rightarrow {}^{70}_{33}As$ (cumulative)	1	b185	
1050	$_{47}Ag \rightarrow {}^{71}_{33}As$ (cumulative)	6	b185	
1051	$_{47}Ag \rightarrow {}^{72}_{33}As$	6	b185	
1052	$_{47}Ag \rightarrow {}^{73}_{33}As$ (cumulative)	2	b185	
1053	$_{47}Ag \rightarrow {}^{74}_{33}As$	7	b186	
1054	$_{47}Ag \rightarrow {}^{72}_{34}Se$	2	b186	
1055	$_{47}Ag \rightarrow {}^{72}_{34}Se$ (cumulative)	2	b186	
1056	$_{47}Ag \rightarrow {}^{73}_{34}Se$ (cumulative)	4	b186	
1057	$_{47}Ag \rightarrow {}^{75}_{34}Se$ (cumulative)	7	b186	
1058	$_{47}Ag \rightarrow {}^{74m}_{35}Br$	1	b186	
1059	$_{47}Ag \rightarrow {}^{75}_{35}Br$ (cumulative)	1	b186	
1060	$_{47}Ag \rightarrow {}^{76}_{35}Br$	4	b187	
1061	$_{47}Ag \rightarrow {}^{77}_{35}Br$ (cumulative)	3	b187	
1062	$_{47}Ag \rightarrow {}^{76}_{36}Kr$ (cumulative)	2	b187	
1063	$_{47}Ag \rightarrow {}^{77}_{36}Kr$ (cumulative)	1	b187	
1064	$_{47}Ag \rightarrow {}^{79}_{36}Kr$ (cumulative)	1	b187	
1065	$_{47}Ag \rightarrow {}^{81}_{36}Kr$	6	b187	
1066	$_{47}Ag \rightarrow {}^{85}_{36}Kr$	5	b187	
1067	$_{47}Ag \rightarrow {}^{79}_{37}Rb$ (cumulative)	2	b187	
1068	$_{47}Ag \rightarrow {}^{81}_{37}Rb$ (cumulative)	4	b188	
1069	$_{47}Ag \rightarrow {}^{82m}_{37}Rb$	4	b188	
1070	$_{47}Ag \rightarrow {}^{83}_{37}Rb$	2	b188	
1071	$_{47}Ag \rightarrow {}^{83}_{37}Rb$ (cumulative)	7	b188	

No.	Excitation function	No. of points	Table on page	Figure on page
1072	$_{47}Ag \rightarrow {}^{84}_{37}Rb$	4	b188	
1073	$_{47}Ag \rightarrow {}^{84m}_{37}Rb$	3	b188	
1074	$_{47}Ag \rightarrow {}^{84m+g}_{37}Rb$	1	b188	
1075	$_{47}Ag \rightarrow {}^{82}_{38}Sr$ (cumulative)	6	b189	
1076	$_{47}Ag \rightarrow {}^{83}_{38}Sr$ (cumulative)	3	b189	
1077	$_{47}Ag \rightarrow {}^{85}_{38}Sr$	2	b189	
1078	$_{47}Ag \rightarrow {}^{84}_{39}Y$	2	b189	
1079	$_{47}Ag \rightarrow {}^{84}_{39}Y$ (cumulative)	1	b189	
1080	$_{47}Ag \rightarrow {}^{85}_{39}Y$ (cumulative)	1	b189	
1081	$_{47}Ag \rightarrow {}^{86m}_{39}Y$	4	b189	
1082	$_{47}Ag \rightarrow {}^{86m+g}_{39}Y$	1	b189	
1083	$_{47}Ag \rightarrow {}^{87}_{39}Y$	1	b189	
1084	$_{47}Ag \rightarrow {}^{87}_{39}Y$ (cumulative)	3	b190	
1085	$_{47}Ag \rightarrow {}^{87m}_{39}Y$ (cumulative)	4	b190	
1086	$_{47}Ag \rightarrow {}^{88}_{39}Y$	3	b190	
1087	$_{47}Ag \rightarrow {}^{90m}_{39}Y$	2	b190	
1088	$_{47}Ag \rightarrow {}^{85}_{40}Zr$ (cumulative)	1	b190	
1089	$_{47}Ag \rightarrow {}^{86}_{40}Zr$ (cumulative)	3	b190	
1090	$_{47}Ag \rightarrow {}^{88}_{40}Zr$ (cumulative)	7	b190	
1091	$_{47}Ag \rightarrow {}^{89}_{40}Zr$ (cumulative)	6	b190	
1092	$_{47}Ag \rightarrow {}^{88}_{41}Nb$ (cumulative)	1	b191	
1093	$_{47}Ag \rightarrow {}^{89}_{41}Nb$	2	b191	
1094	$_{47}Ag \rightarrow {}^{89}_{41}Nb$ (cumulative)	2	b191	
1095	$_{47}Ag \rightarrow {}^{89m}_{41}Nb$	1	b191	
1096	$_{47}Ag \rightarrow {}^{90}_{41}Nb$	1	b191	
1097	$_{47}Ag \rightarrow {}^{90}_{41}Nb$ (cumulative)	3	b191	
1098	$_{47}Ag \rightarrow {}^{92m}_{41}Nb$	2	b191	
1099	$_{47}Ag \rightarrow {}^{95}_{41}Nb$	2	b191	
1100	$_{47}Ag \rightarrow {}^{96}_{41}Nb$	1	b191	
1101	$_{47}Ag \rightarrow {}^{90}_{42}Mo$ (cumulative)	3	b192	
1102	$_{47}Ag \rightarrow {}^{93m}_{42}Mo$	4	b192	
1103	$_{47}Ag \rightarrow {}^{92}_{43}Tc$ (cumulative)	1	b192	
1104	$_{47}Ag \rightarrow {}^{93}_{43}Tc$ (cumulative)	1	b192	
1105	$_{47}Ag \rightarrow {}^{94}_{43}Tc$	2	b192	
1106	$_{47}Ag \rightarrow {}^{94}_{43}Tc$ (cumulative)	3	b192	
1107	$_{47}Ag \rightarrow {}^{94m}_{43}Tc$ (cumulative)	2	b192	
1108	$_{47}Ag \rightarrow {}^{95}_{43}Tc$ (cumulative)	2	b192	
1109	$_{47}Ag \rightarrow {}^{95m}_{43}Tc$	2	b192	
1110	$_{47}Ag \rightarrow {}^{96}_{43}Tc$	3	b193	
1111	$_{47}Ag \rightarrow {}^{95}_{44}Ru$ (cumulative)	3	b193	
1112	$_{47}Ag \rightarrow {}^{97}_{44}Ru$ (cumulative)	6	b193	
1113	$_{47}Ag \rightarrow {}^{103}_{44}Ru$ (cumulative)	6	b193	
1114	$_{47}Ag \rightarrow {}^{96}_{45}Rh$ (cumulative)	1	b193	

No.	Excitation function	No. of points	Table on page	Figure on page
1115	$_{47}Ag \rightarrow {}^{97m}{}_{45}Rh$	1	b193	
1116	$_{47}Ag \rightarrow {}^{99}{}_{45}Rh$	1	b193	
1117	$_{47}Ag \rightarrow {}^{99m}{}_{45}Rh$	1	b193	
1118	$_{47}Ag \rightarrow {}^{100}{}_{45}Rh$	4	b194	
1119	$_{47}Ag \rightarrow {}^{101}{}_{45}Rh$	4	b194	
1120	$_{47}Ag \rightarrow {}^{101m}{}_{45}Rh$ (cumulative)	2	b194	
1121	$_{47}Ag \rightarrow {}^{102}{}_{45}Rh$	3	b194	
1122	$_{47}Ag \rightarrow {}^{102m}{}_{45}Rh$	3	b194	
1123	$_{47}Ag \rightarrow {}^{106m}{}_{45}Rh$ (cumulative)	1	b194	
1124	$_{47}Ag \rightarrow {}^{99}{}_{46}Pd$ (cumulative)	3	b194	
1125	$_{47}Ag \rightarrow {}^{100}{}_{46}Pd$ (cumulative)	6	b194	
1126	$_{47}Ag \rightarrow {}^{101}{}_{46}Pd$ (cumulative)	4	b195	
1127	$_{47}Ag \rightarrow {}^{103}{}_{46}Pd$	2	b195	
1128	$_{47}Ag \rightarrow {}^{103}{}_{46}Pd$ (cumulative)	2	b195	
1129	$_{47}Ag \rightarrow {}^{102}{}_{47}Ag$ (cumulative)	1	b195	
1130	$_{47}Ag \rightarrow {}^{103}{}_{47}Ag$ (cumulative)	6	b195	
1131	$_{47}Ag \rightarrow {}^{104}{}_{47}Ag$	3	b195	
1132	$_{47}Ag \rightarrow {}^{104m}{}_{47}Ag$ (cumulative)	3	b195	
1133	$_{47}Ag \rightarrow {}^{105}{}_{47}Ag$ (cumulative)	7	b195	
1134	$_{47}Ag \rightarrow {}^{106}{}_{47}Ag$	24	b196	b292
1135	$_{47}Ag \rightarrow {}^{106m}{}_{47}Ag$	13	b196	b293
1136	$_{47}Ag \rightarrow {}^{107}{}_{48}Cd$	67	b197	
1137	${}^{107}{}_{47}Ag \rightarrow {}^{107}{}_{48}Cd$	19	b198	b293
1138	${}^{109}{}_{47}Ag \rightarrow {}^{109}{}_{48}Cd$	32	b199	b293
1139	$_{48}Cd \rightarrow {}^{22}{}_{11}Na$	1	b199	
1140	$_{48}Cd \rightarrow {}^{24}{}_{11}Na$	1	b199	
1141	$_{48}Cd \rightarrow {}^{24}{}_{11}Na$ (cumulative)	1	b199	
1142	$_{48}Cd \rightarrow {}^{28}{}_{12}Mg$ (cumulative)	1	b200	
1143	$_{48}Cd \rightarrow {}^{32}{}_{15}P$	1	b200	
1144	$_{48}Cd \rightarrow {}^{33}{}_{15}P$	1	b200	
1145	$_{48}Cd \rightarrow {}^{42}{}_{19}K$	1	b200	
1146	$_{48}Cd \rightarrow {}^{44m}{}_{21}Sc$	1	b200	
1147	$_{48}Cd \rightarrow {}^{48}{}_{21}Sc$	1	b200	
1148	$_{48}Cd \rightarrow {}^{48}{}_{23}V$	1	b200	
1149	$_{48}Cd \rightarrow {}^{52}{}_{25}Mn$	1	b200	
1150	$_{48}Cd \rightarrow {}^{109(m1+m2+g)}{}_{49}In$	34	b200	b293
1151	$_{48}Cd \rightarrow {}^{110m}{}_{49}In$	36	b201	b294
1152	$_{48}Cd \rightarrow {}^{111}{}_{49}In$	1	b202	
1153	$_{48}Cd \rightarrow {}^{111m+g}{}_{49}In$	37	b202	b294
1154	$_{48}Cd \rightarrow {}^{114m}{}_{49}In$	37	b203	b294
1155	${}^{106}{}_{48}Cd \rightarrow {}^{105}{}_{47}Ag$	1	b204	
1156	${}^{106}{}_{48}Cd \rightarrow {}^{105}{}_{48}Cd$	1	b204	

No.	Excitation function	No. of points	Table on page	Figure on page
1157	$^{108}_{48}Cd \rightarrow {}^{107}_{48}Cd$	1	b204	
1158	$^{108}_{48}Cd \rightarrow {}^{107}_{49}In$	1	b204	
1159	$^{110}_{48}Cd \rightarrow {}^{109}_{48}Cd$	4	b204	
1160	$^{110}_{48}Cd \rightarrow {}^{109}_{49}In$	13	b204	b294
1161	$^{110}_{48}Cd \rightarrow {}^{110}_{49}In$	28	b205	b295
1162	$^{110}_{48}Cd \rightarrow {}^{110m}_{49}In$	5	b205	
1163	$^{111}_{48}Cd \rightarrow {}^{108}_{49}In$	4	b205	
1164	$^{111}_{48}Cd \rightarrow {}^{109m+g}_{49}In$	10	b206	b295
1165	$^{111}_{48}Cd \rightarrow {}^{110}_{49}In$	23	b206	b295
1166	$^{111}_{48}Cd \rightarrow {}^{111}_{49}In$	70	b206	b295
1167	$^{112}_{48}Cd \rightarrow {}^{111}_{47}Ag$	1	b208	
1168	$^{112}_{48}Cd \rightarrow {}^{110}_{49}In$	3	b208	
1169	$^{112}_{48}Cd \rightarrow {}^{110m}_{49}In$	3	b208	
1170	$^{112}_{48}Cd \rightarrow {}^{111}_{49}In$	26	b208	b296
1171	$^{112}_{48}Cd \rightarrow {}^{112}_{49}In$	40	b209	b296
1172	$^{112}_{48}Cd \rightarrow {}^{112m}_{49}In$	6	b210	b296
1173	$^{113}_{48}Cd \rightarrow {}^{112}_{47}Ag$	1	b210	
1174	$^{113}_{48}Cd \rightarrow {}^{111}_{49}In$	30	b210	b297
1175	$^{113}_{48}Cd \rightarrow {}^{112}_{49}In$	7	b211	b297
1176	$^{113}_{48}Cd \rightarrow {}^{113m}_{49}In$	15	b211	b297
1177	$^{114}_{48}Cd \rightarrow {}^{113}_{47}Ag$	1	b211	
1178	$^{114}_{48}Cd \rightarrow {}^{111}_{49}In$	29	b212	b297
1179	$^{114}_{48}Cd \rightarrow {}^{113m}_{49}In$	11	b212	b297
1180	$^{114}_{48}Cd \rightarrow {}^{114}_{49}In$	25	b213	b298
1181	$^{114}_{48}Cd \rightarrow {}^{114m}_{49}In$	51	b213	b298
1182	$^{114}_{48}Cd \rightarrow {}^{115m}_{49}In$	6	b214	
1183	$^{116}_{48}Cd \rightarrow {}^{111}_{49}In$	6	b215	b298
1184	$^{116}_{48}Cd \rightarrow {}^{114m}_{49}In$	6	b215	b298
1185	$^{116}_{48}Cd \rightarrow {}^{115m}_{49}In$	7	b215	b299
1186	$^{116}_{48}Cd \rightarrow {}^{116m}_{49}In$	15	b215	b299
1187	$_{49}In \rightarrow {}^{3}_{1}H$	1	b215	
1188	$_{49}In \rightarrow {}^{7}_{4}Be$	1	b216	
1189	$_{49}In \rightarrow {}^{13}_{7}N$	3	b216	
1190	$_{49}In \rightarrow {}^{22}_{11}Na$	1	b216	
1191	$_{49}In \rightarrow {}^{24}_{11}Na$	1	b216	
1192	$_{49}In \rightarrow {}^{32}_{15}P$	1	b216	
1193	$_{49}In \rightarrow {}^{33}_{15}P$	1	b216	
1194	$_{49}In \rightarrow {}^{43}_{21}Sc$	1	b216	
1195	$_{49}In \rightarrow {}^{44}_{21}Sc$	1	b216	
1196	$_{49}In \rightarrow {}^{44m}_{21}Sc$	1	b216	
1197	$_{49}In \rightarrow {}^{46}_{21}Sc$	1	b216	
1198	$_{49}In \rightarrow {}^{47}_{21}Sc$	1	b216	

No.	Excitation function	No. of points	Table on page	Figure on page
1199	$_{49}In \rightarrow {}^{48}{}_{21}Sc$	1	b216	
1200	$_{49}In \rightarrow {}^{65}{}_{28}Ni$ (cumulative)	1	b217	
1201	$_{49}In \rightarrow {}^{66}{}_{28}Ni$ (cumulative)	1	b217	
1202	$_{49}In \rightarrow {}^{67}{}_{29}Cu$ (cumulative)	1	b217	
1203	$_{49}In \rightarrow {}^{72}{}_{30}Zn$ (cumulative)	1	b217	
1204	$_{49}In \rightarrow {}^{66}{}_{31}Ga$	1	b217	
1205	$_{49}In \rightarrow {}^{67}{}_{31}Ga$	1	b217	
1206	$_{49}In \rightarrow {}^{72}{}_{31}Ga$	1	b217	
1207	$_{49}In \rightarrow {}^{73}{}_{31}Ga$ (cumulative)	1	b217	
1208	$_{49}In \rightarrow {}^{66}{}_{32}Ge$ (cumulative)	1	b217	
1209	$_{49}In \rightarrow {}^{67}{}_{32}Ge$ (cumulative)	1	b217	
1210	$_{49}In \rightarrow {}^{71}{}_{33}As$ (cumulative)	1	b217	
1211	$_{49}In \rightarrow {}^{72}{}_{33}As$	1	b217	
1212	$_{49}In \rightarrow {}^{74}{}_{33}As$	1	b218	
1213	$_{49}In \rightarrow {}^{72}{}_{34}Se$	1	b218	
1214	$_{49}In \rightarrow {}^{101}{}_{47}Ag$	2	b218	
1215	$_{49}In \rightarrow {}^{115}{}_{47}Ag$	2	b218	
1216	$_{49}In \rightarrow {}^{107}{}_{48}Cd$	4	b218	
1217	$_{49}In \rightarrow {}^{109}{}_{48}Cd$	4	b218	
1218	$_{49}In \rightarrow {}^{115}{}_{48}Cd$	6	b218	
1219	$_{49}In \rightarrow {}^{115m}{}_{48}Cd$	6	b218	
1220	$_{49}In \rightarrow {}^{109}{}_{49}In$	4	b219	
1221	$_{49}In \rightarrow {}^{110m}{}_{49}In$	4	b219	
1222	$_{49}In \rightarrow {}^{111}{}_{49}In$	6	b219	
1223	$_{49}In \rightarrow {}^{113m}{}_{49}In$	4	b219	
1224	$_{49}In \rightarrow {}^{114m}{}_{49}In$	6	b219	
1225	$_{49}In \rightarrow {}^{115m}{}_{49}In$	6	b219	
1226	$_{49}In \rightarrow {}^{109}{}_{50}Sn$	15	b220	b299
1227	$_{49}In \rightarrow {}^{110}{}_{50}Sn$	23	b220	b299
1228	$_{49}In \rightarrow {}^{111}{}_{50}Sn$	26	b220	b299
1229	$_{49}In \rightarrow {}^{113(0.91m+g)}{}_{50}Sn$	31	b221	b300
1230	${}^{115}{}_{49}In \rightarrow {}^{115}{}_{48}Cd$	12	b222	b300
1231	${}^{115}{}_{49}In \rightarrow {}^{115m}{}_{48}Cd$	12	b222	b300
1232	$_{50}Sn \rightarrow {}^{3}{}_{1}H$	3	b223	
1233	$_{50}Sn \rightarrow {}^{18}{}_{9}F$	1	b223	
1234	$_{50}Sn \rightarrow {}^{22}{}_{11}Na$	1	b223	
1235	$_{50}Sn \rightarrow {}^{24}{}_{11}Na$	2	b223	
1236	$_{50}Sn \rightarrow {}^{32}{}_{15}P$	1	b223	
1237	$_{50}Sn \rightarrow {}^{33}{}_{15}P$	1	b223	
1238	$_{50}Sn \rightarrow {}^{120}{}_{49}In$	1	b223	
1239	$_{50}Sn \rightarrow {}^{121}{}_{49}In$	1	b223	
1240	$_{50}Sn \rightarrow {}^{123}{}_{49}In$	1	b223	

No.	Excitation function	No. of points	Table on page	Figure on page
1241	${}_{50}Sn \rightarrow {}^{124}_{49}In$	1	b223	
1242	${}_{50}Sn \rightarrow {}^{126}_{49}In$	1	b223	
1243	${}^{112}_{50}Sn \rightarrow {}^{113}_{51}Sb$	3	b224	
1244	${}^{116}_{50}Sn \rightarrow {}^{116}_{51}Sb$	10	b224	b300
1245	${}^{117}_{50}Sn \rightarrow {}^{117}_{51}Sb$	55	b224	b300
1246	${}^{118}_{50}Sn \rightarrow {}^{118}_{51}Sb$	35	b225	b301
1247	${}^{119}_{50}Sn \rightarrow {}^{119}_{51}Sb$	40	b226	b301
1248	${}^{120}_{50}Sn \rightarrow {}^{117m}_{49}In$	15	b227	b301
1249	${}^{120}_{50}Sn \rightarrow {}^{120}_{51}Sb$	54	b227	b301
1250	${}^{122}_{50}Sn \rightarrow {}^{122}_{51}Sb$	54	b229	b302
1251	${}^{124}_{50}Sn \rightarrow {}^{124}_{51}Sb$	34	b230	b302
1252	${}_{51}Sb \rightarrow {}^{7}_{4}Be$	1	b231	
1253	${}_{51}Sb \rightarrow {}^{18}_{9}F$	1	b231	
1254	${}_{51}Sb \rightarrow {}^{24}_{11}Na$	1	b231	
1255	${}_{51}Sb \rightarrow {}^{28}_{12}Mg$	1	b231	
1256	${}_{51}Sb \rightarrow {}^{31}_{14}Si$	1	b231	
1257	${}_{51}Sb \rightarrow {}^{32}_{15}P$	1	b231	
1258	${}_{51}Sb \rightarrow {}^{117}_{52}Te$	16	b231	b302
1259	${}_{51}Sb \rightarrow {}^{118}_{52}Te$	35	b231	b303
1260	${}_{51}Sb \rightarrow {}^{119}_{52}Te$	39	b232	b303
1261	${}_{51}Sb \rightarrow {}^{119m}_{52}Te$	39	b233	b303
1262	${}_{51}Sb \rightarrow {}^{121}_{52}Te$	30	b234	b303
1263	${}_{51}Sb \rightarrow {}^{121m}_{52}Te$	30	b235	b303
1264	${}^{121}_{51}Sb \rightarrow {}^{121}_{52}Te$	6	b236	
1265	${}^{121}_{51}Sb \rightarrow {}^{121m}_{52}Te$	7	b236	
1266	${}^{123}_{51}Sb \rightarrow {}^{123}_{52}Te$	6	b236	
1267	${}^{123}_{51}Sb \rightarrow {}^{123m}_{52}Te$	7	b236	
1268	${}_{52}Te \rightarrow {}^{121}_{53}I$	10	b236	b304
1269	${}_{52}Te \rightarrow {}^{123}_{53}I$	23	b237	b304
1270	${}_{52}Te \rightarrow {}^{124}_{53}I$	19	b237	b304
1271	${}_{52}Te \rightarrow {}^{125}_{53}I$	19	b238	b304
1272	${}_{52}Te \rightarrow {}^{126}_{53}I$	10	b238	b304
1273	${}_{52}Te \rightarrow {}^{128}_{53}I$	16	b238	b304
1274	${}_{52}Te \rightarrow {}^{130}_{53}I$	22	b239	b305
1275	${}^{122}_{52}Te \rightarrow {}^{121}_{52}Te$	1	b239	
1276	${}^{122}_{52}Te \rightarrow {}^{121}_{53}I$	1	b239	
1277	${}^{123}_{52}Te \rightarrow {}^{122}_{51}Sb$	1	b240	
1278	${}^{123}_{52}Te \rightarrow {}^{122}_{53}I$	10	b240	b305
1279	${}^{123}_{52}Te \rightarrow {}^{123}_{53}I$	22	b240	b305
1280	${}^{124}_{52}Te \rightarrow {}^{123m}_{52}Te$	1	b240	
1281	${}^{124}_{52}Te \rightarrow {}^{123}_{53}I$	12	b241	b305

No.	Excitation function	No. of points	Table on page	Figure on page
1282	${}^{124}_{52}Te \rightarrow {}^{124}_{53}I$	12	b241	b305
1283	${}^{125}_{52}Te \rightarrow {}^{119}_{51}Sb$	4	b241	
1284	${}^{125}_{52}Te \rightarrow {}^{120}_{51}Sb$	4	b241	
1285	${}^{125}_{52}Te \rightarrow {}^{122}_{51}Sb$	4	b241	
1286	${}^{125}_{52}Te \rightarrow {}^{124}_{51}Sb$	5	b242	
1287	${}^{125}_{52}Te \rightarrow {}^{123}_{53}I$	4	b242	
1288	${}^{125}_{52}Te \rightarrow {}^{124}_{53}I$	6	b242	
1289	${}^{125}_{52}Te \rightarrow {}^{125}_{53}I$	3	b242	
1290	${}^{125}_{52}Te \rightarrow {}^{126}_{53}I$	3	b242	
1291	${}^{126}_{52}Te \rightarrow {}^{119}_{51}Sb$	5	b242	
1292	${}^{126}_{52}Te \rightarrow {}^{120}_{51}Sb$	5	b243	
1293	${}^{126}_{52}Te \rightarrow {}^{122}_{51}Sb$	5	b243	
1294	${}^{126}_{52}Te \rightarrow {}^{125}_{51}Sb$	1	b243	
1295	${}^{126}_{52}Te \rightarrow {}^{125m}_{52}Te$	1	b243	
1296	${}^{126}_{52}Te \rightarrow {}^{123}_{53}I$	6	b243	
1297	${}^{126}_{52}Te \rightarrow {}^{124}_{53}I$	6	b243	
1298	${}^{126}_{52}Te \rightarrow {}^{125}_{53}I$	7	b243	
1299	${}^{126}_{52}Te \rightarrow {}^{126}_{53}I$	6	b244	
1300	${}^{128}_{52}Te \rightarrow {}^{127}_{51}Sb$	1	b244	
1301	${}^{128}_{52}Te \rightarrow {}^{127}_{52}Te$	1	b244	
1302	${}^{128}_{52}Te \rightarrow {}^{127m}_{52}Te$	1	b244	
1303	${}^{128}_{52}Te \rightarrow {}^{128}_{53}I$	12	b244	b306
1304	${}^{130}_{52}Te \rightarrow {}^{129}_{51}Sb$	1	b244	
1305	${}^{130}_{52}Te \rightarrow {}^{129}_{52}Te$	1	b245	
1306	${}^{130}_{52}Te \rightarrow {}^{129m}_{52}Te$	1	b245	
1307	${}^{130}_{52}Te \rightarrow {}^{130}_{53}I$	24	b245	b306
1308	${}^{130}_{52}Te \rightarrow {}^{131}_{53}I$	10	b245	b306

E [MeV]	ΔE [MeV]	σ [mb]	Δσ [mb]	Ref.

3 Excitation functions

3.1 Tables

445 Reaction $_{36}Kr \rightarrow {}^{77}_{35}Br$ (cumulative) (Fig. 329, p. 247)

E [MeV]	ΔE [MeV]	σ [mb]	Δσ [mb]	Ref.
11.50		0.450	0.18	Ste 91
15.20		1.040	0.16	Ste 91
18.80		2.420	0.27	Ste 91
22.00		2.760	0.27	Ste 91
25.70		3.460	0.38	Ste 91
29.00		2.660	0.27	Ste 91
32.80		3.570	0.34	Ste 91
36.80		5.260	0.45	Ste 91
40.90		7.440	0.67	Ste 91
43.70		9.290	0.80	Ste 91
45.90		11.590	0.98	Ste 91
48.40		12.510	1.05	Ste 91
51.00		13.410	1.12	Ste 91
53.30		14.210	1.19	Ste 91
56.00		16.770	1.38	Ste 91
58.30		17.910	1.48	Ste 91
60.80		20.110	1.64	Ste 91
64.40		21.580	1.76	Ste 91
66.70		21.160	1.72	Ste 91
70.00		22.410	1.82	Ste 91
73.20		24.450	1.99	Ste 91
76.20		24.120	1.96	Ste 91
78.60		25.790	2.09	Ste 91
82.20		28.150	2.28	Ste 91
85.20		29.000	2.36	Ste 91
88.90		30.060	2.44	Ste 91
91.90		30.420	2.46	Ste 91
95.80		33.080	2.68	Ste 91
97.80		34.150	2.76	Ste 91

446 Reaction $_{36}Kr \rightarrow {}^{82}_{35}Br$ (Fig. 330, p. 247)

E [MeV]	ΔE [MeV]	σ [mb]	Δσ [mb]	Ref.
22.00		0.600	0.90E-1	Ste 91
25.70		1.760	0.170	Ste 91
29.00		2.560	0.230	Ste 91
32.80		2.840	0.250	Ste 91
36.80		3.000	0.270	Ste 91
40.90		3.540	0.340	Ste 91
43.70		4.200	0.380	Ste 91

E [MeV]	ΔE [MeV]	σ [mb]	Δσ [mb]	Ref.

446 Reaction $_{36}Kr \rightarrow {}^{82}_{35}Br$ (continued)

E [MeV]	ΔE [MeV]	σ [mb]	Δσ [mb]	Ref.
45.90		4.530	0.410	Ste 91
48.40		5.120	0.450	Ste 91
51.00		5.660	0.500	Ste 91
53.30		6.320	0.540	Ste 91
56.00		6.730	0.570	Ste 91
58.30		6.920	0.580	Ste 91
60.80		8.060	0.670	Ste 91
64.40		8.300	0.700	Ste 91
66.70		8.440	0.700	Ste 91
70.00		9.160	0.780	Ste 91
73.20		9.720	0.810	Ste 91
76.20		10.170	0.860	Ste 91
78.60		10.720	0.880	Ste 91
82.20		11.330	0.930	Ste 91
85.20		11.650	0.960	Ste 91
88.90		12.510	1.040	Ste 91
91.90		12.170	1.000	Ste 91
95.80		13.080	1.080	Ste 91
97.80		13.160	1.080	Ste 91

447 Reaction $_{36}Kr \rightarrow {}^{77}_{36}Kr$ (Fig. 331, p. 247)

E [MeV]	ΔE [MeV]	σ [mb]	Δσ [mb]	Ref.
18.80		1.060	0.370	Ste 91
22.00		1.690	0.440	Ste 91
25.70		1.930	0.340	Ste 91
29.00		1.840	0.340	Ste 91
32.80		0.450	0.220	Ste 91
36.80		0.910	0.320	Ste 91
40.90		1.360	0.470	Ste 91
43.70		1.960	0.390	Ste 91
45.90		3.510	0.590	Ste 91
48.40		3.710	0.550	Ste 91
51.00		4.100	0.540	Ste 91
53.30		5.720	0.630	Ste 91
56.00		5.240	0.570	Ste 91
58.30		5.110	0.570	Ste 91
60.80		4.380	0.490	Ste 91
64.40		4.120	0.510	Ste 91
66.70		3.080	0.490	Ste 91
70.00		3.860	0.500	Ste 91
73.20		5.170	0.650	Ste 91
76.20		6.130	0.640	Ste 91
78.60		6.930	0.680	Ste 91
82.20		7.450	0.820	Ste 91
85.20		8.540	0.750	Ste 91
88.90		8.840	0.770	Ste 91

E [MeV]	ΔE [MeV]	σ [mb]	Δσ [mb]	Ref.
447 Reaction $_{36}Kr \rightarrow {}^{77}_{36}Kr$ (continued)				
91.90		8.930	0.760	Ste 91
95.80		10.460	0.910	Ste 91
97.80		10.470	1.120	Ste 91
448 Reaction $_{36}Kr \rightarrow {}^{79}_{36}Kr$ (cumulative) (Fig. 332, p. 247)				
15.20		1.520	0.38	Ste 91
18.80		10.490	1.00	Ste 91
22.00		16.800	1.45	Ste 91
25.70		18.740	1.59	Ste 91
29.00		15.020	1.30	Ste 91
32.80		9.630	0.91	Ste 91
36.80		8.620	0.93	Ste 91
40.90		13.540	1.31	Ste 91
43.70		22.490	1.96	Ste 91
45.90		30.100	2.57	Ste 91
48.40		36.840	3.07	Ste 91
51.00		44.950	3.75	Ste 91
53.30		50.970	4.19	Ste 91
56.00		53.080	4.35	Ste 91
58.30		52.190	4.24	Ste 91
60.80		50.440	4.11	Ste 91
64.40		51.660	4.29	Ste 91
66.70		58.430	4.77	Ste 91
70.00		63.380	5.20	Ste 91
73.20		74.410	6.07	Ste 91
76.20		79.800	6.55	Ste 91
78.60		86.260	6.99	Ste 91
82.20		83.050	0.74	Ste 91
85.20		81.660	6.64	Ste 91
88.90		81.650	6.64	Ste 91
91.90		75.750	6.19	Ste 91
95.80		75.300	6.17	Ste 91
97.80		74.070	6.02	Ste 91
449 Reaction $_{36}Kr \rightarrow {}^{85m}_{36}Kr$ (Fig. 333, p. 248)				
15.20		1.170	0.18	Ste 91
18.80		4.690	0.42	Ste 91
22.00		7.440	0.63	Ste 91
25.70		9.630	0.80	Ste 91
29.00		9.170	0.77	Ste 91
32.80		9.780	0.85	Ste 91
36.80		10.390	0.89	Ste 91
40.90		10.150	0.89	Ste 91
43.70		9.800	0.87	Ste 91

E [MeV]	ΔE [MeV]	σ [mb]	Δσ [mb]	Ref.

449 Reaction $_{36}Kr \rightarrow {}^{85m}{}_{36}Kr$ (continued)

E [MeV]	ΔE [MeV]	σ [mb]	Δσ [mb]	Ref.
45.90		10.750	0.92	Ste 91
48.40		10.640	0.91	Ste 91
51.00		10.640	0.91	Ste 91
53.30		10.600	0.91	Ste 91
56.00		10.920	0.94	Ste 91
58.30		10.680	0.92	Ste 91
60.80		10.530	0.90	Ste 91
64.40		11.070	1.00	Ste 91
66.70		9.890	0.85	Ste 91
70.00		9.840	0.83	Ste 91
73.20		9.660	0.82	Ste 91
76.20		9.920	0.89	Ste 91
78.60		10.030	0.84	Ste 91
82.20		9.920	0.86	Ste 91
85.20		9.180	0.76	Ste 91
88.90		9.330	0.77	Ste 91
91.90		8.830	0.73	Ste 91
95.80		9.340	0.77	Ste 91
97.80		8.820	0.74	Ste 91

450 Reaction $_{36}Kr \rightarrow {}^{79}{}_{37}Rb$ (Fig. 334, p. 248)

E [MeV]	ΔE [MeV]	σ [mb]	Δσ [mb]	Ref.
18.80		2.070	0.45	Ste 91
20.70		5.320	0.46	Ste 91
21.90		6.600	0.56	Ste 91
24.70		6.880	0.57	Ste 91
26.90		5.700	0.48	Ste 91
29.80		3.860	0.40	Ste 91
33.00		1.950	0.25	Ste 91
35.60		1.670	0.22	Ste 91
39.10		1.780	0.16	Ste 91
41.90		2.760	0.25	Ste 91
43.70		4.070	1.02	Ste 91
44.20		4.860	0.43	Ste 91
45.90		6.960	1.10	Ste 91
46.80		7.460	0.66	Ste 91
48.40		8.480	1.63	Ste 91
49.40		9.440	0.83	Ste 91
51.00		10.050	1.29	Ste 91
51.80		9.170	0.80	Ste 91
53.30		10.860	1.52	Ste 91
54.60		9.030	0.79	Ste 91
56.00		9.130	1.26	Ste 91
56.90		9.170	0.80	Ste 91
58.30		8.600	1.09	Ste 91
59.50		8.800	0.77	Ste 91

E [MeV]	ΔE [MeV]	σ [mb]	Δσ [mb]	Ref.
450 Reaction $_{36}$Kr → $^{79}{}_{37}$Rb (continued)				
60.80		7.970	1.04	Ste 91
63.10		6.790	0.59	Ste 91
64.40		7.060	1.36	Ste 91
65.40		8.580	0.76	Ste 91
66.70		8.470	0.87	Ste 91
70.00		8.810	0.92	Ste 91
70.70		10.490	0.92	Ste 91
73.20		9.810	0.98	Ste 91
76.00		11.780	1.02	Ste 91
76.20		10.080	1.02	Ste 91
78.60		10.540	1.05	Ste 91
81.20		11.650	1.00	Ste 91
82.20		10.000	1.13	Ste 91
85.20		8.270	1.07	Ste 91
86.40		10.070	0.86	Ste 91
88.90		8.680	0.96	Ste 91
91.60		8.510	0.74	Ste 91
91.90		8.010	0.94	Ste 91
95.80		7.790	0.88	Ste 91
96.80		7.960	0.68	Ste 91
97.80		7.370	0.98	Ste 91
102.0		6.390	0.55	Ste 91
107.1		6.310	0.58	Ste 91
112.3		5.370	0.46	Ste 91
115.6		5.360	0.49	Ste 91
451 Reaction $_{36}$Kr → $^{81}{}_{37}$Rb (cumulative) **(Fig. 335, p. 248)**				
14.30		1.000	0.140	Ste 91
14.80		0.950		Kov 91*
15.20		14.090	1.190	Ste 91
16.50		20.000		Kov 91*
16.70		41.770	3.350	Ste 91
18.50		50.000		Kov 91*
18.80		66.610	5.340	Ste 91
18.80		73.050	5.870	Ste 91
19.80		55.000		Kov 91*
20.30		60.000		Kov 91*
20.70		81.660	6.510	Ste 91
21.90		84.310	6.760	Ste 91
22.00		60.000		Kov 91*
22.00		85.920	6.900	Ste 91
22.50		55.000		Kov 91*
24.70		86.580	6.940	Ste 91
25.00		50.000		Kov 91*
25.70		99.500	7.990	Ste 91

451 Reaction $_{36}Kr \rightarrow {}^{81}{}_{37}Rb$ (continued)

E [MeV]	ΔE [MeV]	σ [mb]	Δσ [mb]	Ref.
26.90		97.220	7.790	Ste 91
27.30		65.000		Kov 91*
29.00		98.390	7.900	Ste 91
29.60		70.000		Kov 91*
29.80		94.630	7.590	Ste 91
32.80		85.260	6.850	Ste 91
33.00		81.940	6.570	Ste 91
35.60		81.100	6.500	Ste 91
36.80		85.050	6.830	Ste 91
39.10		106.590	8.590	Ste 91
40.90		131.090	10.580	Ste 91
41.90		145.280	11.700	Ste 91
43.70		155.120	12.500	Ste 91
44.20		170.810	11.730	Ste 91
45.90		181.240	14.590	Ste 91
46.80		177.750	14.280	Ste 91
48.40		181.340	14.580	Ste 91
49.40		194.200	15.590	Ste 91
51.00		165.580	13.310	Ste 91
51.80		170.650	13.690	Ste 91
53.30		148.410	11.920	Ste 91
54.60		140.420	11.260	Ste 91
56.00		127.410	10.210	Ste 91
56.90		120.830	9.690	Ste 91
58.30		106.380	8.550	Ste 91
59.50		98.600	7.900	Ste 91
60.80		90.720	7.300	Ste 91
63.10		69.210	5.550	Ste 91
64.40		74.110	5.970	Ste 91
65.40		71.310	5.900	Ste 91
66.70		67.370	5.440	Ste 91
70.00		57.840	4.660	Ste 91
70.70		61.820	5.090	Ste 91
73.20		59.260	4.710	Ste 91
76.00		54.630	4.480	Ste 91
76.20		52.820	4.280	Ste 91
78.60		18.480	3.910	Ste 91
81.20		48.370	3.950	Ste 91
82.20		46.350	3.730	Ste 91
85.20		43.250	3.480	Ste 91
86.40		42.610	3.460	Ste 91
88.90		40.540	3.260	Ste 91
91.60		38.100	3.090	Ste 91
91.90		36.930	2.970	Ste 91
95.80		35.380	2.850	Ste 91
96.80		33.000	2.670	Ste 91

E [MeV]	ΔE [MeV]	σ [mb]	Δσ [mb]	Ref.

451 Reaction $_{36}Kr \rightarrow {}^{81}{}_{37}Rb$ (continued)

E [MeV]	ΔE [MeV]	σ [mb]	Δσ [mb]	Ref.
97.80		35.230	2.810	Ste 91
102.00		29.890	2.410	Ste 91
107.00		27.860	2.240	Ste 91
112.30		26.490	2.130	Ste 91
115.60		24.910	2.000	Ste 91

452 Reaction $_{36}Kr \rightarrow {}^{81m}{}_{37}Rb$ (Fig. 336, p.248)

E [MeV]	ΔE [MeV]	σ [mb]	Δσ [mb]	Ref.
16.70		19.210	2.01	Ste 91
18.80		29.820	2.83	Ste 91
20.70		34.240	2.86	Ste 91
21.90		35.440	3.04	Ste 91
24.70		36.900	3.10	Ste 91
26.90		46.960	3.92	Ste 91
29.80		54.500	4.63	Ste 91
33.00		47.900	3.93	Ste 91
35.60		45.970	3.87	Ste 91
39.10		64.380	5.22	Ste 91
41.90		92.620	7.50	Ste 91
44.20		112.650	9.14	Ste 91
46.80		112.260	9.08	Ste 91
49.40		128.770	10.41	Ste 91
51.80		108.260	8.72	Ste 91
54.60		91.970	7.41	Ste 91
56.90		81.310	6.59	Ste 91
59.50		68.390	5.55	Ste 91
63.10		49.860	4.11	Ste 91
65.40		46.180	3.93	Ste 91
70.70		39.710	3.74	Ste 91
76.00		35.640	3.12	Ste 91
81.20		34.080	3.09	Ste 91
86.40		27.080	2.40	Ste 91
91.60		26.090	2.36	Ste 91
96.80		19.480	1.80	Ste 91
102.0		19.290	1.80	Ste 91
107.1		18.400	1.76	Ste 91
112.3		16.740	1.64	Ste 91
115.6		15.910	1.58	Ste 91

453 Reaction $_{36}Kr \rightarrow {}^{82m}{}_{37}Rb$ (Fig. 337, p.248)

E [MeV]	ΔE [MeV]	σ [mb]	Δσ [mb]	Ref.
5.00		4.800		Kov 91*
5.20		2.610	0.29	Ste 91
7.70		15.320	1.29	Ste 91
8.00		12.000		Kov 91*
8.50		32.290	2.62	Ste 91

453 Reaction $_{36}Kr \rightarrow {}^{82m}{}_{37}Rb$ (continued)

E [MeV]	ΔE [MeV]	σ [mb]	Δσ [mb]	Ref.
10.80		26.000		Kov 91*
11.50		36.940	3.03	Ste 91
11.70		39.450	3.19	Ste 91
12.90		40.000		Kov 91*
14.00		50.000		Kov 91*
14.30		57.470	4.64	Ste 91
14.50		62.000		Kov 91*
15.20		73.220	5.94	Ste 91
16.30		55.000		Kov 91*
16.70		81.630	6.56	Ste 91
18.20		70.000		Kov 91*
18.80		76.980	6.23	Ste 91
19.50		60.000		Kov 91*
20.10		60.000		Kov 91*
20.70		74.560	5.99	Ste 91
21.80		60.000		Kov 91*
21.90		72.770	5.86	Ste 91
22.00		74.230	6.01	Ste 91
22.50		60.000		Kov 91*
24.70		68.640	5.53	Ste 91
25.00		50.000		Kov 91*
25.70		82.210	6.64	Ste 91
26.90		104.000	8.35	Ste 91
27.30		90.000		Kov 91*
29.00		159.410	12.82	Ste 91
29.60		160.000		Kov 91*
29.80		185.550	14.89	Ste 91
32.80		227.980	18.30	Ste 91
33.00		223.670	17.93	Ste 91
35.60		222.430	17.83	Ste 91
36.80		230.100	18.48	Ste 91
39.10		214.030	17.27	Ste 91
40.90		163.300	13.24	Ste 91
41.90		155.810	12.56	Ste 91
43.70		112.350	9.13	Ste 91
44.20		112.060	9.03	Ste 91
45.90		95.800	7.80	Ste 91
46.80		86.380	6.97	Ste 91
48.40		76.300	6.24	Ste 91
49.40		75.180	6.06	Ste 91
51.00		65.790	5.39	Ste 91
51.80		68.520	5.52	Ste 91
53.30		60.350	4.94	Ste 91
54.60		62.760	5.05	Ste 91
56.00		62.600	5.13	Ste 91
56.90		63.440	5.11	Ste 91

E [MeV]	ΔE [MeV]	σ [mb]	Δσ [mb]	Ref.
453 Reaction $_{36}Kr \rightarrow {}^{82m}{}_{37}Rb$ (continued)				
58.30		58.970	4.83	Ste 91
59.50		59.840	4.81	Ste 91
60.80		60.220	4.93	Ste 91
63.10		46.320	3.74	Ste 91
64.40		50.550	4.15	Ste 91
65.40		48.690	4.05	Ste 91
66.70		45.160	3.72	Ste 91
70.00		38.580	3.19	Ste 91
73.20		36.160	3.00	Ste 91
76.00		32.790	2.71	Ste 91
76.20		32.520	2.72	Ste 91
78.60		29.530	2.46	Ste 91
81.20		29.170	2.42	Ste 91
82.20		27.640	2.26	Ste 91
85.20		27.330	2.26	Ste 91
86.40		25.520	2.11	Ste 91
88.90		24.510	2.02	Ste 91
91.60		22.960	1.89	Ste 91
91.90		22.640	1.90	Ste 91
95.80		21.340	1.77	Ste 91
96.80		20.570	1.69	Ste 91
97.80		20.050	1.67	Ste 91
102.0		18.370	1.51	Ste 91
107.1		17.880	1.46	Ste 91
112.3		17.110	1.41	Ste 91
115.6		16.010	1.31	Ste 91
454 Reaction $_{36}Kr \rightarrow {}^{83}{}_{37}Rb$ (Fig. 338, p. 249)				
5.00		8.400		Kov 91*
5.20		17.610	2.04	Ste 91
7.70		55.650	5.24	Ste 91
8.00		33.000		Kov 91*
8.50		82.260	6.89	Ste 91
10.80		58.000		Kov 91*
11.50		100.020	9.38	Ste 91
11.70		99.120	8.15	Ste 91
12.90		85.000		Kov 91*
14.00		180.000		Kov 91*
14.30		260.850	21.08	Ste 91
14.50		235.000		Kov 91*
15.20		341.660	27.78	Ste 91
16.30		290.000		Kov 91*
16.70		449.470	36.21	Ste 91
18.20		390.000		Kov 91*
18.80		489.010	39.50	Ste 91

454 Reaction $_{36}$Kr → $^{83}_{37}$Rb (continued)

E [MeV]	ΔE [MeV]	σ [mb]	Δσ [mb]	Ref.
19.50		340.000		Kov 91*
20.10		395.000		Kov 91*
20.70		510.090	40.95	Ste 91
21.80		375.000		Kov 91*
21.90		490.300	39.44	Ste 91
22.00		500.860	40.43	Ste 91
22.50		390.000		Kov 91*
24.70		473.800	38.14	Ste 91
25.00		310.000		Kov 91*
25.70		462.780	37.40	Ste 91
26.90		365.940	29.48	Ste 91
27.30		275.000		Kov 91*
29.00		249.410	20.32	Ste 91
29.60		200.000		Kov 91*
29.80		203.980	16.58	Ste 91
32.80		131.440	10.95	Ste 91
33.00		120.140	9.87	Ste 91
35.60		109.000	8.84	Ste 91
36.80		113.440	9.56	Ste 91
39.10		126.860	10.51	Ste 91
40.90		122.480	10.48	Ste 91
41.90		128.370	10.51	Ste 91
43.70		124.100	10.47	Ste 91
44.20		124.980	10.27	Ste 91
45.90		122.180	10.34	Ste 91
46.80		119.030	9.75	Ste 91
48.40		118.010	10.15	Ste 91
49.40		114.040	9.38	Ste 91
51.00		98.210	8.45	Ste 91
51.80		103.850	8.61	Ste 91
53.30		88.220	7.70	Ste 91
54.60		82.520	6.85	Ste 91
56.00		75.890	6.74	Ste 91
56.90		74.620	6.20	Ste 91
58.30		69.940	6.69	Ste 91
59.50		65.510	5.55	Ste 91
60.80		57.250	5.57	Ste 91
63.10		54.370	5.01	Ste 91
64.40		56.160	5.58	Ste 91
65.40		48.020	4.11	Ste 91
66.70		42.870	4.07	Ste 91
70.00		34.850	3.44	Ste 91
70.70		45.160	4.04	Ste 91
73.20		33.850	3.61	Ste 91
76.00		39.920	3.60	Ste 91
76.20		31.100	3.22	Ste 91

E [MeV]	ΔE [MeV]	σ [mb]	Δσ [mb]	Ref.

454 Reaction $_{36}Kr \rightarrow {}^{83}_{37}Rb$ (continued)

E [MeV]	ΔE [MeV]	σ [mb]	Δσ [mb]	Ref.
78.60		32.660	3.50	Ste 91
81.20		33.600	3.48	Ste 91
82.20		27.530	3.50	Ste 91
85.20		26.540	3.91	Ste 91
86.40		30.080	2.91	Ste 91
88.90		25.530	3.44	Ste 91
91.60		29.660	2.77	Ste 91
91.90		26.220	3.26	Ste 91
95.80		23.320	3.33	Ste 91
96.80		27.180	2.70	Ste 91
97.80		23.790	3.17	Ste 91
102.0		23.080	2.25	Ste 91
107.1		25.600	2.54	Ste 91
112.3		21.710	2.31	Ste 91
115.6		21.280	2.11	Ste 91

455 Reaction $_{36}Kr \rightarrow {}^{84}_{37}Rb$ (cumulative) **(Fig. 339, p. 249)**

E [MeV]	ΔE [MeV]	σ [mb]	Δσ [mb]	Ref.
5.00		30.000		Kov 91*
5.20		81.120	6.60	Ste 91
7.70		254.350	20.51	Ste 91
8.00		150.000		Kov 91*
8.50		425.810	34.15	Ste 91
10.80		295.000		Kov 91*
11.50		434.830	34.99	Ste 91
11.70		448.670	35.97	Ste 91
12.90		280.000		Kov 91*
14.00		205.000		Kov 91*
14.30		313.210	25.13	Ste 91
14.50		220.000		Kov 91*
15.20		210.590	16.98	Ste 91
16.30		80.000		Kov 91*
16.70		135.640	10.92	Ste 91
18.20		60.000		Kov 91*
18.80		54.580	4.45	Ste 91
19.50		30.000		Kov 91*
20.10		35.000		Kov 91*
20.70		47.100	3.84	Ste 91
21.80		32.000		Kov 91*
21.90		43.200	3.51	Ste 91
22.00		40.800	3.36	Ste 91
22.50		30.000		Kov 91*
24.70		69.530	5.62	Ste 91
25.00		55.000		Kov 91*
25.70		100.200	8.12	Ste 91
26.90		117.300	9.44	Ste 91

455 Reaction $_{36}$Kr → $^{84}_{37}$Rb (continued)

E [MeV]	ΔE [MeV]	σ [mb]	Δσ [mb]	Ref.
27.30		80.000		Kov 91*
29.00		130.050	10.50	Ste 91
29.60		105.000		Kov 91*
29.80		141.980	11.42	Ste 91
32.80		140.580	11.33	Ste 91
33.00		140.450	11.30	Ste 91
35.60		117.300	9.43	Ste 91
36.80		108.490	8.77	Ste 91
39.10		100.230	8.11	Ste 91
40.90		72.010	5.90	Ste 91
41.90		71.140	5.77	Ste 91
43.70		53.250	4.42	Ste 91
44.20		55.460	4.51	Ste 91
45.90		49.160	4.08	Ste 91
46.80		46.420	3.78	Ste 91
48.40		41.590	3.52	Ste 91
49.40		41.090	3.35	Ste 91
51.00		35.530	3.01	Ste 91
51.80		38.920	3.18	Ste 91
53.30		31.370	2.66	Ste 91
54.60		33.100	2.71	Ste 91
56.00		31.160	2.65	Ste 91
56.90		30.980	2.53	Ste 91
58.30		25.600	2.23	Ste 91
59.50		28.930	2.38	Ste 91
60.80		24.500	2.18	Ste 91
63.10		21.430	1.78	Ste 91
64.40		23.090	2.00	Ste 91
65.40		23.730	2.02	Ste 91
66.70		21.550	1.83	Ste 91
70.00		19.120	1.63	Ste 91
70.70		21.600	1.83	Ste 91
73.20		18.950	1.64	Ste 91
76.00		19.390	1.65	Ste 91
76.20		17.320	1.51	Ste 91
78.60		17.420	1.54	Ste 91
81.20		18.130	1.54	Ste 91
82.20		16.830	1.69	Ste 91
85.20		15.170	1.42	Ste 91
86.40		16.290	1.39	Ste 91
88.90		15.070	1.43	Ste 91
91.60		15.040	1.28	Ste 91
91.90		13.100	1.29	Ste 91
95.80		12.380	1.29	Ste 91
96.80		13.700	1.17	Ste 91
97.80		12.710	1.25	Ste 91

E [MeV]	ΔE [MeV]	σ [mb]	Δσ [mb]	Ref.

455 Reaction $_{36}Kr \rightarrow {}^{84}{}_{37}Rb$ (continued)

E [MeV]	ΔE [MeV]	σ [mb]	Δσ [mb]	Ref.
102.0		12.670	1.08	Ste 91
107.1		12.320	1.05	Ste 91
112.3		11.780	1.01	Ste 91
115.6		11.030	0.95	Ste 91

456 Reaction $_{36}Kr \rightarrow {}^{84m}{}_{37}Rb$ (Fig. 340, p. 249)

E [MeV]	ΔE [MeV]	σ [mb]	Δσ [mb]	Ref.
5.20		8.010	0.68	Ste 91
7.70		35.880	2.91	Ste 91
8.50		43.200	3.47	Ste 91
11.50		96.020	7.74	Ste 91
11.70		98.040	7.87	Ste 91
14.30		88.710	7.12	Ste 91
15.20		66.040	5.36	Ste 91
16.70		38.030	3.05	Ste 91
18.80		23.550	1.89	Ste 91
20.70		16.050	1.29	Ste 91
21.90		13.170	1.06	Ste 91
22.00		13.140	1.13	Ste 91
24.70		22.010	1.77	Ste 91
25.70		38.190	6.64	Ste 91
26.90		42.080	3.37	Ste 91
29.00		57.830	4.67	Ste 91
29.80		56.610	4.54	Ste 91
32.80		64.490	5.20	Ste 91
33.00		60.280	4.83	Ste 91
35.60		54.000	4.32	Ste 91
36.80		51.860	4.21	Ste 91
39.10		34.130	2.92	Ste 91
40.90		34.660	2.93	Ste 91
41.90		29.850	2.54	Ste 91
43.70		22.390	1.91	Ste 91
44.20		24.830	2.10	Ste 91
45.90		18.830	1.74	Ste 91
46.80		17.360	1.47	Ste 91
48.40		15.430	1.46	Ste 91
49.40		16.550	1.39	Ste 91
51.00		14.170	1.40	Ste 91
51.80		13.550	1.14	Ste 91
53.30		10.980	1.26	Ste 91
54.60		10.840	0.91	Ste 91
56.00		10.000	1.12	Ste 91
56.90		10.820	0.91	Ste 91
58.30		9.580	1.02	Ste 91
59.50		9.460	0.79	Ste 91
60.80		9.250	1.12	Ste 91

E [MeV]	ΔE [MeV]	σ [mb]	Δσ [mb]	Ref.

456 Reaction $_{36}Kr \rightarrow {}^{84m}{}_{37}Rb$ (continued)

E [MeV]	ΔE [MeV]	σ [mb]	Δσ [mb]	Ref.
63.10		8.790	0.73	Ste 91
64.40		7.760	0.92	Ste 91
65.40		7.530	0.63	Ste 91
66.70		6.910	0.68	Ste 91
70.00		6.520	0.64	Ste 91
70.70		7.690	0.64	Ste 91
73.20		6.370	0.68	Ste 91
76.00		6.850	0.57	Ste 91
76.20		5.670	0.59	Ste 91
78.60		5.500	0.59	Ste 91
81.20		6.630	0.55	Ste 91
82.20		4.600	0.69	Ste 91
85.20		4.640	0.69	Ste 91
86.40		5.750	0.48	Ste 91
88.90		4.720	0.72	Ste 91
91.60		5.330	0.44	Ste 91
91.90		4.080	0.49	Ste 91
95.80		4.540	0.65	Ste 91
96.80		4.760	0.39	Ste 91
97.80		4.120	0.61	Ste 91
102.0		4.510	0.37	Ste 91
107.1		4.500	0.37	Ste 91
112.3		3.950	0.33	Ste 91
115.6		3.800	0.31	Ste 91

457 Reaction $_{36}Kr \rightarrow {}^{86}{}_{37}Rb$ (Fig. 341, p. 249)

E [MeV]	ΔE [MeV]	σ [mb]	Δσ [mb]	Ref.
5.20		23.720	2.57	Ste 91
7.70		82.650	7.35	Ste 91
11.50		87.740	7.63	Ste 91
15.20		18.140	2.24	Ste 91
18.80		11.410	1.61	Ste 91
22.00		6.310	1.23	Ste 91
25.70		6.440	1.43	Ste 91
29.00		6.710	1.63	Ste 91
32.80		7.040	1.68	Ste 91
36.80		5.950	1.41	Ste 91

458 Reaction ${}^{78}{}_{36}Kr \rightarrow {}^{75}{}_{35}Br$ (Fig. 342, p. 249)

E [MeV]	ΔE [MeV]	σ [mb]	Δσ [mb]	Ref.
13.00		20.000		Hel 91*
16.00		50.000		Hel 91*
18.00		75.000		Hel 91*
20.00		95.000		Hel 91*
21.00		110.000		Hel 91*
23.00		105.000		Hel 91*

E [MeV]	ΔE [MeV]	σ [mb]	Δσ [mb]	Ref.

458 Reaction $^{78}_{36}Kr \rightarrow ^{75}_{35}Br$ (continued)

E [MeV]	ΔE [MeV]	σ [mb]	Δσ [mb]	Ref.
25.20		75.000		Hel 91*
26.00		60.000		Hel 91*
30.00		45.000		Hel 91*

459 Reaction $^{78}_{36}Kr \rightarrow ^{77}_{36}Kr$ (Fig. 343, p. 250)

E [MeV]	ΔE [MeV]	σ [mb]	Δσ [mb]	Ref.
15.50		50.000		Hel 91*
17.50		950.000		Hel 91*
19.50		1300.000		Hel 91*
21.00		1600.000		Hel 91*
23.00		1250.000		Hel 91*
25.00		1100.000		Hel 91*
26.00		850.000		Hel 91*
29.00		830.000		Hel 91*

460 Reaction $^{78}_{36}Kr \rightarrow ^{77}_{37}Rb$ (Fig. 344, p. 250)

E [MeV]	ΔE [MeV]	σ [mb]	Δσ [mb]	Ref.
21.00		10.000		Hel 91*
22.20		180.000		Hel 91*
24.60		700.000		Hel 91*
26.20		950.000		Hel 91*
27.80		1200.000		Hel 91*
29.20		1400.000		Hel 91*

461 Reaction $^{80}_{36}Kr \rightarrow ^{76}_{35}Br$

E [MeV]	ΔE [MeV]	σ [mb]	Δσ [mb]	Ref.
23.00		9.500		Hel 91*
26.20		95.000		Hel 91*
29.00		200.000		Hel 91*
29.20		210.000		Hel 91*

462 Reaction $^{80}_{36}Kr \rightarrow ^{77}_{35}Br$ (Fig. 345, p. 250)

E [MeV]	ΔE [MeV]	σ [mb]	Δσ [mb]	Ref.
12.50		7.000		Hel 91*
15.20		15.000		Hel 91*
17.30		31.000		Hel 91*
19.20		49.000		Hel 91*
23.00		74.000		Hel 91*
26.00		70.000		Hel 91*
29.20		62.000		Hel 91*

463 Reaction $^{80}_{36}Kr \rightarrow ^{79}_{37}Rb$

E [MeV]	ΔE [MeV]	σ [mb]	Δσ [mb]	Ref.
19.20		3.000		Hel 91*
21.00		92.000		Hel 91*
22.80		240.000		Hel 91*

E [MeV]	ΔE [MeV]	σ [mb]	Δσ [mb]	Ref.

463 Reaction $^{80}_{36}Kr \rightarrow ^{79}_{37}Rb$ (continued)

E [MeV]	ΔE [MeV]	σ [mb]	Δσ [mb]	Ref.
26.20		420.000		Hel 91*
29.20		400.000		Hel 91*

464 Reaction $^{82}_{36}Kr \rightarrow ^{81}_{37}Rb$ (Fig. 346, p. 250)

E [MeV]	ΔE [MeV]	σ [mb]	Δσ [mb]	Ref.
15.00		34.000		Kov 91*
17.00		250.000		Kov 91*
18.80		380.000		Kov 91*
20.00		405.000		Kov 91*
21.30		380.000		Kov 91*
22.80		390.000		Kov 91*
23.00		400.000		Kov 91*
25.00		300.000		Kov 91*
27.50		280.000		Kov 91*
29.60		275.000		Kov 91*

465 Reaction $^{82}_{36}Kr \rightarrow ^{82m}_{37}Rb$ (Fig. 347, p. 250)

E [MeV]	ΔE [MeV]	σ [mb]	Δσ [mb]	Ref.
5.50		10.500		Kov 91*
7.50		110.000		Kov 91*
10.90		300.000		Kov 91*
11.80		225.000		Kov 91*
12.80		310.000		Kov 91*
13.40		275.000		Kov 91*
15.00		300.000		Kov 91*
17.00		205.000		Kov 91*
18.80		185.000		Kov 91*
20.00		160.000		Kov 91*
21.20		110.000		Kov 91*
22.90		120.000		Kov 91*
25.00		80.000		Kov 91*
27.50		65.000		Kov 91*
29.60		55.000		Kov 91*

466 Reaction $^{82}_{36}Kr \rightarrow ^{83}_{37}Rb$ (Fig. 348, p. 251)

E [MeV]	ΔE [MeV]	σ [mb]	Δσ [mb]	Ref.
5.50		20.000		Kov 91*
7.50		65.000		Kov 91*
10.90		110.000		Kov 91*
11.70		105.000		Kov 91*
12.80		120.000		Kov 91*
13.40		95.000		Kov 91*
15.00		80.000		Kov 91*
17.00		40.000		Kov 91*
18.70		22.000		Kov 91*
20.00		16.000		Kov 91*

E [MeV]	ΔE [MeV]	σ [mb]	Δσ [mb]	Ref.
466 Reaction $^{82}_{36}Kr \rightarrow {}^{83}_{37}Rb$ (continued)				
21.20		13.000		Kov 91*
22.90		11.000		Kov 91*
23.00		13.000		Kov 91*
25.10		7.000		Kov 91*
27.50		6.500		Kov 91*
29.60		7.000		Kov 91*
467 Reaction $^{83}_{36}Kr \rightarrow {}^{81}_{37}Rb$ (Fig. 349, p. 251)				
14.90		3.800		Kov 91*
15.00		9.000		Kov 91*
17.00		45.000		Kov 91*
18.90		75.000		Kov 91*
20.80		90.000		Kov 91*
22.00		100.000		Kov 91*
23.00		105.000		Kov 91*
25.50		125.000		Kov 91*
27.50		200.000		Kov 91*
29.60		220.000		Kov 91*
468 Reaction $^{83}_{36}Kr \rightarrow {}^{82m}_{37}Rb$ (Fig. 350, p. 251)				
7.00		20.000		Kov 91*
9.00		40.000		Kov 91*
10.50		65.000		Kov 91*
12.80		90.000		Kov 91*
14.90		190.000		Kov 91*
15.00		160.000		Kov 91*
17.00		265.000		Kov 91*
18.90		295.000		Kov 91*
20.50		300.000		Kov 91*
20.60		305.000		Kov 91*
22.00		305.000		Kov 91*
23.00		310.000		Kov 91*
25.50		290.000		Kov 91*
27.50		200.000		Kov 91*
29.60		120.000		Kov 91*
469 Reaction $^{83}_{36}Kr \rightarrow {}^{83}_{37}Rb$ (Fig. 351, p. 251)				
7.00		200.000		Kov 91*
9.00		310.000		Kov 91*
10.50		410.000		Kov 91*
12.80		430.000		Kov 91*
14.90		350.000		Kov 91*
15.00		310.000		Kov 91*

E [MeV]	ΔE [MeV]	σ [mb]	Δσ [mb]	Ref.

469 Reaction $^{83}_{36}$Kr → $^{83}_{37}$Rb (continued)

E [MeV]	ΔE [MeV]	σ [mb]	Δσ [mb]	Ref.
17.00		210.000		Kov 91*
18.90		35.000		Kov 91*
20.50		70.000		Kov 91*
22.00		70.000		Kov 91*
23.00		65.000		Kov 91*
25.50		55.000		Kov 91*
27.50		45.000		Kov 91*
29.60		40.000		Kov 91*

470 Reaction $^{83}_{36}$Kr → $^{84}_{37}$Rb (Fig. 352, p. 251)

E [MeV]	ΔE [MeV]	σ [mb]	Δσ [mb]	Ref.
7.00		10.500		Kov 91*
9.00		19.000		Kov 91*
10.50		25.000		Kov 91*
12.80		23.000		Kov 91*
14.90		13.000		Kov 91*
15.00		14.000		Kov 91*
17.00		7.000		Kov 91*
18.90		3.400		Kov 91*
20.50		2.900		Kov 91*
22.00		2.800		Kov 91*
23.00		2.300		Kov 91*
25.50		1.800		Kov 91*
27.50		1.800		Kov 91*
29.60		1.300		Kov 91*

471 Reaction $_{37}$Rb → $^{74}_{33}$As

E [MeV]	ΔE [MeV]	σ [mb]	Δσ [mb]	Ref.
1000		10.60	0.60	Ale 84

472 Reaction $_{37}$Rb → $^{77}_{35}$Br (cumulative)

E [MeV]	ΔE [MeV]	σ [mb]	Δσ [mb]	Ref.
1000		18.30	0.60	Ale 84

473 Reaction $_{37}$Rb → $^{76}_{36}$Kr (cumulative)

E [MeV]	ΔE [MeV]	σ [mb]	Δσ [mb]	Ref.
1000		1.000	0.20	Ale 84

474 Reaction $_{37}$Rb → $^{77}_{36}$Kr (cumulative)

E [MeV]	ΔE [MeV]	σ [mb]	Δσ [mb]	Ref.
1000		4.500	0.70	Ale 84

475 Reaction $_{37}$Rb → $^{79}_{36}$Kr (cumulative)

E [MeV]	ΔE [MeV]	σ [mb]	Δσ [mb]	Ref.
1000		14.000	1.20	Ale 84

E [MeV]	ΔE [MeV]	σ [mb]	Δσ [mb]	Ref.

476 Reaction $_{37}Rb \rightarrow {}^{84}_{37}Rb$ (Fig. 353, p. 252)

E [MeV]	ΔE [MeV]	σ [mb]	Δσ [mb]	Ref.
12.00		1.200		Sak 85*
15.00		5.800		Sak 85*
18.00		30.000		Sak 85*
21.00		70.000		Sak 85*
23.00		100.000		Sak 85*
30.00		130.000		Sak 85*
36.00		110.000		Sak 85*
41.00		120.000		Sak 85*
52.00		140.000		Sak 85*
1000		42.000	2.00	Ale 84

477 Reaction $_{37}Rb \rightarrow {}^{86}_{37}Rb$ (Fig. 354, p. 252)

E [MeV]	ΔE [MeV]	σ [mb]	Δσ [mb]	Ref.
14.50		2.300		Sak 85*
17.50		10.000		Sak 85*
21.00		18.000		Sak 85*
23.00		30.000		Sak 85*
30.00		40.000		Sak 85*
36.00		45.000		Sak 85*
41.00		40.000		Sak 85*
52.00		35.000		Sak 85*

478 Reaction $_{37}Rb \rightarrow {}^{82}_{38}Sr$ (Fig. 355, p. 252)

E [MeV]	ΔE [MeV]	σ [mb]	Δσ [mb]	Ref.
39.50		30.000		Zai 91*
41.50		45.000		Zai 91*
44.00		85.000		Zai 91*
45.50		90.000		Zai 91*
46.50		120.000		Zai 91*
49.00		180.000		Zai 91*
51.00		190.000		Zai 91*
52.50		160.000		Zai 91*
54.00		80.000		Zai 91*
56.50		75.000		Zai 91*
58.00		60.000		Zai 91*
60.00		60.000		Zai 91*
63.00		60.000		Zai 91*
68.00		70.000		Zai 91*
70.00		60.000		Zai 91*
72.00		50.000		Zai 91*
74.00		55.000		Zai 91*
76.00		50.000		Zai 91*
77.00		45.000		Zai 91*
81.00		45.000		Zai 91*
84.50		45.000		Zai 91*
85.50		50.000		Zai 91*

478 Reaction $_{37}Rb \rightarrow {}^{82}{}_{38}Sr$ (continued)

E [MeV]	ΔE [MeV]	σ [mb]	Δσ [mb]	Ref.
88.00		45.000		Zai 91*
89.50		45.000		Zai 91*
91.50		35.000		Zai 91*
92.50		40.000		Zai 91*
94.00		40.000		Zai 91*
95.00		40.000		Zai 91*
98.00		35.000		Zai 91*

479 Reaction $_{37}Rb \rightarrow {}^{83m+g}{}_{38}Sr$ (Fig. 356, p. 252)

E [MeV]	ΔE [MeV]	σ [mb]	Δσ [mb]	Ref.
23.00		1.000		Sak 85*
30.00		180.000		Sak 85*
33.00		280.000		Sak 85*
36.00		280.000		Sak 85*
39.00		200.000		Sak 85*
41.00		180.000		Sak 85*
44.00		120.000		Sak 85*
52.00		70.000		Sak 85*

480 Reaction $_{37}Rb \rightarrow {}^{85m}{}_{38}Sr$

E [MeV]	ΔE [MeV]	σ [mb]	Δσ [mb]	Ref.
18.00		39.000		Sak 85*
33.00		85.000		Sak 85*
38.50		80.000		Sak 85*
44.00		38.000		Sak 85*
52.00		45.000		Sak 85*

481 Reaction $_{37}Rb \rightarrow {}^{85m+g}{}_{38}Sr$ (Fig. 357, p. 252)

E [MeV]	ΔE [MeV]	σ [mb]	Δσ [mb]	Ref.
8.00		220.000		Sak 85*
12.00		310.000		Sak 85*
15.00		280.000		Sak 85*
18.00		105.000		Sak 85*
21.00		60.000		Sak 85*
23.00		60.000		Sak 85*
30.00		180.000		Sak 85*
36.00		180.000		Sak 85*

482 Reaction $_{37}Rb \rightarrow {}^{87m}{}_{38}Sr$ (Fig. 358, p. 253)

E [MeV]	ΔE [MeV]	σ [mb]	Δσ [mb]	Ref.
8.00		32.000		Sak 85*
12.00		30.000		Sak 85*
15.00		12.000		Sak 85*
18.00		5.300		Sak 85*
21.00		3.800		Sak 85*
23.00		3.400		Sak 85*

E [MeV]	ΔE [MeV]	σ [mb]	Δσ [mb]	Ref.

482 Reaction $_{37}$Rb → $^{87m}_{38}$Sr (continued)

E [MeV]	ΔE [MeV]	σ [mb]	Δσ [mb]	Ref.
33.00		1.600		Sak 85*
39.00		1.300		Sak 85*
44.00		1.200		Sak 85*
52.00		1.100		Sak 85*

483 Reaction $^{85}_{37}$Rb → $^{79}_{36}$Kr (Fig. 359, p. 253)

E [MeV]	ΔE [MeV]	σ [mb]	Δσ [mb]	Ref.
42.00		0.200		Hor 80*
45.00		7.300		Hor 80*
47.00		18.000		Hor 80*
48.00		25.000		Hor 80*
52.00		40.000		Hor 80*
53.00		50.000		Hor 80*
55.00		45.000		Hor 80*
58.00		50.000		Hor 80*
59.00		45.000		Hor 80*
61.00		50.000		Hor 80*
62.00		40.000		Hor 80*
63.00		35.000		Hor 80*
66.00		30.000		Hor 80*
67.00		35.000		Hor 80*
68.00		30.000		Hor 80*
69.00		25.000		Hor 80*
70.00		30.000		Hor 80*

484 Reaction $^{85}_{37}$Rb → $^{81}_{37}$Rb (Fig. 360, p. 253)

E [MeV]	ΔE [MeV]	σ [mb]	Δσ [mb]	Ref.
46.00		0.600		Hor 80*
48.00		1.400		Hor 80*
49.00		1.700		Hor 80*
50.00		3.800		Hor 80*
53.00		11.000		Hor 80*
54.00		18.000		Hor 80*
55.00		25.000		Hor 80*
58.00		40.000		Hor 80*
59.00		40.000		Hor 80*
60.00		50.000		Hor 80*
62.00		60.000		Hor 80*
63.00		60.000		Hor 80*
65.00		65.000		Hor 80*
66.00		80.000		Hor 80*
67.00		75.000		Hor 80*
68.00		90.000		Hor 80*
70.00		85.000		Hor 80*

E [MeV]	ΔE [MeV]	σ [mb]	Δσ [mb]	Ref.

485 Reaction $^{85}{}_{37}Rb \rightarrow {}^{81m}{}_{37}Rb$ (Fig. 361, p. 253)

E [MeV]	ΔE [MeV]	σ [mb]	Δσ [mb]	Ref.
48.00		1.300		Hor 80*
49.00		2.200		Hor 80*
50.00		4.200		Hor 80*
53.00		12.000		Hor 80*
54.00		25.000		Hor 80*
55.00		40.000		Hor 80*
58.00		60.000		Hor 80*
59.00		65.000		Hor 80*
60.00		80.000		Hor 80*
62.00		85.000		Hor 80*
63.00		95.000		Hor 80*
65.00		110.000		Hor 80*
66.00		125.000		Hor 80*
68.00		120.000		Hor 80*
70.00		115.000		Hor 80*

486 Reaction $^{85}{}_{37}Rb \rightarrow {}^{82m}{}_{37}Rb$ (Fig. 362, p. 253)

E [MeV]	ΔE [MeV]	σ [mb]	Δσ [mb]	Ref.
36.00		4.200		Hor 80*
41.00		17.000		Hor 80*
45.00		85.000		Hor 80*
47.00		110.000		Hor 80*
48.00		110.000		Hor 80*
49.00		120.000		Hor 80*
50.00		150.000		Hor 80*
52.00		170.000		Hor 80*
53.00		180.000		Hor 80*
55.00		170.000		Hor 80*
58.00		170.000		Hor 80*
59.00		160.000		Hor 80*
61.00		150.000		Hor 80*
62.00		140.000		Hor 80*
65.00		130.000		Hor 80*
68.00		120.000		Hor 80*
70.00		130.000		Hor 80*

487 Reaction $^{85}{}_{37}Rb \rightarrow {}^{83}{}_{37}Rb$ (Fig. 363, p. 254)

E [MeV]	ΔE [MeV]	σ [mb]	Δσ [mb]	Ref.
22.00		3.900		Hor 80*
24.80		12.000		Lev 91
25.70		35.000		Lev 91
26.00		30.000		Hor 80*
26.60		64.000		Lev 91
27.60		98.000		Lev 91
28.50		144.000		Lev 91
29.00		115.000		Hor 80*

E [MeV]	ΔE [MeV]	σ [mb]	Δσ [mb]	Ref.

487 Reaction $^{85}_{37}$Rb → $^{83}_{37}$Rb (continued)

E [MeV]	ΔE [MeV]	σ [mb]	Δσ [mb]	Ref.
29.50		186.000		Lev 91
34.00		220.000		Hor 80*
38.00		320.000		Hor 80*
41.00		350.000		Hor 80*
45.00		360.000		Hor 80*
47.00		330.000		Hor 80*
48.00		260.000		Hor 80*
50.00		280.000		Hor 80*
52.00		220.000		Hor 80*
53.00		260.000		Hor 80*
55.00		210.000		Hor 80*
58.00		200.000		Hor 80*
59.00		200.000		Hor 80*
61.00		190.000		Hor 80*
62.00		180.000		Hor 80*
67.00		190.000		Hor 80*
68.00		180.000		Hor 80*
70.00		170.000		Hor 80*

488 Reaction $^{85}_{37}$Rb → $^{84}_{37}$Rb (Fig. 364, p. 254)

E [MeV]	ΔE [MeV]	σ [mb]	Δσ [mb]	Ref.
14.00		3.500		Hor 80*
22.00		160.000		Hor 80*
26.00		280.000		Hor 80*
30.00		270.000		Hor 80*
32.00		280.000		Hor 80*
34.00		260.000		Hor 80*
38.00		220.000		Hor 80*
41.00		140.000		Hor 80*
45.00		160.000		Hor 80*
47.00		130.000		Hor 80*
49.00		120.000		Hor 80*
53.00		125.000		Hor 80*
56.00		125.000		Hor 80*
58.00		125.000		Hor 80*
59.00		130.000		Hor 80*
62.00		120.000		Hor 80*
65.00		110.000		Hor 80*
67.00		130.000		Hor 80*
68.00		120.000		Hor 80*
70.00		120.000		Hor 80*

489 Reaction $^{85}_{37}$Rb → $^{84m}_{37}$Rb (Fig. 365, p. 254)

E [MeV]	ΔE [MeV]	σ [mb]	Δσ [mb]	Ref.
13.80		0.800		Lev 91
14.00		4.700		Hor 80*

489 Reaction $^{85}_{37}Rb \rightarrow {}^{84m}_{37}Rb$ (continued)

E [MeV]	ΔE [MeV]	σ [mb]	Δσ [mb]	Ref.
14.80		1.600		Lev 91
15.80		6.300		Lev 91
16.70		15.000		Lev 91
17.50		30.000		Lev 91
18.30		36.000		Lev 91
19.30		44.000		Lev 91
20.30		70.000		Lev 91
21.40		83.000		Lev 91
22.00		130.000		Hor 80*
22.20		98.000		Lev 91
23.10		107.000		Lev 91
24.00		135.000		Lev 91
24.80		130.000		Lev 91
25.70		144.000		Lev 91
26.00		180.000		Hor 80*
26.60		146.000		Lev 91
27.60		156.000		Lev 91
28.50		142.000		Lev 91
29.50		144.000		Lev 91
30.00		190.000		Hor 80*
32.00		140.000		Hor 80*
34.00		150.000		Hor 80*
39.00		110.000		Hor 80*
41.00		100.000		Hor 80*
45.00		115.000		Hor 80*
47.00		85.000		Hor 80*
49.00		75.000		Hor 80*
53.00		85.000		Hor 80*
56.00		80.000		Hor 80*
58.00		85.000		Hor 80*
59.00		80.000		Hor 80*
62.00		80.000		Hor 80*
65.00		80.000		Hor 80*
67.00		80.000		Hor 80*
68.00		75.000		Hor 80*
70.00		80.000		Hor 80*

490 Reaction $^{85}_{37}Rb \rightarrow {}^{84m+g}_{37}Rb$ (Fig. 366, p. 254)

E [MeV]	ΔE [MeV]	σ [mb]	Δσ [mb]	Ref.
12.80		27.000		Lev 91
13.80		30.000		Lev 91
14.80		43.000		Lev 91
15.80		58.000		Lev 91
16.70		82.000		Lev 91
17.50		92.000		Lev 91
18.30		119.000		Lev 91

E [MeV]	ΔE [MeV]	σ [mb]	Δσ [mb]	Ref.

490 Reaction $^{85}_{37}Rb \rightarrow ^{84m+g}_{37}Rb$ (continued)

E [MeV]	ΔE [MeV]	σ [mb]	Δσ [mb]	Ref.
19.30		140.000		Lev 91
20.30		180.000		Lev 91
21.40		220.000		Lev 91
22.20		154.000		Lev 91
23.10		280.000		Lev 91
24.00		285.000		Lev 91
24.80		320.000		Lev 91
25.70		325.000		Lev 91
26.60		335.000		Lev 91
27.60		328.000		Lev 91
28.50		320.000		Lev 91
29.50		351.000		Lev 91

491 Reaction $^{85}_{37}Rb \rightarrow ^{81}_{38}Sr$ (Fig. 367, p. 255)

E [MeV]	ΔE [MeV]	σ [mb]	Δσ [mb]	Ref.
48.00		0.130		Hor 80*
53.00		1.300		Hor 80*
54.00		2.200		Hor 80*
56.00		6.000		Hor 80*
58.00		12.000		Hor 80*
59.00		13.000		Hor 80*
61.00		14.000		Hor 80*
63.00		20.000		Hor 80*
64.00		20.000		Hor 80*
65.00		25.000		Hor 80*
66.00		30.000		Hor 80*
68.00		30.000		Hor 80*
70.00		30.000		Hor 80*

492 Reaction $^{85}_{37}Rb \rightarrow ^{82}_{38}Sr$ (Fig. 368, p. 255)

E [MeV]	ΔE [MeV]	σ [mb]	Δσ [mb]	Ref.
34.00		12.000		Hor 80*
38.00		12.000		Hor 80*
41.00		50.000		Hor 80*
46.00		170.000		Hor 80*
47.00		220.000		Hor 80*
48.00		180.000		Hor 80*
49.00		175.000		Hor 80*
50.00		230.000		Hor 80*
53.00		180.000		Hor 80*
54.00		210.000		Hor 80*
56.00		170.000		Hor 80*
58.00		150.000		Hor 80*
59.00		125.000		Hor 80*
61.00		110.000		Hor 80*

E [MeV]	ΔE [MeV]	σ [mb]	Δσ [mb]	Ref.

492 Reaction $^{85}_{37}Rb \rightarrow ^{82}_{38}Sr$ (continued)

E [MeV]	ΔE [MeV]	σ [mb]	Δσ [mb]	Ref.
63.00		95.000		Hor 80*
65.00		85.000		Hor 80*
66.00		90.000		Hor 80*
68.00		70.000		Hor 80*
70.00		60.000		Hor 80*

493 Reaction $^{85}_{37}Rb \rightarrow ^{83}_{38}Sr$ (Fig. 369, p. 255)

E [MeV]	ΔE [MeV]	σ [mb]	Δσ [mb]	Ref.
22.00		0.270		Hor 80*
23.10		4.000		Lev 91
24.00		12.300		Lev 91
24.80		39.000		Lev 91
25.70		102.000		Lev 91
26.00		65.000		Hor 80*
26.60		178.000		Lev 91
27.60		270.000		Lev 91
28.50		347.000		Lev 91
29.50		468.000		Lev 91
30.00		230.000		Hor 80*
34.00		400.000		Hor 80*
39.00		410.000		Hor 80*
42.00		290.000		Hor 80*
46.00		210.000		Hor 80*
48.00		125.000		Hor 80*
49.00		90.000		Hor 80*
53.00		70.000		Hor 80*
56.00		60.000		Hor 80*
58.00		55.000		Hor 80*
60.00		50.000		Hor 80*
61.00		50.000		Hor 80*
63.00		50.000		Hor 80*
64.00		40.000		Hor 80*
65.00		45.000		Hor 80*
66.00		50.000		Hor 80*
68.00		45.000		Hor 80*
69.00		40.000		Hor 80*
70.00		45.000		Hor 80*

494 Reaction $^{85}_{37}Rb \rightarrow ^{85m}_{38}Sr$ (Fig. 370, p. 255)

E [MeV]	ΔE [MeV]	σ [mb]	Δσ [mb]	Ref.
7.70		153.000		Lev 91
8.60		201.000		Lev 91
9.50		270.000		Lev 91
10.40		260.000		Lev 91
11.30		272.000		Lev 91
12.10		249.000		Lev 91

E [MeV]	ΔE [MeV]	σ [mb]	Δσ [mb]	Ref.
494 Reaction $^{85}_{37}$Rb → $^{85m}_{38}$Sr (continued)				
12.80		215.000		Lev 91
13.80		134.000		Lev 91
14.80		71.000		Lev 91
15.80		49.000		Lev 91
16.70		37.000		Lev 91
17.50		32.000		Lev 91
18.30		26.000		Lev 91
19.30		19.000		Lev 91
20.30		18.000		Lev 91
21.40		16.000		Lev 91
22.20		16.000		Lev 91
23.10		14.000		Lev 91
24.00		15.000		Lev 91
24.80		13.200		Lev 91
25.70		12.700		Lev 91
26.60		13.500		Lev 91
27.60		11.500		Lev 91
28.50		10.500		Lev 91
29.50		11.000		Lev 91
495 Reaction $^{85}_{37}$Rb → $^{85m+g}_{38}$Sr (Fig. 371, p. 256)				
7.70		389.000		Lev 91
8.60		575.000		Lev 91
9.50		825.000		Lev 91
10.40		830.000		Lev 91
11.30		860.000		Lev 91
12.10		890.000		Lev 91
12.80		794.000		Lev 91
13.80		664.000		Lev 91
14.80		447.000		Lev 91
15.80		357.000		Lev 91
16.70		288.000		Lev 91
17.50		234.000		Lev 91
496 Reaction $^{87}_{37}$Rb → $^{87}_{38}$Sr				
2.60		0.400		Bla 51*
3.10		2.000		Bla 51*
3.70		7.000		Bla 51*
4.25		20.000		Bla 51*
4.80		40.000		Bla 51*
5.25		70.000		Bla 51*
5.75		120.000		Bla 51*
6.20		180.000		Bla 51*
6.70		200.000		Bla 51*

E [MeV]	ΔE [MeV]	σ [mb]	Δσ [mb]	Ref.
497 Reaction $_{38}Sr \rightarrow {}^{48}{}_{23}V$				
1000		1.100	0.17	Ale 91
498 Reaction $_{38}Sr \rightarrow {}^{51}{}_{24}Cr$ (cumulative)				
1000		3.600	0.57	Ale 91
499 Reaction $_{38}Sr \rightarrow {}^{52m+g}{}_{25}Mn$				
1000		1.100	0.18	Ale 91
500 Reaction $_{38}Sr \rightarrow {}^{54}{}_{25}Mn$				
1000		3.900	0.65	Ale 91
501 Reaction $_{38}Sr \rightarrow {}^{59}{}_{26}Fe$				
1000		0.850	0.14	Ale 91
502 Reaction $_{38}Sr \rightarrow {}^{56}{}_{27}Co$				
1000		1.100	0.18	Ale 91
503 Reaction $_{38}Sr \rightarrow {}^{57}{}_{27}Co$				
1000		2.400	0.40	Ale 91
504 Reaction $_{38}Sr \rightarrow {}^{58m+g}{}_{27}Co$				
1000		6.400	1.10	Ale 91
505 Reaction $_{38}Sr \rightarrow {}^{60m+g}{}_{27}Co$				
1000		3.000	0.52	Ale 91
506 Reaction $_{38}Sr \rightarrow {}^{65}{}_{30}Zn$ (cumulative)				
1000		13.800	2.40	Ale 91
507 Reaction $_{38}Sr \rightarrow {}^{71m+g}{}_{30}Zn$				
1000		0.300	0.50E-1	Ale 91
508 Reaction $_{38}Sr \rightarrow {}^{67}{}_{31}Ga$ (cumulative)				
1000		6.100	1.10	Ale 91

E [MeV]	ΔE [MeV]	σ [mb]	Δσ [mb]	Ref.

509 Reaction $_{38}Sr \rightarrow {}^{68}_{32}Ge$ (cumulative)

E [MeV]	ΔE [MeV]	σ [mb]	Δσ [mb]	Ref.
1000		9.000	2.00	Ale 91

510 Reaction $_{38}Sr \rightarrow {}^{69}_{32}Ge$ (cumulative)

E [MeV]	ΔE [MeV]	σ [mb]	Δσ [mb]	Ref.
1000		12.000	1.90	Ale 91

511 Reaction $_{38}Sr \rightarrow {}^{71}_{33}As$ (cumulative)

E [MeV]	ΔE [MeV]	σ [mb]	Δσ [mb]	Ref.
1000		9.700	1.80	Ale 91

512 Reaction $_{38}Sr \rightarrow {}^{74}_{33}As$

E [MeV]	ΔE [MeV]	σ [mb]	Δσ [mb]	Ref.
600.0		10.600	1.60	Fel 83
1000		7.600	1.20	Ale 91
1000		8.600	0.90	Ale 84

513 Reaction $_{38}Sr \rightarrow {}^{72}_{34}Se$ (cumulative)

E [MeV]	ΔE [MeV]	σ [mb]	Δσ [mb]	Ref.
1000		7.000	2.30	Ale 91

514 Reaction $_{38}Sr \rightarrow {}^{73m+g}_{34}Se$

E [MeV]	ΔE [MeV]	σ [mb]	Δσ [mb]	Ref.
1000		13.000	2.30	Ale 91

515 Reaction $_{38}Sr \rightarrow {}^{75}_{34}Se$

E [MeV]	ΔE [MeV]	σ [mb]	Δσ [mb]	Ref.
1000		18.000	2.80	Ale 91

516 Reaction $_{38}Sr \rightarrow {}^{75}_{35}Br$

E [MeV]	ΔE [MeV]	σ [mb]	Δσ [mb]	Ref.
1000		12.000	2.10	Ale 91

517 Reaction $_{38}Sr \rightarrow {}^{77m+g}_{35}Br$ (cumulative)

E [MeV]	ΔE [MeV]	σ [mb]	Δσ [mb]	Ref.
1000		17.000	2.70	Ale 91

518 Reaction $_{38}Sr \rightarrow {}^{82m+g}_{35}Br$

E [MeV]	ΔE [MeV]	σ [mb]	Δσ [mb]	Ref.
1000		2.700	0.45	Ale 91

519 Reaction $_{38}Sr \rightarrow {}^{76}_{36}Kr$ (cumulative)

E [MeV]	ΔE [MeV]	σ [mb]	Δσ [mb]	Ref.
1000		1.600	0.40	Ale 84

E [MeV]	ΔE [MeV]	σ [mb]	Δσ [mb]	Ref.

520 Reaction $_{38}Sr \rightarrow {}^{77}{}_{36}Kr$ (cumulative)

E [MeV]	ΔE [MeV]	σ [mb]	Δσ [mb]	Ref.
1000		6.500	1.20	Ale 84
1000		5.700	1.10	Ale 91

521 Reaction $_{38}Sr \rightarrow {}^{79m+g}{}_{36}Kr$ (cumulative)

E [MeV]	ΔE [MeV]	σ [mb]	Δσ [mb]	Ref.
1000		17.500	1.90	Ale 84
1000		17.000	3.10	Ale 91

522 Reaction $_{38}Sr \rightarrow {}^{85m}{}_{36}Kr$ (cumulative)

E [MeV]	ΔE [MeV]	σ [mb]	Δσ [mb]	Ref.
1000		1.950	0.32	Ale 91

523 Reaction $_{38}Sr \rightarrow {}^{81}{}_{37}Rb$ (cumulative)

E [MeV]	ΔE [MeV]	σ [mb]	Δσ [mb]	Ref.
1000		24.000	3.80	Ale 91

524 Reaction $_{38}Sr \rightarrow {}^{82m}{}_{37}Rb$

E [MeV]	ΔE [MeV]	σ [mb]	Δσ [mb]	Ref.
600.0		17.500	1.90	Fel 83
1000		23.000	3.80	Ale 91

525 Reaction $_{38}Sr \rightarrow {}^{83}{}_{37}Rb$

E [MeV]	ΔE [MeV]	σ [mb]	Δσ [mb]	Ref.
1000		25.000	4.50	Ale 91

526 Reaction $_{38}Sr \rightarrow {}^{84}{}_{37}Rb$

E [MeV]	ΔE [MeV]	σ [mb]	Δσ [mb]	Ref.
600.0		21.200	3.30	Fel 83
1000		20.500	4.50	Ale 84
1000		15.000	2.30	Ale 91

527 Reaction $_{38}Sr \rightarrow {}^{84m}{}_{37}Rb$

E [MeV]	ΔE [MeV]	σ [mb]	Δσ [mb]	Ref.
1000		13.700	2.40	Ale 91

528 Reaction $_{38}Sr \rightarrow {}^{86m+g}{}_{37}Rb$

E [MeV]	ΔE [MeV]	σ [mb]	Δσ [mb]	Ref.
1000		25.000	4.20	Ale 91

529 Reaction $_{38}Sr \rightarrow {}^{81}{}_{38}Sr$

E [MeV]	ΔE [MeV]	σ [mb]	Δσ [mb]	Ref.
1000		11.900	2.70	Ale 91

530 Reaction $_{38}Sr \rightarrow {}^{85m}{}_{38}Sr$

E [MeV]	ΔE [MeV]	σ [mb]	Δσ [mb]	Ref.
1000		7.700	1.20	Ale 91

E [MeV]	ΔE [MeV]	σ [mb]	Δσ [mb]	Ref.

531 Reaction $_{38}Sr \rightarrow {}^{87m}{}_{38}Sr$

600.0		14.400	4.60	Fel 83
1000		25.000	4.50	Ale 91

532 Reaction $_{38}Sr \rightarrow {}^{87}{}_{39}Y$ (cumulative)

600.0		2.000	0.40	Fel 83
1000		5.100	0.80	Ale 91

533 Reaction $_{38}Sr \rightarrow {}^{87m}{}_{39}Y$ (cumulative)

1000		24.000	4.30	Ale 91

534 Reaction $_{38}Sr \rightarrow {}^{88}{}_{39}Y$

600.0		1.500	0.30	Fel 83
1000		2.200	0.40	Ale 91

535 Reaction ${}^{86}{}_{38}Sr \rightarrow {}^{82m}{}_{37}Rb$

23.10		7.000		Lev 91
24.00		11.000		Lev 91
24.80		16.000		Lev 91
25.70		20.000		Lev 91
26.60		28.000		Lev 91
27.60		32.000		Lev 91
28.50		36.000		Lev 91
29.50		39.000		Lev 91

536 Reaction ${}^{86}{}_{38}Sr \rightarrow {}^{83}{}_{37}Rb$ (Fig. 372, p. 256)

14.80		3.500		Lev 91
16.70		6.000		Lev 91
17.50		13.000		Lev 91
18.30		19.000		Lev 91
19.30		31.000		Lev 91
20.30		32.000		Lev 91
21.40		36.000		Lev 91
22.20		35.000		Lev 91
23.10		34.000		Lev 91
24.00		33.000		Lev 91
24.80		30.000		Lev 91
25.70		23.000		Lev 91
26.60		21.000		Lev 91
27.60		18.000		Lev 91
28.50		15.000		Lev 91
29.50		14.000		Lev 91

E [MeV]	ΔE [MeV]	σ [mb]	Δσ [mb]	Ref.

537 Reaction $^{86}_{38}Sr \rightarrow {}^{85m+g}_{38}Sr$ (Fig. 373, p. 256)

E [MeV]	ΔE [MeV]	σ [mb]	Δσ [mb]	Ref.
17.50		52.000		Lev 91
18.30		155.000		Lev 91
19.30		300.000		Lev 91
20.30		300.000		Lev 91
21.40		400.000		Lev 91
22.20		498.000		Lev 91
23.10		562.000		Lev 91
24.00		613.000		Lev 91
24.80		646.000		Lev 91
25.70		525.000		Lev 91
26.60		560.000		Lev 91
27.60		549.000		Lev 91
28.50		479.000		Lev 91
29.50		617.000		Lev 91

538 Reaction $^{86}_{38}Sr \rightarrow {}^{85}_{39}Y$ (Fig. 374, p. 256)

E [MeV]	ΔE [MeV]	σ [mb]	Δσ [mb]	Ref.
16.70		66.000		Lev 91
17.50		87.000		Lev 91
18.30		182.000		Lev 91
19.30		247.000		Lev 91
20.30		282.000		Lev 91
21.40		320.000		Lev 91
22.20		320.000		Lev 91
23.10		327.000		Lev 91
24.00		327.000		Lev 91
24.80		303.000		Lev 91
25.70		311.000		Lev 91
26.60		253.000		Lev 91
27.60		260.000		Lev 91
28.50		238.000		Lev 91
29.50		209.000		Lev 91

539 Reaction $^{86}_{38}Sr \rightarrow {}^{85m}_{39}Y$ (Fig. 375, p. 256)

E [MeV]	ΔE [MeV]	σ [mb]	Δσ [mb]	Ref.
15.80		22.000		Lev 91
16.70		95.000		Lev 91
17.50		163.000		Lev 91
18.30		200.000		Lev 91
19.30		245.000		Lev 91
20.30		323.000		Lev 91
21.40		349.000		Lev 91
22.20		395.000		Lev 91
23.10		396.000		Lev 91
24.00		371.000		Lev 91
24.80		367.000		Lev 91

E [MeV]	ΔE [MeV]	σ [mb]	Δσ [mb]	Ref.

539 Reaction $^{86}_{38}Sr \rightarrow {}^{85m}_{39}Y$ (continued)

E [MeV]	ΔE [MeV]	σ [mb]	Δσ [mb]	Ref.
25.70		394.000		Lev 91
26.60		323.000		Lev 91
27.60		340.000		Lev 91
28.50		338.000		Lev 91
29.50		309.000		Lev 91

540 Reaction $^{86}_{38}Sr \rightarrow {}^{86}_{39}Y$ (Fig. 376, p. 256)

E [MeV]	ΔE [MeV]	σ [mb]	Δσ [mb]	Ref.
7.70		267.000		Lev 91
8.60		612.000		Lev 91
9.50		700.000		Lev 91
10.40		712.000		Lev 91
11.20		738.000	17.00	Del 63
11.30		800.000		Lev 91
12.10		760.000		Lev 91
12.80		795.000		Lev 91
13.80		790.000		Lev 91
14.80		810.000		Lev 91
15.80		732.000		Lev 91
16.70		562.000		Lev 91
17.50		432.000		Lev 91
18.30		350.000		Lev 91
19.30		257.000		Lev 91
20.30		159.000		Lev 91
21.40		135.000		Lev 91
22.20		117.000		Lev 91
23.10		96.000		Lev 91
24.00		86.000		Lev 91
24.80		76.000		Lev 91
25.70		73.000		Lev 91
26.60		72.000		Lev 91
27.60		62.000		Lev 91
28.50		64.000		Lev 91
29.50		62.000		Lev 91

541 Reaction $^{86}_{38}Sr \rightarrow {}^{86m}_{39}Y$ (Fig. 377, p. 257)

E [MeV]	ΔE [MeV]	σ [mb]	Δσ [mb]	Ref.
7.70		35.000		Lev 91
8.60		100.000		Lev 91
9.50		121.000		Lev 91
10.40		122.000		Lev 91
11.30		139.000		Lev 91
12.10		143.000		Lev 91
12.80		162.000		Lev 91
13.80		182.000		Lev 91
14.80		191.000		Lev 91

E [MeV]	ΔE [MeV]	σ [mb]	Δσ [mb]	Ref.

541 Reaction $^{86}_{38}Sr \rightarrow {}^{86m}_{39}Y$ (continued)

E [MeV]	ΔE [MeV]	σ [mb]	Δσ [mb]	Ref.
15.80		199.000		Lev 91
16.70		170.000		Lev 91
17.50		137.000		Lev 91
18.30		123.000		Lev 91
19.30		99.000		Lev 91
20.30		68.000		Lev 91
21.40		58.000		Lev 91
22.20		49.000		Lev 91
23.10		43.000		Lev 91
24.00		38.000		Lev 91
24.80		33.000		Lev 91
25.70		32.000		Lev 91
26.60		29.000		Lev 91
27.60		28.000		Lev 91
28.50		25.000		Lev 91
29.50		27.000		Lev 91

542 Reaction $^{87}_{38}Sr \rightarrow {}^{83}_{37}Rb$

E [MeV]	ΔE [MeV]	σ [mb]	Δσ [mb]	Ref.
21.40		6.000		Lev 91
22.20		14.000		Lev 91
23.10		20.000		Lev 91
24.00		22.000		Lev 91
24.80		28.000		Lev 91
25.70		35.000		Lev 91
26.60		42.000		Lev 91
27.60		45.000		Lev 91
28.50		49.000		Lev 91
29.50		48.000		Lev 91

543 Reaction $^{87}_{38}Sr \rightarrow {}^{84}_{37}Rb$

E [MeV]	ΔE [MeV]	σ [mb]	Δσ [mb]	Ref.
11.20		0.200E-1		Del 63

544 Reaction $^{87}_{38}Sr \rightarrow {}^{84m}_{37}Rb$ (Fig. 378, p. 257)

E [MeV]	ΔE [MeV]	σ [mb]	Δσ [mb]	Ref.
12.10		1.200		Lev 91
12.80		1.500		Lev 91
13.80		3.000		Lev 91
14.80		3.600		Lev 91
15.80		5.500		Lev 91
16.70		6.400		Lev 91
17.50		9.100		Lev 91
18.30		9.100		Lev 91
19.30		9.700		Lev 91
20.30		10.500		Lev 91

E [MeV]	ΔE [MeV]	σ [mb]	Δσ [mb]	Ref.
544 Reaction $^{87}_{38}Sr \rightarrow ^{84m}_{37}Rb$ (continued)				
21.40		9.700		Lev 91
22.20		8.600		Lev 91
23.10		8.900		Lev 91
24.00		7.800		Lev 91
545 Reaction $^{87}_{38}Sr \rightarrow ^{84m+g}_{37}Rb$ (Fig. 379, p. 257)				
12.10		2.200		Lev 91
12.80		3.300		Lev 91
13.80		6.000		Lev 91
14.80		8.000		Lev 91
15.80		9.500		Lev 91
16.70		13.000		Lev 91
17.50		15.000		Lev 91
18.30		17.000		Lev 91
19.30		19.000		Lev 91
20.30		21.000		Lev 91
21.40		22.000		Lev 91
22.20		21.000		Lev 91
23.10		19.000		Lev 91
24.00		17.000		Lev 91
24.80		13.000		Lev 91
25.70		12.000		Lev 91
26.60		12.000		Lev 91
27.60		10.800		Lev 91
28.50		9.700		Lev 91
29.50		10.300		Lev 91
546 Reaction $^{87}_{38}Sr \rightarrow ^{86}_{39}Y$ (Fig. 380, p. 257)				
14.80		10.000		Lev 91
15.80		38.000		Lev 91
16.70		170.000		Lev 91
17.50		248.000		Lev 91
18.30		449.000		Lev 91
19.30		492.000		Lev 91
20.30		522.000		Lev 91
21.40		562.000		Lev 91
22.20		610.000		Lev 91
23.10		590.000		Lev 91
24.00		600.000		Lev 91
24.80		480.000		Lev 91
25.70		415.000		Lev 91
26.60		440.000		Lev 91
27.60		352.000		Lev 91
28.50		335.000		Lev 91
29.50		292.000		Lev 91

E [MeV]	ΔE [MeV]	σ [mb]	Δσ [mb]	Ref.

547 Reaction $^{87}_{38}Sr \rightarrow ^{86m}_{39}Y$ (Fig. 381, p. 257)

E [MeV]	ΔE [MeV]	σ [mb]	Δσ [mb]	Ref.
14.80		2.200		Lev 91
15.80		20.000		Lev 91
16.70		114.000		Lev 91
17.50		187.000		Lev 91
18.30		300.000		Lev 91
19.30		342.000		Lev 91
20.30		362.000		Lev 91
21.40		390.000		Lev 91
22.20		410.000		Lev 91
23.10		422.000		Lev 91
24.00		427.000		Lev 91
24.80		412.000		Lev 91
25.70		410.000		Lev 91
26.60		380.000		Lev 91
27.60		350.000		Lev 91
28.50		335.000		Lev 91
29.50		295.000		Lev 91

548 Reaction $^{87}_{38}Sr \rightarrow ^{87}_{39}Y$ (Fig. 382, p. 258)

E [MeV]	ΔE [MeV]	σ [mb]	Δσ [mb]	Ref.
2.70		0.300		Bla 51*
3.10		1.600		Bla 51*
3.50		5.000		Bla 51*
3.90		11.000		Bla 51*
4.30		20.000		Bla 51*
4.70		40.000		Bla 51*
5.30		90.000		Bla 51*
5.60		120.000		Bla 51*
5.80		170.000		Bla 51*
6.10		210.000		Bla 51*
6.40		240.000		Bla 51*
6.60		290.000		Bla 51*
6.80		320.000		Bla 51*
7.70		120.000		Lev 91
8.60		220.000		Lev 91
9.50		240.000		Lev 91
10.40		240.000		Lev 91
11.30		250.000		Lev 91
12.10		270.000		Lev 91
12.80		250.000		Lev 91
13.80		220.000		Lev 91
14.80		190.000		Lev 91
15.80		145.000		Lev 91
16.70		98.000		Lev 91
17.50		84.000		Lev 91
18.30		67.000		Lev 91

E [MeV]	ΔE [MeV]	σ [mb]	Δσ [mb]	Ref.

549 Reaction $^{87}_{38}Sr \rightarrow {}^{87m}_{39}Y$ (Fig. 383, p. 258)

E [MeV]	ΔE [MeV]	σ [mb]	Δσ [mb]	Ref.
7.70		415.000		Lev 91
8.60		765.000		Lev 91
9.50		880.000		Lev 91
10.40		880.000		Lev 91
11.30		955.000		Lev 91
12.10		990.000		Lev 91
12.80		1020.000		Lev 91
13.80		977.000		Lev 91
14.80		905.000		Lev 91
15.80		685.000		Lev 91
16.70		517.000		Lev 91
17.50		435.000		Lev 91
18.30		327.000		Lev 91
19.30		252.000		Lev 91
20.30		160.000		Lev 91
21.40		115.000		Lev 91
22.20		110.000		Lev 91
23.10		110.000		Lev 91
24.00		107.000		Lev 91
24.80		87.000		Lev 91
25.70		78.000		Lev 91
26.60		85.000		Lev 91
27.60		60.000		Lev 91
28.50		80.000		Lev 91
29.50		70.000		Lev 91

550 Reaction $^{88}_{38}Sr \rightarrow {}^{83}_{37}Rb$ (Fig. 384, p. 258)

E [MeV]	ΔE [MeV]	σ [mb]	Δσ [mb]	Ref.
33.00		1.500	0.20	Sac 67
36.00		11.000	1.20	Sac 67
40.00		35.000	3.90	Sac 67
42.00		44.000	4.80	Sac 67
48.00		73.000	8.00	Sac 67
54.00		38.000	4.20	Sac 67
60.00		37.000	4.10	Sac 67
66.00		33.000	3.60	Sac 67
72.00		38.000	4.20	Sac 67
78.00		41.000	4.50	Sac 67
85.00		57.000	6.30	Sac 67

551 Reaction $^{88}_{38}Sr \rightarrow {}^{84}_{37}Rb$ (Fig. 385, p. 258)

E [MeV]	ΔE [MeV]	σ [mb]	Δσ [mb]	Ref.
25.00		3.000	0.30	Sac 67
29.00		14.000	1.50	Sac 67
33.00		27.000	3.00	Sac 67
36.00		32.000	3.50	Sac 67

E [MeV]	ΔE [MeV]	σ [mb]	Δσ [mb]	Ref.
551 Reaction $^{88}_{38}Sr \rightarrow {}^{84}_{37}Rb$ (continued)				
40.00		24.000	2.60	Sac 67
42.00		18.000	2.00	Sac 67
48.00		12.000	1.30	Sac 67
54.00		10.000	1.10	Sac 67
60.00		24.000	2.60	Sac 67
66.00		42.000	4.60	Sac 67
72.00		49.000	5.40	Sac 67
78.00		46.000	5.10	Sac 67
85.00		43.000	4.70	Sac 67
552 Reaction $^{88}_{38}Sr \rightarrow {}^{84m}_{37}Rb$				
23.10		2.400		Lev 91
24.00		3.300		Lev 91
24.80		5.500		Lev 91
25.70		7.000		Lev 91
26.60		8.000		Lev 91
27.60		10.000		Lev 91
28.50		11.000		Lev 91
29.50		12.000		Lev 91
553 Reaction $^{88}_{38}Sr \rightarrow {}^{84m+g}_{37}Rb$				
22.20		4.400		Lev 91
23.10		6.600		Lev 91
24.00		10.000		Lev 91
24.80		13.000		Lev 91
25.70		18.000		Lev 91
26.60		25.000		Lev 91
27.60		28.000		Lev 91
28.50		33.000		Lev 91
29.50		37.000		Lev 91
554 Reaction $^{88}_{38}Sr \rightarrow {}^{86}_{37}Rb$ (Fig. 386, p. 258)				
33.00		1.100	0.11	Sac 67
36.00		2.300	0.23	Sac 67
40.00		5.400	0.54	Sac 67
42.00		8.800	0.88	Sac 67
48.00		14.000	1.40	Sac 67
54.00		16.000	1.60	Sac 67
60.00		30.000	3.00	Sac 67
66.00		32.000	3.20	Sac 67
72.00		30.000	3.00	Sac 67
78.00		25.000	2.50	Sac 67
85.00		22.000	2.20	Sac 67

E [MeV]	ΔE [MeV]	σ [mb]	Δσ [mb]	Ref.

555 Reaction $^{88}_{38}Sr \rightarrow ^{85}_{38}Sr$ (Fig. 387, p. 259)

E [MeV]	ΔE [MeV]	σ [mb]	Δσ [mb]	Ref.
42.00		26.000	3.10	Sac 67
48.00		140.000	17.00	Sac 67
54.00		230.000	28.00	Sac 67
60.00		160.000	19.00	Sac 67
66.00		150.000	18.00	Sac 67
72.00		130.000	16.00	Sac 67
78.00		120.000	14.00	Sac 67
85.00		100.000	12.00	Sac 67

556 Reaction $^{88}_{38}Sr \rightarrow ^{85m}_{38}Sr$ (Fig. 388, p. 259)

E [MeV]	ΔE [MeV]	σ [mb]	Δσ [mb]	Ref.
42.00		11.000	2.00	Sac 67
48.00		41.000	8.00	Sac 67
54.00		25.000	5.00	Sac 67
60.00		28.000	6.00	Sac 67
66.00		28.000	6.00	Sac 67
72.00		26.000	5.00	Sac 67
78.00		29.000	6.00	Sac 67
85.00		25.000	5.00	Sac 67

557 Reaction $^{88}_{38}Sr \rightarrow ^{85m+g}_{38}Sr$ (Fig. 389, p. 259)

E [MeV]	ΔE [MeV]	σ [mb]	Δσ [mb]	Ref.
42.00		37.000	6.00	Sac 67
48.00		180.000	30.00	Sac 67
54.00		250.000	40.00	Sac 67
60.00		190.000	30.00	Sac 67
66.00		180.000	30.00	Sac 67
72.00		160.000	25.00	Sac 67
78.00		150.000	24.00	Sac 67
85.00		130.000	20.00	Sac 67

558 Reaction $^{88}_{38}Sr \rightarrow ^{84}_{39}Y$ (Fig. 390, p. 259)

E [MeV]	ΔE [MeV]	σ [mb]	Δσ [mb]	Ref.
60.00		11.000	2.00	Sac 67
66.00		24.000	4.00	Sac 67
69.00		26.000	5.00	Sac 67
72.00		24.000	4.00	Sac 67
78.00		18.000	3.00	Sac 67
85.00		12.000	2.00	Sac 67

559 Reaction $^{88}_{38}Sr \rightarrow ^{85}_{39}Y$ (Fig. 391, p. 259)

E [MeV]	ΔE [MeV]	σ [mb]	Δσ [mb]	Ref.
42.00		2.500	0.25	Sac 67
45.00		69.000	6.90	Sac 67
48.00		95.000	9.50	Sac 67
54.00		140.000	14.00	Sac 67

E [MeV]	ΔE [MeV]	σ [mb]	Δσ [mb]	Ref.

559 Reaction $^{88}_{38}Sr \rightarrow {}^{85}_{39}Y$ (continued)

E [MeV]	ΔE [MeV]	σ [mb]	Δσ [mb]	Ref.
60.00		91.000	9.10	Sac 67
66.00		59.000	5.90	Sac 67
72.00		34.000	3.40	Sac 67
78.00		27.000	2.70	Sac 67
85.00		23.000	2.30	Sac 67

560 Reaction $^{88}_{38}Sr \rightarrow {}^{85m}_{39}Y$ (Fig. 392, p. 260)

E [MeV]	ΔE [MeV]	σ [mb]	Δσ [mb]	Ref.
42.00		22.000	2.20	Sac 67
45.00		48.000	4.80	Sac 67
48.00		59.000	5.90	Sac 67
54.00		94.000	9.40	Sac 67
60.00		53.000	5.30	Sac 67
66.00		46.000	4.60	Sac 67
72.00		28.000	2.80	Sac 67
78.00		23.000	2.30	Sac 67
85.00		19.000	1.90	Sac 67

561 Reaction $^{88}_{38}Sr \rightarrow {}^{85m+g}_{39}Y$ (Fig. 393, p. 260)

E [MeV]	ΔE [MeV]	σ [mb]	Δσ [mb]	Ref.
42.00		25.000	3.00	Sac 67
45.00		120.000	16.00	Sac 67
48.00		150.000	20.00	Sac 67
54.00		230.000	30.00	Sac 67
60.00		140.000	18.00	Sac 67
66.00		100.000	13.00	Sac 67
72.00		62.000	8.00	Sac 67
78.00		50.000	7.00	Sac 67
85.00		42.000	6.00	Sac 67

562 Reaction $^{88}_{38}Sr \rightarrow {}^{86}_{39}Y$ (Fig. 394, p. 260)

E [MeV]	ΔE [MeV]	σ [mb]	Δσ [mb]	Ref.
27.60		18.000		Lev 91
28.50		57.000		Lev 91
29.00		28.000		Sac 67
29.50		117.000		Lev 91
33.00		190.000		Sac 67
36.00		310.000		Sac 67
40.00		290.000		Sac 67
42.00		260.000		Sac 67
48.00		110.000		Sac 67
54.00		58.000		Sac 67
60.00		27.000		Sac 67
66.00		31.000		Sac 67
72.00		17.000		Sac 67
78.00		23.000		Sac 67
85.00		20.000		Sac 67

E [MeV]	ΔE [MeV]	σ [mb]	Δσ [mb]	Ref.

563 Reaction $^{88}{}_{38}Sr \rightarrow {}^{86m}{}_{39}Y$ (Fig. 395, p. 260)

E [MeV]	ΔE [MeV]	σ [mb]	Δσ [mb]	Ref.
27.60		11.000		Lev 91
28.50		35.000		Lev 91
29.50		70.000		Lev 91
33.00		77.000	14.00	Sac 67
36.00		130.000	23.00	Sac 67
40.00		180.000	32.00	Sac 67
42.00		180.000	32.00	Sac 67
48.00		79.000	14.00	Sac 67
54.00		59.000	11.00	Sac 67
60.00		42.000	8.00	Sac 67
66.00		33.000	6.00	Sac 67
72.00		26.000	5.00	Sac 67
78.00		18.000	3.00	Sac 67
85.00		17.000	3.00	Sac 67

564 Reaction $^{88}{}_{38}Sr \rightarrow {}^{86m+g}{}_{39}Y$ (Fig. 396, p. 260)

E [MeV]	ΔE [MeV]	σ [mb]	Δσ [mb]	Ref.
29.00		28.000	7.00	Sac 67
33.00		270.000	70.00	Sac 67
36.00		440.000	110.00	Sac 67
40.00		470.000	120.00	Sac 67
42.00		440.000	110.00	Sac 67
48.00		190.000	50.00	Sac 67
54.00		120.000	30.00	Sac 67
60.00		69.000	17.00	Sac 67
66.00		64.000	16.00	Sac 67
72.00		43.000	11.00	Sac 67
78.00		41.000	10.00	Sac 67
85.00		37.000	9.00	Sac 67

565 Reaction $^{88}{}_{38}Sr \rightarrow {}^{87}{}_{39}Y$ (Fig. 397, p. 260)

E [MeV]	ΔE [MeV]	σ [mb]	Δσ [mb]	Ref.
13.80		35.000		Lev 91
14.80		88.000		Lev 91
15.00		130.000	14.00	Sac 67
15.80		235.000		Lev 91
16.70		310.000		Lev 91
17.50		362.000		Lev 91
18.00		250.000	28.00	Sac 67
18.30		387.000		Lev 91
19.30		415.000		Lev 91
20.30		380.000		Lev 91
21.00		270.000	30.00	Sac 67
21.40		402.000		Lev 91
22.20		390.000		Lev 91
23.10		382.000		Lev 91

E [MeV]	ΔE [MeV]	σ [mb]	Δσ [mb]	Ref.

565 Reaction $^{88}_{38}Sr \rightarrow ^{87}_{39}Y$ (continued)

E [MeV]	ΔE [MeV]	σ [mb]	Δσ [mb]	Ref.
24.00		320.000		Lev 91
24.00		510.000	56.00	Sac 67
24.80		325.000		Lev 91
25.00		520.000	57.00	Sac 67
25.70		290.000		Lev 91
26.00		420.000	46.00	Sac 67
26.60		250.000		Lev 91
27.60		270.000		Lev 91
28.50		230.000		Lev 91
29.50		185.000		Lev 91
30.00		330.000	50.00	Sac 67
33.00		180.000	27.00	Sac 67

566 Reaction $^{88}_{38}Sr \rightarrow ^{87m}_{39}Y$ (Fig. 398, p. 261)

E [MeV]	ΔE [MeV]	σ [mb]	Δσ [mb]	Ref.
13.80		42.000		Lev 91
14.80		97.000		Lev 91
15.00		77.000	9.00	Sac 67
15.80		302.000		Lev 91
16.70		487.000		Lev 91
17.50		567.000		Lev 91
18.00		220.000	25.00	Sac 67
18.30		580.000		Lev 91
19.30		590.000		Lev 91
20.30		622.000		Lev 91
21.00		330.000	35.00	Sac 67
21.40		685.000		Lev 91
22.20		700.000		Lev 91
23.10		727.000		Lev 91
24.00		650.000	70.00	Sac 67
24.00		787.000		Lev 91
24.80		770.000		Lev 91
25.00		690.000	80.00	Sac 67
25.70		750.000		Lev 91
26.00		650.000	70.00	Sac 67
26.60		748.000		Lev 91
27.60		740.000		Lev 91
28.50		662.000		Lev 91
29.50		580.000		Lev 91
30.00		580.000	90.00	Sac 67
33.00		320.000	50.00	Sac 67

567 Reaction $^{88}_{38}Sr \rightarrow ^{87m+g}_{39}Y$ (Fig. 399, p. 261)

E [MeV]	ΔE [MeV]	σ [mb]	Δσ [mb]	Ref.
15.00		210.000	30.00	Sac 67
18.00		470.000	70.00	Sac 67

E [MeV]	ΔE [MeV]	σ [mb]	Δσ [mb]	Ref.

567 Reaction $^{88}_{38}Sr \rightarrow {}^{87m+g}_{39}Y$ (continued)

E [MeV]	ΔE [MeV]	σ [mb]	Δσ [mb]	Ref.
21.00		600.000	90.00	Sac 67
24.00		1160.000	170.00	Sac 67
25.00		1210.000	180.00	Sac 67
26.00		1070.000	160.00	Sac 67
30.00		910.000	140.00	Sac 67
33.00		500.000	75.00	Sac 67

568 Reaction $^{88}_{38}Sr \rightarrow {}^{88}_{39}Y$ (Fig. 400, p. 261)

E [MeV]	ΔE [MeV]	σ [mb]	Δσ [mb]	Ref.
4.80		0.320		Bla 51*
4.90		5.500		Bla 51*
5.30		30.000		Bla 51*
5.60		40.000		Bla 51*
5.80		65.000		Bla 51*
6.10		95.000		Bla 51*
6.40		120.000		Bla 51*
6.60		130.000		Bla 51*
6.80		170.000		Bla 51*
7.00		260.000	30.00	Sac 67
7.70		640.000		Lev 91
8.60		741.000		Lev 91
9.00		630.000	75.00	Sac 67
9.50		912.000		Lev 91
10.40		890.000		Lev 91
11.20		959.000	84.00	Del 63
11.30		1010.000		Lev 91
12.00		940.000	110.00	Sac 67
12.10		1030.000		Lev 91
12.80		1070.000		Lev 91
13.80		1040.000		Lev 91
14.80		1023.000		Lev 91
15.00		960.000	115.00	Sac 67
15.80		692.000		Lev 91
16.70		501.000		Lev 91
17.50		389.000		Lev 91
18.00		580.000	70.00	Sac 67
18.30		347.000		Lev 91
19.30		275.000		Lev 91
20.30		250.000		Lev 91
21.00		300.000	35.00	Sac 67
21.40		155.000		Lev 91
22.20		129.000		Lev 91
23.10		117.000		Lev 91
24.00		107.000		Lev 91
24.80		100.000		Lev 91
25.00		180.000	20.00	Sac 67

568 Reaction $^{88}_{38}Sr \rightarrow ^{88}_{39}Y$ (continued)

E [MeV]	ΔE [MeV]	σ [mb]	Δσ [mb]	Ref.
25.70		96.000		Lev 91
26.60		100.000		Lev 91
27.60		95.000		Lev 91
28.50		80.000		Lev 91
29.00		69.000	8.00	Sac 67
29.50		78.000		Lev 91
33.00		59.000	7.00	Sac 67
36.00		52.000	6.00	Sac 67
40.00		41.000	5.00	Sac 67
42.00		40.000	5.00	Sac 67
48.00		33.000	4.00	Sac 67
54.00		27.000	3.00	Sac 67
60.00		25.000	3.00	Sac 67
66.00		24.000	3.00	Sac 67
72.00		17.000	2.00	Sac 67
78.00		16.000	2.00	Sac 67
85.00		16.000	2.00	Sac 67

569 Reaction $_{39}Y \rightarrow ^{7}_{4}Be$

E [MeV]	ΔE [MeV]	σ [mb]	Δσ [mb]	Ref.
600.0		1.950	0.20	Mic 89
1005		10.000	2.00	Reg 82
2500		12.000	2.40	Reg 82

570 Reaction $_{39}Y \rightarrow ^{22}_{11}Na$

E [MeV]	ΔE [MeV]	σ [mb]	Δσ [mb]	Ref.
400.0		0.254E-1	0.36E-2	Kor 67
2500		1.600	0.30	Reg 82
24000		2.200	0.40	Reg 82

571 Reaction $_{39}Y \rightarrow ^{39}_{18}Ar$

E [MeV]	ΔE [MeV]	σ [mb]	Δσ [mb]	Ref.
1050		0.540	0.15	Reg 82
2500		2.700	0.50	Reg 82
24000		3.700	0.60	Reg 82

572 Reaction $_{39}Y \rightarrow ^{46}_{21}Sc$

E [MeV]	ΔE [MeV]	σ [mb]	Δσ [mb]	Ref.
2500		3.400	0.40	Reg 82
24000		3.500	0.40	Reg 82

573 Reaction $_{39}Y \rightarrow ^{48}_{23}V$

E [MeV]	ΔE [MeV]	σ [mb]	Δσ [mb]	Ref.
1005		1.100	0.20	Reg 82
2500		3.500	0.40	Reg 82

E [MeV]	ΔE [MeV]	σ [mb]	Δσ [mb]	Ref.
574 Reaction $_{39}Y \rightarrow {}^{51}_{24}Cr$				
1005		6.000	2.00	Reg 82
2500		9.400	3.50	Reg 82
575 Reaction $_{39}Y \rightarrow {}^{52}_{25}Mn$				
1005		2.500	0.80	Reg 82
2500		2.200	0.40	Reg 82
576 Reaction $_{39}Y \rightarrow {}^{54}_{25}Mn$				
1005		11.500	6.00	Reg 82
2500		14.000	7.00	Reg 82
24000		6.200	3.00	Reg 82
577 Reaction $_{39}Y \rightarrow {}^{59}_{26}Fe$				
1005		4.300	1.00	Reg 82
25000		5.400	1.20	Reg 82
578 Reaction $_{39}Y \rightarrow {}^{56}_{27}Co$				
1005		1.900	0.40	Reg 82
2500		1.900	0.40	Reg 82
24000		1.700	0.30	Reg 82
579 Reaction $_{39}Y \rightarrow {}^{58}_{27}Co$				
1005		14.300	3.60	Reg 82
2500		14.200	2.30	Reg 82
25000		10.300	2.00	Reg 82
580 Reaction $_{39}Y \rightarrow {}^{60}_{27}Co$				
25000		2.700	0.80	Reg 82
581 Reaction $_{39}Y \rightarrow {}^{65}_{30}Zn$				
600.0		8.200	0.50	Mic 89
1005		22.400	3.00	Reg 82
2500		15.000	4.00	Reg 82
24000		11.000	3.00	Reg 82
582 Reaction $_{39}Y \rightarrow {}^{67}_{31}Ga$				
600.0		12.200	0.70	Mic 89

E [MeV]	ΔE [MeV]	σ [mb]	Δσ [mb]	Ref.

583 Reaction $_{39}Y \rightarrow {}^{69}_{32}Ge$

E [MeV]	ΔE [MeV]	σ [mb]	Δσ [mb]	Ref.
600.0		11.500	0.60	Mic 89

584 Reaction $_{39}Y \rightarrow {}^{74}_{33}As$ (cumulative) **(Fig. 401, p. 261)**

E [MeV]	ΔE [MeV]	σ [mb]	Δσ [mb]	Ref.
100.0		0.270E-1	0.10E-1	Car 59
150.0		0.310	0.10	Car 59
180.0		0.700	0.20	Car 59
240.0		0.770	0.30	Car 56
593.0		6.900		Gru 82
600.0		7.580	0.36	Mic 89
1000		6.600	0.50	Ale 84
1000		9.500	1.90	Reg 82
2500		5.600	1.10	Reg 82

585 Reaction $_{39}Y \rightarrow {}^{72}_{34}Se$

E [MeV]	ΔE [MeV]	σ [mb]	Δσ [mb]	Ref.
100.0		0.700E-3	0.20E-3	Car 59
150.0		0.270E-1	0.70E-2	Car 59
180.0		0.390	0.10	Car 59
240.0		0.960	0.30	Car 56
593.0		6.050		Gru 82

586 Reaction $_{39}Y \rightarrow {}^{73m}_{34}Se$

E [MeV]	ΔE [MeV]	σ [mb]	Δσ [mb]	Ref.
100.0		0.100E-2	0.300E-3	Car 59
150.0		0.250	0.700E-1	Car 59
180.0		0.290	0.700	Car 59
240.0		3.500	1.000	Car 56
593.0		0.660		Gru 82

587 Reaction $_{39}Y \rightarrow {}^{75}_{34}Se$ (cumulative)

E [MeV]	ΔE [MeV]	σ [mb]	Δσ [mb]	Ref.
168.0		3.800	0.80	Reg 82
200.0		8.500	1.70	Reg 82
593.0		18.700		Reg 82
600.0		34.900	1.30	Mic 89
1000		46.000	9.00	Reg 82
2500		21.300	4.20	Reg 82
24000		16.100	3.20	Reg 82

588 Reaction $_{39}Y \rightarrow {}^{77}_{35}Br$ (cumulative)

E [MeV]	ΔE [MeV]	σ [mb]	Δσ [mb]	Ref.
1000		19.400	1.40	Ale 84

589 Reaction $_{39}Y \rightarrow {}^{76}_{36}Kr$ (cumulative)

E [MeV]	ΔE [MeV]	σ [mb]	Δσ [mb]	Ref.
1000		1.900	0.50	Ale 84

E [MeV]	ΔE [MeV]	σ [mb]	Δσ [mb]	Ref.

590 Reaction $_{39}Y \rightarrow {}^{77}_{36}Kr$ (cumulative)

E [MeV]	ΔE [MeV]	σ [mb]	Δσ [mb]	Ref.
1000		7.000	1.50	Ale 84

591 Reaction $_{39}Y \rightarrow {}^{79}_{36}Kr$ (cumulative)

E [MeV]	ΔE [MeV]	σ [mb]	Δσ [mb]	Ref.
1000		17.400	3.20	Ale 84

592 Reaction $_{39}Y \rightarrow {}^{81}_{36}Kr$ (Fig. 402, p. 262)

E [MeV]	ΔE [MeV]	σ [mb]	Δσ [mb]	Ref.
59.00		1.000	0.10	Reg 82
75.00		12.500	1.20	Reg 82
80.00		12.800	4.10	Reg 82
150.0		44.700	4.60	Reg 82
168.0		55.800	5.60	Reg 82
200.0		74.000	7.50	Reg 82
1050		53.000	5.30	Reg 82
2500		27.300	2.70	Reg 82
24000		24.000	5.80	Reg 82

593 Reaction $_{39}Y \rightarrow {}^{85}_{36}Kr$ (Fig. 403, p. 262)

E [MeV]	ΔE [MeV]	σ [mb]	Δσ [mb]	Ref.
75.00		0.390E-2	0.700E-3	Reg 82
80.00		0.710E-2	0.230E-2	Reg 82
150.0		0.350E-1	0.400E-2	Reg 82
168.0		0.600E-1	0.600E-2	Reg 82
200.0		0.950E-1	0.100E-1	Reg 82
1050		0.350	0.350E-1	Reg 82
2500		0.220	0.800E-1	Reg 82
24000		0.234	0.640E-1	Reg 82

594 Reaction $_{39}Y \rightarrow {}^{83}_{37}Rb$ (cumulative) (Fig. 404, p. 262)

E [MeV]	ΔE [MeV]	σ [mb]	Δσ [mb]	Ref.
75.00		29.000	6.00	Reg 82
168.0		54.000	11.00	Reg 82
200.0		55.000	11.00	Reg 82
600.0		93.000	4.00	Lag 79a
600.0		67.500	3.50	Mic 89
1000		45.000	9.00	Reg 82
2500		23.000	9.00	Reg 82
10500		39.000	2.00	Lag 79a
24000		19.200	3.80	Reg 82

595 Reaction $_{39}Y \rightarrow {}^{84}_{37}Rb$ (Fig. 405, p. 262)

E [MeV]	ΔE [MeV]	σ [mb]	Δσ [mb]	Ref.
75.00		5.400	1.10	Reg 82
168.0		13.300	2.60	Reg 82
200.0		14.400	2.80	Reg 82

595 Reaction $_{39}Y \rightarrow {}^{84}{}_{37}Rb$ (continued)

E [MeV]	ΔE [MeV]	σ [mb]	Δσ [mb]	Ref.
600.0		11.300	1.50	Lag 79a
600.0		16.300	0.70	Mic 89
1000		14.800	3.00	Reg 82
1000		11.300	0.40	Ale 84
2500		12.000	2.40	Reg 82
10500		10.900	0.80	Reg 82
24000		11.800	2.30	Reg 82

596 Reaction $_{39}Y \rightarrow {}^{84m+g}{}_{37}Rb$

E [MeV]	ΔE [MeV]	σ [mb]	Δσ [mb]	Ref.
60.00		3.900	0.20	Car 59
100.0		9.400	1.00	Car 59
150.0		27.000	5.00	Car 59
180.0		68.000	12.00	Car 59
240.0		71.400	12.00	Car 56

597 Reaction $_{39}Y \rightarrow {}^{86}{}_{37}Rb$ (Fig. 406, p. 262)

E [MeV]	ΔE [MeV]	σ [mb]	Δσ [mb]	Ref.
100.0		0.200	0.20E-1	Car 59
150.0		0.410	0.10	Car 59
180.0		1.700	0.50	Car 59
240.0		4.300	1.00	Car 56
593.0		4.000		Gru 82
600.0		3.500		Lag 79a
1000		6.900	1.60	Reg 82
2500		4.800	1.10	Reg 82

598 Reaction $_{39}Y \rightarrow {}^{82}{}_{38}Sr$

E [MeV]	ΔE [MeV]	σ [mb]	Δσ [mb]	Ref.
100.0		2.000	0.50	Car 59
150.0		7.000	1.00	Car 59
180.0		11.000	3.00	Car 59
240.0		9.300	5.00	Car 56
593.0		19.700		Gru 82

599 Reaction $_{39}Y \rightarrow {}^{83}{}_{38}Sr$ (Fig. 407, p. 263)

E [MeV]	ΔE [MeV]	σ [mb]	Δσ [mb]	Ref.
60.00		3.500	0.80	Car 59
100.0		8.900	1.00	Car 59
150.0		11.000	2.00	Car 59
180.0		18.500	3.00	Car 59
240.0		27.500	4.00	Car 56
593.0		31.500		Gru 82

600 Reaction $_{39}Y \rightarrow {}^{85}_{38}Sr$ (Fig. 408, p. 263)

E [MeV]	ΔE [MeV]	σ [mb]	Δσ [mb]	Ref.
75.00		17.000	7.00	Reg 82
168.0		39.000	8.00	Reg 82
200.0		46.000	9.00	Reg 82
240.0		59.500	10.00	Car 56
593.0		51.500		Gru 82
600.0		57.000	3.70	Mic 89
1000		50.000	22.00	Reg 82
2500		34.500	9.30	Reg 82
24000		18.200	3.60	Reg 82

601 Reaction $_{39}Y \rightarrow {}^{85m+g}_{38}Sr$ (Fig. 409, p. 263)

E [MeV]	ΔE [MeV]	σ [mb]	Δσ [mb]	Ref.
18.30		10.000		Lev 91
19.30		12.000		Lev 91
20.30		16.000		Lev 91
21.40		16.000		Lev 91
22.20		20.000		Lev 91
23.10		24.000		Lev 91
24.00		30.000		Lev 91
24.80		36.000		Lev 91
25.70		41.000		Lev 91
26.60		45.000		Lev 91
27.60		51.000		Lev 91
28.50		52.000		Lev 91
29.50		59.000		Lev 91

602 Reaction $_{39}Y \rightarrow {}^{87m}_{38}Sr$

E [MeV]	ΔE [MeV]	σ [mb]	Δσ [mb]	Ref.
60.00		54.000	8.00	Car 59
100.0		18.400	3.00	Car 59
180.0		6.000	0.50	Car 59
240.0		5.000	0.50	Car 56
593.0		8.500		Gru 82

603 Reaction $_{39}Y \rightarrow {}^{84}_{39}Y$

E [MeV]	ΔE [MeV]	σ [mb]	Δσ [mb]	Ref.
66.00		0.370	0.74E-1	Sah 66
78.00		11.000	2.20	Sah 66
85.00		24.000	4.80	Sah 66

604 Reaction $_{39}Y \rightarrow {}^{85}_{39}Y$

E [MeV]	ΔE [MeV]	σ [mb]	Δσ [mb]	Ref.
60.00		4.000	1.00	Car 59
100.0		30.000	5.00	Car 59
180.0		17.500	4.00	Car 59
240.0		15.800	3.00	Car 56
593.0		6.100		Gru 82

E [MeV]	ΔE [MeV]	σ [mb]	Δσ [mb]	Ref.

605 Reaction $_{39}Y \rightarrow {}^{85m+g}{}_{39}Y$

E [MeV]	ΔE [MeV]	σ [mb]	Δσ [mb]	Ref.
60.00		57.600		Sah 66
66.00		155.000		Sah 66
72.00		141.400		Sah 66
78.00		133.000		Sah 66
85.00		109.000		Sah 66

606 Reaction $_{39}Y \rightarrow {}^{86m+g}{}_{39}Y$ (Fig. 410, p. 263)

E [MeV]	ΔE [MeV]	σ [mb]	Δσ [mb]	Ref.
36.80		10.700		Sah 66
42.00		19.500		Sah 66
45.00		61.500		Sah 66
48.00		126.500		Sah 66
54.00		208.000		Sah 66
57.00		209.000		Sah 66
60.00		237.000		Sah 66
66.00		161.000		Sah 66
72.00		121.000		Sah 66
78.00		117.500		Sah 66
85.00		105.600		Sah 66

607 Reaction $_{39}Y \rightarrow {}^{87}{}_{39}Y$ (Fig. 411, p. 263)

E [MeV]	ΔE [MeV]	σ [mb]	Δσ [mb]	Ref.
60.0		82.000	15.00	Car 59
100.0		41.000	7.00	Car 59
150.0		49.000	7.00	Car 59
180.0		35.000	5.00	Car 59
240.0		23.000	4.00	Car 56
593.0		53.500		Gru 82
600.0		60.900	2.30	Mic 89

608 Reaction $_{39}Y \rightarrow {}^{87m+g}{}_{39}Y$ (Fig. 412, p. 264)

E [MeV]	ΔE [MeV]	σ [mb]	Δσ [mb]	Ref.
24.80		1.400		Sah 66
27.00		3.500		Sah 66
30.50		95.700		Sah 66
33.50		229.000		Sah 66
36.80		284.000		Sah 66
40.00		320.000		Sah 66
42.00		370.000		Sah 66
45.00		395.000		Sah 66
48.00		340.000		Sah 66
54.00		263.000		Sah 66
60.00		218.000		Sah 66
66.00		164.000		Sah 66
72.00		158.600		Sah 66
78.00		155.000		Sah 66
85.00		128.000		Sah 66

609 Reaction $_{39}Y \rightarrow {}^{88}{}_{39}Y$ (Figs. 413, 414, p. 264, 265)

E [MeV]	ΔE [MeV]	σ [mb]	Δσ [mb]	Ref.
11.20		≤ 1.000		Del 63
11.97	0.17	0.250E-2	0.26E-2	Mus 88
12.70	0.14	0.123E-1	0.30E-2	Mus 88
13.30	0.11	0.960E-1	0.16E-1	Mus 88
13.40	0.11	0.151	0.38E-1	Mus 88
13.80	0.10	0.272	0.11	Mus 88
14.19	0.10	1.210	0.13	Mus 88
14.68	0.26	5.310	0.52	Mus 88
14.95	0.18	8.390	0.21	Mus 88
15.00		3.600	0.50	Sah 66
15.41	0.37	17.870	0.21	Mus 88
15.80		50.000		Lev 91
15.95	0.18	31.200	0.30	Mus 88
16.57	0.30	45.200	0.40	Mus 88
16.70		80.000		Lev 91
17.39	0.32	64.800	0.30	Mus 88
17.50		110.000		Lev 91
18.30		151.000		Lev 91
18.31	0.24	85.900	1.50	Mus 88
18.50		38.000	5.00	Sah 66
19.21	0.26	115.200	0.80	Mus 88
19.30		166.000		Lev 91
19.89	0.44	128.400	0.40	Mus 88
19.92	0.17	129.800	2.10	Mus 88
20.30		214.000		Lev 91
20.87	0.31	146.400	0.80	Mus 88
20.91	0.19	156.600	0.70	Mus 88
21.40		251.000		Lev 91
21.44	0.11	163.900	0.40	Mus 88
21.50		141.000	18.00	Sah 66
22.00		36.000	25.00	Gus 62
22.20		263.000		Lev 91
22.38	0.26	184.100	0.40	Mus 88
22.98	0.37	200.000	0.50	Mus 88
23.10		280.000		Lev 91
23.85	0.21	212.200	0.60	Mus 88
24.00		301.000		Lev 91
24.44	0.13	226.100	1.10	Mus 88
24.80		257.000	33.00	Sah 66
24.80		339.000		Lev 91
25.25	0.33	234.100	1.30	Mus 88
25.70		347.000		Lev 91
26.60		386.000		Lev 91
26.86	0.27	263.100	0.50	Mus 88
27.00		220.000	35.00	Gus 62
27.50		304.000	40.00	Sah 66

609 Reaction $_{39}Y \rightarrow {}^{88}{}_{39}Y$ (continued)

E [MeV]	ΔE [MeV]	σ [mb]	Δσ [mb]	Ref.
27.60		380.000		Lev 91
28.30	0.51	298.000	5.70	Mus 88
28.37	0.20	283.300	0.60	Mus 88
28.50		396.000		Lev 91
29.50		355.000		Lev 91
30.10	0.48	286.100	4.00	Mus 88
30.50		365.000	47.00	Sah 66
31.00		139.000	30.00	Gus 62
31.21	0.34	282.300	2.00	Mus 88
31.82	0.41	264.900	4.00	Mus 88
32.86	0.28	242.800	2.50	Mus 88
33.46	0.38	249.800	4.00	Mus 88
33.50		310.000	40.00	Sah 66
34.44	0.20	231.800	1.80	Mus 88
35.04	0.22	237.100	3.20	Mus 88
36.00		172.000	33.00	Gus 62
36.47	0.43	212.500	2.70	Mus 88
36.72	0.34	218.700	0.60	Mus 88
36.80		283.000	37.00	Sah 66
37.93	0.38	208.800	2.80	Mus 88
38.17	0.28	206.700	1.00	Mus 88
39.35	0.24	202.600	2.70	Mus 88
39.56	0.19	198.700	0.60	Mus 88
42.00		231.000	30.00	Sah 66
48.00		228.000	30.00	Sah 66
48.50		202.000	35.00	Gus 62
54.00		186.000	24.00	Sah 66
59.00		195.000	42.00	Reg 82
60.00		199.000	26.00	Sah 66
60.00		420.000	50.00	Car 59
61.00		201.000	35.00	Gus 62
66.00		177.000	23.00	Sah 66
71.00		148.000	30.00	Gus 62
72.00		175.000	23.00	Sah 66
74.00		124.000	25.00	Gus 62
75.00		165.000	37.00	Reg 82
78.00		162.000	21.00	Sah 66
85.00		144.000	19.00	Sah 66
87.00		126.000	25.00	Gus 62
100.0		150.000	20.00	Car 59
103.0		143.000	20.00	Gus 62
117.0		119.000	18.00	Gus 62
120.0		95.500	3.00	Lev 66
134.0		109.000	18.00	Gus 62
150.0		94.000	12.00	Car 59
152.0		120.000	20.00	Gus 62

E [MeV]	ΔE [MeV]	σ [mb]	Δσ [mb]	Ref.

609 Reaction $_{39}Y \rightarrow {}^{88}_{39}Y$ (continued)

E [MeV]	ΔE [MeV]	σ [mb]	Δσ [mb]	Ref.
168.0		110.000	24.00	Reg 82
180.0		120.000	20.00	Car 59
200.0		115.000	25.00	Reg 82
200.0		73.700	2.50	Lev 66
240.0		93.000	15.00	Car 56
300.0		67.700	1.00	Lev 66
400.0		66.100	2.80	Lev 66
500.0		66.100	1.80	Lev 66
593.0		68.500		Gru 82
600.0		63.000	1.40	Lev 66
600.0		88.300	3.40	Mic 89
760.0		62.000	0.90	Lev 66
2500		63.000	13.00	Reg 82
24000		61.000	12.00	Reg 82

610 Reaction $_{39}Y \rightarrow {}^{86}_{40}Zr$ (Fig. 415, p. 265)

E [MeV]	ΔE [MeV]	σ [mb]	Δσ [mb]	Ref.
45.00		26.600	4.30	Sah 66
48.00		57.000	9.10	Sah 66
54.00		78.000	12.50	Sah 66
57.00		81.500	13.00	Sah 66
60.00		63.000	10.00	Sah 66
60.00		72.000	20.00	Car 59
66.00		42.000	6.70	Sah 66
72.00		32.000	5.10	Sah 66
78.00		27.700	4.40	Sah 66
85.00		22.000	3.50	Sah 66
100.0		2.900	0.50	Car 59
150.0		2.000	0.50	Car 59
180.0		2.000	0.50	Car 59
240.0		0.800	0.20	Car 56
400.0		2.390	0.38	Chu 69
593.0		1.970		Gru 82

611 Reaction $_{39}Y \rightarrow {}^{87}_{40}Zr$ (Fig. 416, p. 265)

E [MeV]	ΔE [MeV]	σ [mb]	Δσ [mb]	Ref.
27.60		16.000		Lev 91
28.50		58.000		Lev 91
29.50		125.000		Lev 91
30.50		55.000	6.60	Sah 66
33.50		118.000	14.00	Sah 66
36.80		313.000	38.00	Sah 66
40.00		385.000	46.00	Sah 66
42.00		341.000	42.00	Sah 66
45.00		299.000	36.00	Sah 66
48.00		168.000	20.00	Sah 66

E [MeV]	ΔE [MeV]	σ [mb]	Δσ [mb]	Ref.
611 Reaction $_{39}Y \rightarrow {}^{87}_{40}Zr$ (continued)				
54.00		91.000	11.00	Sah 66
60.00		55.400	6.60	Sah 66
66.00		54.000	6.50	Sah 66
72.00		47.500	5.70	Sah 66
78.00		36.500	4.40	Sah 66
85.00		31.500	3.80	Sah 66
100.0		44.000	5.00	Car 59
150.0		15.000	3.00	Car 59
240.0		9.200	1.00	Car 56
400.0		3.850	0.70	Chu 69
612 Reaction $_{39}Y \rightarrow {}^{88}_{40}Zr$ (Figs. 417, 418, p. 266)				
13.30	0.11	3.620	0.90E-1	Mus 88
13.40	0.11	6.490	0.11	Mus 88
13.80	0.10	34.880	0.26	Mus 88
14.19	0.10	69.980	0.48	Mus 88
14.68	0.26	125.200	0.50	Mus 88
14.80		288.000		Lev 91
14.95	0.18	165.900	0.90	Mus 88
15.00		68.000	8.80	Sah 66
15.41	0.37	229.600	2.40	Mus 88
15.80		363.000		Lev 91
15.95	0.18	324.800	1.50	Mus 88
16.57	0.30	418.000	2.30	Mus 88
16.70		537.000		Lev 91
17.39	0.32	533.900	1.60	Mus 88
17.50		575.000		Lev 91
18.30		794.000		Lev 91
18.31	0.24	643.900	10.70	Mus 88
18.50		352.000	46.00	Sah 66
19.21	0.26	711.100	4.00	Mus 88
19.30		891.000		Lev 91
19.89	0.44	725.800	1.40	Mus 88
19.92	0.17	765.000	7.40	Mus 88
20.30		1020.000		Lev 91
20.87	0.31	746.500	5.00	Mus 88
20.91	0.19	774.700	2.40	Mus 88
21.40		1020.000		Lev 91
21.44	0.11	765.400	1.40	Mus 88
21.50		495.000	64.00	Sah 66
22.20		1050.000		Lev 91
22.38	0.26	784.000	4.00	Mus 88
22.98	0.37	786.900	1.50	Mus 88
23.10		1050.000		Lev 91
23.85	0.21	770.000	5.00	Mus 88

612 Reaction $_{39}Y \rightarrow {}^{88}_{40}Zr$ (continued)

E [MeV]	ΔE [MeV]	σ [mb]	Δσ [mb]	Ref.
24.00		1000.000		Lev 91
24.44	0.13	771.200	1.10	Mus 88
24.80		1020.000		Lev 91
24.80		1252.000	163.00	Sah 66
25.25	0.33	766.000	2.00	Mus 88
25.70		1000.000		Lev 91
26.60		933.000		Lev 91
26.86	0.27	736.300	1.60	Mus 88
27.00		1248.000	162.00	Sah 66
27.50		1318.000	171.00	Sah 66
27.60		851.000		Lev 91
28.30	0.51	629.300	6.00	Mus 88
28.37	0.20	669.000	1.10	Mus 88
28.50		759.000		Lev 91
29.50		631.000		Lev 91
30.10	0.48	469.400	6.00	Mus 88
30.50		862.000	116.00	Sah 66
31.21	0.34	439.200	3.00	Mus 88
31.82	0.41	336.300	6.00	Mus 88
32.86	0.28	291.000	0.70	Mus 88
33.46	0.38	247.400	6.00	Mus 88
33.50		506.000	66.00	Sah 66
34.44	0.20	220.600	0.90	Mus 88
35.04	0.22	190.700	4.00	Mus 88
36.47	0.43	153.800	0.30	Mus 88
36.72	0.34	155.300	0.50	Mus 88
36.80		329.000	43.00	Sah 66
37.93	0.38	129.100	0.30	Mus 88
38.17	0.28	125.600	0.40	Mus 88
39.35	0.24	113.500	0.30	Mus 88
39.56	0.19	114.800	0.50	Mus 88
42.00		172.000	22.00	Sah 66
48.00		112.000	15.00	Sah 66
54.00		82.000	11.00	Sah 66
59.00		126.000	25.00	Reg 82
60.00		140.000	20.00	Car 59
60.00		73.400	9.50	Sah 66
66.00		60.000	7.80	Sah 66
72.00		54.000	7.00	Sah 66
75.00		55.000	11.00	Reg 82
78.00		46.000	6.00	Sah 66
85.00		41.000	5.30	Sah 66
100.0		71.000	8.00	Car 59
120.0		15.000	1.00	Lev 66
150.0		41.000	5.00	Car 59
168.0		18.300	3.70	Reg 82

612 Reaction $_{39}Y \rightarrow {}^{88}{}_{40}Zr$ (continued)

E [MeV]	ΔE [MeV]	σ [mb]	Δσ [mb]	Ref.
180.0		38.000	4.00	Car 59
200.0		19.200	3.80	Reg 82
200.0		11.000	0.20	Lev 66
240.0		25.600	2.00	Car 56
300.0		7.700	0.20	Lev 66
400.0		4.510	0.52	Chu 69
400.0		5.900	0.20	Lev 66
500.0		4.800	0.20	Lev 66
593.0		4.950		Gru 82
600.0		3.200	0.10	Lev 66
600.0		5.650	0.41	Mic 89
760.0		3.100	0.20	Lev 66
2500		5.900	1.50	Reg 82
24000		4.200	0.80	Reg 82

613 Reaction $_{39}Y \rightarrow {}^{89}{}_{40}Zr$ (Figs. 419, 420, 421, p. 267, 268)

E [MeV]	ΔE [MeV]	σ [mb]	Δσ [mb]	Ref.
3.50		0.700E-2		Bla 51*
3.65		0.220		Joh 68*
3.70		0.400		Joh 68*
3.80		0.900		Joh 68*
3.80		0.490		Bla 51*
4.00		100.000		Bir 73*
4.10		1.600		Joh 68*
4.20		3.200		Joh 68*
4.25		3.800		Joh 68*
4.30		9.000		Joh 68*
4.30		3.900		Bla 51*
4.35		11.000		Joh 68*
4.40		12.500		Joh 68*
4.50		15.000		Joh 68*
4.59	0.23	17.150	0.50E-1	Mus 88
4.60		19.000		Joh 68*
4.70		21.000		Joh 68*
4.78	0.44	26.300	0.10	Mus 88
4.80		50.000		Joh 68*
4.80		18.000		Bla 51*
4.90		32.000		Joh 68*
5.00		63.000		Joh 68*
5.00		50.000	5.50	Sah 66
5.05		100.000		Joh 68*
5.10		80.000		Joh 68*
5.10		40.000		Bla 51*
5.15		45.000		Joh 68*
5.20		50.000		Joh 68*
5.30	0.41	51.200	0.20	Mus 88

613 Reaction $_{39}Y \rightarrow {}^{89}_{40}Zr$ (continued)

E [MeV]	ΔE [MeV]	σ [mb]	Δσ [mb]	Ref.
5.30		52.000		Joh 68*
5.40		60.000		Joh 68*
5.50		70.000		Joh 68*
5.50		55.000		Bla 51*
5.51	0.19	70.500	0.20	Mus 88
5.60		80.000		Joh 68*
5.70		90.000		Joh 68*
5.80		100.000		Joh 68*
5.80		90.000		Bla 51*
5.81	0.38	98.200	0.20	Mus 88
5.85		110.000		Joh 68*
6.20		130.000		Bla 51*
6.31	0.19	142.400	0.30	Mus 88
6.50		150.000		Bla 51*
6.62	0.49	200.400	1.40	Mus 88
6.80		210.000		Bla 51*
6.83	0.35	224.300	0.50	Mus 88
7.00		380.000		Bir 73*
7.36	0.29	330.200	0.60	Mus 88
7.70		339.000		Lev 91
7.90	0.37	391.100	0.90	Mus 88
8.38	0.38	467.100	1.80	Mus 88
8.50		352.000	39.00	Sah 66
8.60		490.000		Lev 91
8.98	0.32	520.700	1.20	Mus 88
9.18	0.29	572.500	2.20	Mus 88
9.50		530.000		Bir 73*
9.50		549.000		Lev 91
9.88	0.41	616.600	8.00	Mus 88
9.94	0.33	636.000	2.40	Mus 88
10.40		776.000		Lev 91
10.87	0.26	674.500	1.80	Mus 88
10.88	0.25	674.400	1.60	Mus 88
11.00		650.000		Bir 73*
11.20		514.000	9.00	Del 63
11.30		880.000		Lev 91
11.97	0.17	741.500	4.70	Mus 88
12.00		750.000	100.00	Blo 55
12.00		720.000	79.00	Sah 66
12.10		912.000		Lev 91
12.46	0.33	750.800	1.50	Mus 88
12.70	0.14	749.000	2.00	Mus 88
12.80		933.000		Lev 91
13.30	0.11	756.200	2.10	Mus 88
13.40	0.11	769.000	2.50	Mus 88
13.50		700.000		Bir 73*

613 Reaction $_{39}Y \rightarrow {}^{89}_{40}Zr$ (continued)

E [MeV]	ΔE [MeV]	σ [mb]	Δσ [mb]	Ref.
13.80		871.000		Lev 91
13.80	0.10	787.300	2.60	Mus 88
14.19	0.10	764.900	2.60	Mus 88
14.68	0.26	700.000	1.40	Mus 88
14.80		891.000		Lev 91
14.95	0.18	664.400	2.20	Mus 88
15.00		712.000	78.00	Sah 66
15.41	0.37	599.200	7.30	Mus 88
15.80		741.000		Lev 91
15.95	0.18	536.000	1.70	Mus 88
16.00		530.000		Bir 73*
16.57	0.30	475.400	0.90	Mus 88
16.70		500.000		Lev 91
17.39	0.32	320.800	2.30	Mus 88
17.50		339.000		Lev 91
18.00		270.000		Bir 73*
18.30		251.000		Lev 91
18.31	0.24	241.900	0.30	Mus 88
18.50		552.000	61.00	Sah 66
19.21	0.26	169.200	1.60	Mus 88
19.30		186.000		Lev 91
19.89	0.44	135.300	0.20	Mus 88
19.92	0.17	140.800	0.50	Mus 88
20.00		130.000		Bir 73*
20.30		135.000		Lev 91
20.87	0.31	106.200	0.10	Mus 88
20.91	0.19	102.000	0.40	Mus 88
21.40		107.000		Lev 91
21.44	0.11	89.400	0.10	Mus 88
21.50		85.000		Bir 73*
21.50		395.000	43.00	Sah 66
22.20		95.000		Lev 91
22.38	0.26	78.200	0.10	Mus 88
22.98	0.37	69.700	0.10	Mus 88
23.00		55.000		Bir 73*
23.10		83.000		Lev 91
23.85	0.21	62.500	0.10	Mus 88
24.00		87.000		Lev 91
24.44	0.13	58.200	0.10	Mus 88
24.80		76.000		Lev 91
24.80		194.000	21.00	Sah 66
25.00		40.000		Bir 73*
25.25	0.33	57.000	0.10	Mus 88
25.70		68.000		Lev 91
26.60		69.000		Lev 91
26.86	0.27	50.400	0.10	Mus 88

613 Reaction $_{39}Y \rightarrow {}^{89}{}_{40}Zr$ (continued)

E [MeV]	ΔE [MeV]	σ [mb]	Δσ [mb]	Ref.
27.50		103.000	11.00	Sah 66
27.60		59.000		Lev 91
28.00		35.000		Bir 73*
28.30	0.51	47.700	1.00	Mus 88
28.37	0.20	46.600	0.10	Mus 88
28.50		57.000		Lev 91
29.50		30.000		Bir 73*
29.50		50.000		Lev 91
30.10	0.48	44.500	1.00	Mus 88
30.50		61.400	6.80	Sah 66
31.21	0.34	41.800	0.10	Mus 88
31.82	0.41	41.300	1.00	Mus 88
32.00		30.000		Bir 73*
32.86	0.28	37.500	0.10	Mus 88
33.46	0.38	38.500	1.00	Mus 88
33.50		43.000	4.70	Sah 66
34.00		30.000		Bir 73*
34.44	0.20	35.300	0.10	Mus 88
35.04	0.22	35.900	1.00	Mus 88
35.50		20.000		Bir 73*
36.47	0.43	33.400	0.10	Mus 88
36.72	0.34	33.500	0.10	Mus 88
36.80		45.000	5.00	Sah 66
37.50		20.000		Bir 73*
37.93	0.38	31.500	0.10	Mus 88
38.17	0.28	30.600	0.10	Mus 88
39.35	0.24	29.600	0.10	Mus 88
39.56	0.19	30.200	0.10	Mus 88
40.00		20.000		Bir 73*
41.50		20.000		Bir 73*
42.00		37.000	4.10	Sah 66
45.00		20.000		Bir 73*
48.00		29.400	3.20	Sah 66
54.00		29.700	3.30	Sah 66
60.00		44.000	4.00	Car 59
60.00		23.300	2.60	Sah 66
66.00		19.700	2.20	Sah 66
72.00		17.000	1.90	Sah 66
78.00		14.400	1.60	Sah 66
85.00		12.000	1.30	Sah 66
100.0		19.300	2.00	Car 59
120.0		5.800	0.30	Lev 66
150.0		9.100	0.90	Car 59
180.0		9.300	0.90	Car 59
200.0		4.200	0.10	Lev 66
240.0		6.100	0.70	Car 56

E [MeV]	ΔE [MeV]	σ [mb]	Δσ [mb]	Ref.
613 Reaction $_{39}Y \rightarrow {}^{89}_{40}Zr$ (continued)				
300.0		3.050	0.10	Lev 66
400.0		1.850	0.27	Chu 69
400.0		2.350	0.50E-1	Lev 66
500.0		2.050	0.20	Lev 66
593.0		2.740		Gru 82
600.0		1.500	0.50E-1	Lev 66
600.0		3.230	0.18	Mic 89
760.0		1.500	0.10	Lev 66
614 Reaction $_{40}Zr \rightarrow {}^{7}_{4}Be$				
800		2.320	0.31	Dit 90
1200		3.700	0.32	Dit 90
2500		10.000	4.50	Reg 82
2600		6.410	0.52	Dit 90
615 Reaction $_{40}Zr \rightarrow {}^{22}_{11}Na$				
400.0		0.124E-1	0.10E-2	Kor 67
1200		0.191	0.32E-1	Dit 90
24000		1.60	0.30	Reg 82
616 Reaction $_{40}Zr \rightarrow {}^{39}_{18}Ar$				
1005		1.000	0.20	Reg 82
2500		2.500	0.30	Reg 82
24000		3.300	0.40	Reg 82
617 Reaction $_{40}Zr \rightarrow {}^{42}_{18}Ar$				
2500		0.200	0.60E-1	Reg 82
24000		0.170	0.30E-1	Reg 82
618 Reaction $_{40}Zr \rightarrow {}^{46}_{21}Sc$				
2500		3.100	0.40	Reg 82
24000		3.100	0.40	Reg 82
619 Reaction $_{40}Zr \rightarrow {}^{48}_{23}V$				
1005		0.950	0.15	Reg 82
2500		3.300	0.40	Reg 82
620 Reaction $_{40}Zr \rightarrow {}^{51}_{24}Cr$				
1005		3.800	1.30	Reg 82
2500		8.600	2.80	Reg 82

E [MeV]	ΔE [MeV]	σ [mb]	Δσ [mb]	Ref.
621 Reaction $_{40}Zr \rightarrow {}^{54}_{25}Mn$				
1005		9.500	5.00	Reg 82
2500		14.000	7.00	Reg 82
24000		6.000	3.00	Reg 82
622 Reaction $_{40}Zr \rightarrow {}^{56}_{27}Co$				
1005		1.200	0.20	Reg 82
2500		2.300	0.50	Reg 82
24000		1.500	0.20	Reg 82
623 Reaction $_{40}Zr \rightarrow {}^{58}_{27}Co$				
1005		10.300	3.00	Reg 82
2500		12.200	1.50	Reg 82
24000		10.700	2.00	Reg 82
624 Reaction $_{40}Zr \rightarrow {}^{65}_{30}Zn$				
600.0		5.220	0.57	Mic 89
1005		17.200	2.50	Reg 82
2500		16.000	2.00	Reg 82
24000		10.000	1.50	Reg 82
625 Reaction $_{40}Zr \rightarrow {}^{67}_{31}Ga$				
600.0		7.510	0.35	Mic 89
626 Reaction $_{40}Zr \rightarrow {}^{74}_{33}As$				
600.0		5.150	0.22	Mic 89
1000		6.000	0.90	Ale 84
1000		7.400	1.50	Reg 82
2500		4.800	1.00	Reg 82
627 Reaction $_{40}Zr \rightarrow {}^{75}_{34}Se$ (cumulative)				
168.0		1.700	0.80	Reg 82
200.0		3.900	0.80	Reg 82
600.0		28.000	1.10	Mic 89
1000		37.000	7.00	Reg 82
2500		22.000	4.00	Reg 82
24000		14.500	2.90	Reg 82
628 Reaction $_{40}Zr \rightarrow {}^{77}_{35}Br$ (cumulative)				
1000		19.000	1.20	Ale 84

E [MeV]	ΔE [MeV]	σ [mb]	Δσ [mb]	Ref.

629 Reaction $_{40}Zr \rightarrow {}^{76}_{36}Kr$ (cumulative)

E [MeV]	ΔE [MeV]	σ [mb]	Δσ [mb]	Ref.
1000		1.600	0.60	Ale 84

630 Reaction $_{40}Zr \rightarrow {}^{77}_{36}Kr$ (cumulative)

E [MeV]	ΔE [MeV]	σ [mb]	Δσ [mb]	Ref.
1000		7.600	1.50	Ale 84

631 Reaction $_{40}Zr \rightarrow {}^{79}_{36}Kr$ (cumulative)

E [MeV]	ΔE [MeV]	σ [mb]	Δσ [mb]	Ref.
1000		19.600	3.50	Ale 84

632 Reaction $_{40}Zr \rightarrow {}^{81}_{36}Kr$ (Fig. 422, p. 267)

E [MeV]	ΔE [MeV]	σ [mb]	Δσ [mb]	Ref.
59.00		1.330	0.13	Reg 82
75.00		3.540	0.21	Reg 82
168.0		30.800	3.10	Reg 82
200.0		41.700	4.20	Reg 82
1000		46.700	4.60	Reg 82
2500		29.800	3.00	Reg 82
24000		21.000	2.10	Reg 82

633 Reaction $_{40}Zr \rightarrow {}^{85}_{36}Kr$ (Fig. 423, p. 268)

E [MeV]	ΔE [MeV]	σ [mb]	Δσ [mb]	Ref.
59.00		<0.100E-2		Reg 82
75.00		0.390E-2	0.18E-2	Reg 82
168.0		0.480E-1	0.11E-1	Reg 82
200.0		0.880E-1	0.10E-1	Reg 82
1000		0.240	0.60E-1	Reg 82
2500		0.190	0.20E-1	Reg 82
24000		0.280	0.30E-1	Reg 82

634 Reaction $_{40}Zr \rightarrow {}^{83}_{37}Rb$ (Fig. 424, p. 268)

E [MeV]	ΔE [MeV]	σ [mb]	Δσ [mb]	Ref.
35.00		0.190E-1	3.00E-2	Kon 91
36.00		0.280E-1	4.00E-2	Kon 91
37.50		0.750E-1	0.14E-1	Kon 91
39.00		0.356	0.19E-1	Kon 91
40.00		0.461	0.52E-1	Kon 91
42.50		0.520	0.70E-1	Kon 91
44.00		0.580	0.80E-1	Kon 91
45.00		0.510	0.60E-1	Kon 91
46.00		0.530	0.60E-1	Kon 91
47.50		0.400	0.70E-1	Kon 91
49.00		0.390	0.50E-1	Kon 91
50.00		0.400	0.50E-1	Kon 91
51.00		0.420	0.60E-1	Kon 91
52.00		0.380	0.50E-1	Kon 91

E [MeV]	ΔE [MeV]	σ [mb]	Δσ [mb]	Ref.

634 Reaction $_{40}Zr \rightarrow {}^{83}_{37}Rb$ (continued)

E [MeV]	ΔE [MeV]	σ [mb]	Δσ [mb]	Ref.
53.00		0.390	0.60E-1	Kon 91
55.00		0.450	0.40E-1	Kon 91
56.50		0.420	0.70E-1	Kon 91
57.50		0.460	0.70E-1	Kon 91
60.00		0.390	0.70E-1	Kon 91
62.00		0.460	0.60E-1	Kon 91
64.00		0.510	0.80E-1	Kon 91
65.00		0.630	0.90E-1	Kon 91
67.00		0.760	0.11E-1	Kon 91
68.00		0.980	0.16E-1	Kon 91

635 Reaction $_{40}Zr \rightarrow {}^{83}_{37}Rb$ (cumulative) (Fig. 425, p. 268)

E [MeV]	ΔE [MeV]	σ [mb]	Δσ [mb]	Ref.
59.00		2.900		Reg 82
75.00		27.000		Reg 82
168.0		37.000		Reg 82
200.0		38.000		Reg 82
600.0		55.200	2.90	Mic 89
1000		34.000		Reg 82
2500		21.000		Reg 82
24000		16.000		Reg 82

636 Reaction $_{40}Zr \rightarrow {}^{84}_{37}Rb$ (Fig. 426, p. 269)

E [MeV]	ΔE [MeV]	σ [mb]	Δσ [mb]	Ref.
47.50		0.285E-1	0.17E-2	Kon 91
49.00		0.440E-1	0.25E-2	Kon 91
50.00		0.490E-1	0.32E-2	Kon 91
51.00		0.480E-1	0.70E-2	Kon 91
52.00		0.760E-1	0.12E-1	Kon 91
53.00		0.150	0.12E-1	Kon 91
55.00		0.250	0.25E-1	Kon 91
56.50		0.260	0.40E-1	Kon 91
60.00		0.560	0.70E-1	Kon 91
62.00		0.740	0.12	Kon 91
64.00		1.050	0.16	Kon 91
65.00		1.090	0.14	Kon 91
67.00		1.230	0.13	Kon 91
68.00		1.110	0.11	Kon 91
75.00		2.100	0.40	Reg 82
200.0		4.500	1.60	Reg 82
600.0		9.500	0.40	Mic 89
1000		8.900	1.80	Reg 82
1000		7.600	0.40	Ale 84
2500		6.000	1.20	Reg 82
24000		5.800	1.20	Reg 82

E [MeV]	ΔE [MeV]	σ [mb]	Δσ [mb]	Ref.

637 Reaction $_{40}Zr \rightarrow {}^{86}_{37}Rb$

E [MeV]	ΔE [MeV]	σ [mb]	Δσ [mb]	Ref.
1000		5.000	1.00	Reg 82
2500		2.900	0.60	Reg 82

638 Reaction $_{40}Zr \rightarrow {}^{85}_{38}Sr$ (cumulative)

E [MeV]	ΔE [MeV]	σ [mb]	Δσ [mb]	Ref.
75.00		22.000	7.00	Reg 82
168.0		28.000	12.00	Reg 82
200.0		32.000	8.00	Reg 82
600.0		102.000	9.70	Mic 89
1000		30.000	13.00	Reg 82
2500		22.000	8.00	Reg 82
24000		18.800	5.40	Reg 82

639 Reaction $_{40}Zr \rightarrow {}^{86}_{39}Y$ (Fig. 427, p. 269)

E [MeV]	ΔE [MeV]	σ [mb]	Δσ [mb]	Ref.
14.90		3.500	0.30	Kon 91
17.00		9.100	0.80	Kon 91
20.00		11.300	1.20	Kon 91
21.00		16.000	1.10	Kon 91
22.50		15.000	1.80	Kon 91
24.00		19.100	2.10	Kon 91
25.00		19.800	1.80	Kon 91
27.00		25.900	2.00	Kon 91
27.50		30.000	2.60	Kon 91
30.00		29.500	2.60	Kon 91
32.50		20.800	2.50	Kon 91
34.00		18.300	1.90	Kon 91
36.00		16.700	1.60	Kon 91
39.00		13.600	1.30	Kon 91
41.00		14.900	1.31	Kon 91
44.00		15.100	1.50	Kon 91
46.00		19.000	0.80	Kon 91
50.00		20.600	0.90	Kon 91
53.00		21.300	0.90	Kon 91
56.00		21.800	0.80	Kon 91
60.00		29.500	0.90	Kon 91
62.50		31.200	1.10	Kon 91
66.00		32.000	1.20	Kon 91
68.00		30.500	0.90	Kon 91

640 Reaction $_{40}Zr \rightarrow {}^{87}_{39}Y$ (Fig. 428, p. 269)

E [MeV]	ΔE [MeV]	σ [mb]	Δσ [mb]	Ref.
9.60		0.910	0.10	Kon 91
12.40		1.400	0.23	Kon 91
14.90		3.600	0.47	Kon 91
17.00		6.100	0.55	Kon 91

E [MeV]	ΔE [MeV]	σ [mb]	Δσ [mb]	Ref.

640 Reaction $_{40}Zr \rightarrow {}^{87}_{39}Y$ (continued)

E [MeV]	ΔE [MeV]	σ [mb]	Δσ [mb]	Ref.
20.00		9.500	0.76	Kon 91
22.50		6.600	0.63	Kon 91
25.00		7.900	0.84	Kon 91
27.00		8.900	0.68	Kon 91
27.50		7.300	0.75	Kon 91
30.00		7.500	0.80	Kon 91
32.50		7.200	0.65	Kon 91
36.00		6.100	0.60	Kon 91
39.00		8.120	0.71	Kon 91
41.00		10.210	0.96	Kon 91
44.00		13.600	1.10	Kon 91
46.00		19.200	1.70	Kon 91
50.00		29.600	2.40	Kon 91
53.00		43.400	3.80	Kon 91
56.00		66.800	6.70	Kon 91
60.00		69.500	6.50	Kon 91
62.50		56.300	5.80	Kon 91
66.00		55.200	5.40	Kon 91
68.00		57.300	5.50	Kon 91
600.0		66.900	2.50	Mic 89

641 Reaction $_{40}Zr \rightarrow {}^{88}_{39}Y$

E [MeV]	ΔE [MeV]	σ [mb]	Δσ [mb]	Ref.
168.0		52.000	10.00	Reg 82
200.0		56.000	11.00	Reg 82
600.0		51.100	2.70	Mic 89
1000		27.000	5.40	Reg 82
2500		20.000	4.00	Reg 82

642 Reaction $_{40}Zr \rightarrow {}^{91}_{39}Y$

E [MeV]	ΔE [MeV]	σ [mb]	Δσ [mb]	Ref.
600.0		19.400	1.50	Mic 89

643 Reaction $_{40}Zr \rightarrow {}^{88}_{40}Zr$ (cumulative) (Fig. 429, p. 269)

E [MeV]	ΔE [MeV]	σ [mb]	Δσ [mb]	Ref.
59.00		456.000	91.00	Reg 82
75.00		232.000	46.00	Reg 82
168.0		80.000	16.00	Reg 82
200.0		79.000	16.00	Reg 82
600.0		44.800	2.40	Mic 89
1000		39.000	8.00	Reg 82
2500		28.000	6.00	Reg 82
24000		24.500	5.00	Reg 82

644 Reaction $_{40}Zr \rightarrow {}^{89}_{40}Zr$

E [MeV]	ΔE [MeV]	σ [mb]	Δσ [mb]	Ref.
600.0		58.600	2.20	Mic 89

E [MeV]	ΔE [MeV]	σ [mb]	Δσ [mb]	Ref.

645 Reaction $_{40}Zr \rightarrow {}^{95}_{40}Zr$ (cumulative) (Fig. 430, p. 270)

E [MeV]	ΔE [MeV]	σ [mb]	Δσ [mb]	Ref.
59.00		10.000	2.00	Reg 82
75.00		5.200	1.00	Reg 82
168.0		2.300	0.90	Reg 82
200.0		2.400	0.60	Reg 82
600.0		2.420	0.15	Mic 89
1000		2.800	0.60	Reg 82
2500		2.600	0.50	Reg 82
24000		2.800	0.50	Reg 82

646 Reaction $_{40}Zr \rightarrow {}^{90}_{41}Nb$ (Fig. 431, p. 270)

E [MeV]	ΔE [MeV]	σ [mb]	Δσ [mb]	Ref.
9.60		45.000	4.00	Kon 91
12.40		170.000	8.00	Kon 91
14.90		280.000	10.00	Kon 91
17.00		340.000	21.00	Kon 91
19.00		278.000	17.00	Kon 91
21.00		250.000	15.00	Kon 91
22.50		154.000	10.00	Kon 91
24.00		120.000	10.00	Kon 91
26.00		103.000	9.00	Kon 91
27.00		95.000	8.00	Kon 91
28.00		72.000	7.00	Kon 91
30.00		67.000	6.00	Kon 91
31.50		72.000	4.00	Kon 91
32.50		70.000	4.00	Kon 91
33.50		72.000	4.00	Kon 91
35.00		67.000	4.00	Kon 91
36.00		67.000	4.00	Kon 91
37.50		53.000	4.00	Kon 91
39.00		47.000	4.00	Kon 91
40.00		43.000	3.00	Kon 91
42.50		41.000	4.00	Kon 91
44.00		39.000	3.00	Kon 91
45.00		39.000	3.00	Kon 91
46.00		43.000	4.00	Kon 91
47.50		40.000	3.00	Kon 91
49.00		41.000	3.00	Kon 91
50.00		38.000	3.00	Kon 91
51.00		39.000	3.00	Kon 91
52.00		41.000	3.00	Kon 91
53.00		36.000	3.00	Kon 91
55.00		40.000	3.00	Kon 91
56.50		33.000	2.00	Kon 91
57.50		30.000	2.00	Kon 91
60.00		32.000	2.00	Kon 91
62.00		28.000	2.00	Kon 91

E [MeV]	ΔE [MeV]	σ [mb]	Δσ [mb]	Ref.
646 Reaction $_{40}Zr \rightarrow {}^{90}_{41}Nb$ (continued)				
64.00		25.000	2.00	Kon 91
65.00		31.000	2.00	Kon 91
67.00		27.000	2.00	Kon 91
68.00		26.000	2.00	Kon 91
647 Reaction $_{40}Zr \rightarrow {}^{92m}_{41}Nb$ (cumulative)				
600.0		1.000	0.10E-1	Mic 89
1000		0.700	0.14	Reg 82
2500		1.200	0.24	Reg 82
648 Reaction $^{90}_{40}Zr \rightarrow {}^{71}_{33}As$ (cumulative)				
1000		14.300	1.00	Bel 78
649 Reaction $^{90}_{40}Zr \rightarrow {}^{72}_{33}As$				
1000		16.900	1.40	Bel 78
650 Reaction $^{90}_{40}Zr \rightarrow {}^{74}_{33}As$				
1000		4.200	0.30	Bel 78
651 Reaction $^{90}_{40}Zr \rightarrow {}^{72}_{34}Se$ (cumulative)				
1000		5.500	0.50	Bel 78
652 Reaction $^{90}_{40}Zr \rightarrow {}^{73}_{34}Se$ (cumulative)				
1000		12.900	0.80	Bel 78
653 Reaction $^{90}_{40}Zr \rightarrow {}^{75}_{34}Se$ (cumulative)				
1000		23.900	1.60	Bel 78
654 Reaction $^{90}_{40}Zr \rightarrow {}^{77}_{35}Br$ (cumulative)				
1000		20.700	1.50	Bel 78
655 Reaction $^{90}_{40}Zr \rightarrow {}^{74}_{37}Rb$				
1000		0.600E-2	0.60E-2	Bel 78
656 Reaction $^{90}_{40}Zr \rightarrow {}^{75}_{37}Rb$				
1000		0.130E-1	0.50E-2	Bel 78

E [MeV]	ΔE [MeV]	σ [mb]	Δσ [mb]	Ref.

657 Reaction $^{90}_{40}Zr \rightarrow {}^{76}_{37}Rb$

1000		0.125	0.31E-1	Bel 78

658 Reaction $^{90}_{40}Zr \rightarrow {}^{77}_{37}Rb$

1000		1.190	0.40E-1	Bel 78

659 Reaction $^{90}_{40}Zr \rightarrow {}^{78}_{37}Rb$

1000		4.400	0.20	Bel 78

660 Reaction $^{90}_{40}Zr \rightarrow {}^{79}_{37}Rb$

1000		13.200	0.40	Bel 78

661 Reaction $^{90}_{40}Zr \rightarrow {}^{80}_{37}Rb$

1000		17.700	0.90	Bel 78

662 Reaction $^{90}_{40}Zr \rightarrow {}^{81}_{37}Rb$ (cumulative)

1000		20.300	1.20	Bel 78
1000		26.400	0.90	Bel 78

663 Reaction $^{90}_{40}Zr \rightarrow {}^{82}_{37}Rb$

1000		12.200	1.00	Bel 78
1000		15.700	0.40	Bel 78

664 Reaction $^{90}_{40}Zr \rightarrow {}^{82m}_{37}Rb$ (Fig. 432, p. 270)

44.00		0.134	0.130E-1	Kan 76
50.00		0.970	0.700E-1	Kan 76
56.00		2.000	0.100	Kan 76
62.00		1.800	0.100	Kan 76
68.00		1.600	0.100	Kan 76
74.00		0.950	0.500E-1	Kan 76
80.00		1.900	0.100	Kan 76
86.00		6.500	0.400	Kan 76

665 Reaction $^{90}_{40}Zr \rightarrow {}^{83}_{37}Rb$ (Fig. 433, p. 270)

38.00		0.280	2.00E-1	Kan 76
44.00		0.770	5.00E-1	Kan 76
50.00		0.510	3.00E-1	Kan 76
56.00		0.370	2.00E-1	Kan 76
62.00		0.260	5.00E-1	Kan 76

E [MeV]	ΔE [MeV]	σ [mb]	Δσ [mb]	Ref.
665 Reaction $^{90}_{40}Zr \rightarrow ^{83}_{37}Rb$ (continued)				
68.00		1.800	0.11	Kan 76
74.00		5.500	0.40	Kan 76
80.00		7.600	0.90	Kan 76
86.00		11.600	0.70	Kan 76
1000		13.900	2.50	Bel 78
1000		12.500	0.50	Bel 78
666 Reaction $^{90}_{40}Zr \rightarrow ^{84}_{37}Rb$ (Fig. 434, p. 270)				
50.00		0.240E-1	0.30E-2	Kan 76
56.00		0.380E-1	0.20E-1	Kan 76
62.00		1.000	0.60E-1	Kan 76
68.00		2.100	0.10	Kan 76
74.00		2.500	0.20	Kan 76
80.00		2.400	0.20	Kan 76
86.00		2.200	0.10	Kan 76
1000		4.600	0.40	Bel 78
1000		4.300	0.30	Bel 78
667 Reaction $^{90}_{40}Zr \rightarrow ^{86}_{37}Rb$				
1000		0.450	0.50E-1	Bel 78
668 Reaction $^{90}_{40}Zr \rightarrow ^{82}_{38}Sr$ (cumulative)				
1000		15.100	1.20	Bel 78
669 Reaction $^{90}_{40}Zr \rightarrow ^{83}_{38}Sr$ (cumulative) (Fig. 435, p. 271)				
60.00		0.320E-1		Uns 61
62.00		1.200	0.10	Kan 76
68.00		13.400	0.80	Kan 76
74.00		38.000	2.00	Kan 76
80.00		45.000	3.00	Kan 76
86.00		48.000	3.00	Kan 76
100.0		12.000		Uns 61
150.0		11.300		Uns 61
240.0		16.900		Uns 61
1000		20.100	1.50	Bel 78
670 Reaction $^{90}_{40}Zr \rightarrow ^{85}_{38}Sr$ (Fig. 436, p. 271)				
38.00		7.400	0.40	Kan 76
44.00		18.900	0.90	Kan 76
50.00		27.000	2.00	Kan 76
56.00		25.000	2.00	Kan 76

E [MeV]	ΔE [MeV]	σ [mb]	Δσ [mb]	Ref.

670 Reaction $^{90}_{40}Zr \rightarrow ^{85}_{38}Sr$ (continued)

E [MeV]	ΔE [MeV]	σ [mb]	Δσ [mb]	Ref.
62.00		16.500	0.80	Kan 76
68.00		12.400	0.70	Kan 76
74.00		15.700	1.10	Kan 76
80.00		18.000	1.00	Kan 76
86.00		34.000	2.00	Kan 76

671 Reaction $^{90}_{40}Zr \rightarrow ^{83}_{39}Y$

E [MeV]	ΔE [MeV]	σ [mb]	Δσ [mb]	Ref.
60.00		0.210E-2		Uns 61
100.0		0.910E-2		Uns 61
150.0		0.660E-2		Uns 61
240.0		0.780E-2		Uns 61

672 Reaction $^{90}_{40}Zr \rightarrow ^{84}_{39}Y$ (Fig. 437, p. 271)

E [MeV]	ΔE [MeV]	σ [mb]	Δσ [mb]	Ref.
56.00		7.700	1.00	Kan 76
62.00		19.300	1.80	Kan 76
68.00		23.000	2.00	Kan 76
74.00		15.100	2.00	Kan 76
80.00		10.200	1.00	Kan 76
86.00		9.600	1.00	Kan 76

673 Reaction $^{90}_{40}Zr \rightarrow ^{85}_{39}Y$ (Fig. 438, p. 271)

E [MeV]	ΔE [MeV]	σ [mb]	Δσ [mb]	Ref.
38.00		3.600	0.30	Kan 76
44.00		25.000	1.00	Kan 76
50.00		45.000	3.00	Kan 76
56.00		29.000	2.00	Kan 76
62.00		15.400	0.90	Kan 76
68.00		12.900	0.60	Kan 76
74.00		16.300	0.50	Kan 76
80.00		34.000	2.00	Kan 76
86.00		57.000	4.00	Kan 76

674 Reaction $^{90}_{40}Zr \rightarrow ^{86}_{39}Y$ (Fig. 439, p. 271)

E [MeV]	ΔE [MeV]	σ [mb]	Δσ [mb]	Ref.
24.00		2.100	0.10	Kan 76
30.00		41.000	2.00	Kan 76
38.00		57.000	2.00	Kan 76
44.00		28.000	1.00	Kan 76
50.00		14.000	1.00	Kan 76
56.00		13.300	0.80	Kan 76
62.00		29.000	2.00	Kan 76
68.00		82.000	5.00	Kan 76
74.00		117.000	7.00	Kan 76
80.00		122.000	9.00	Kan 76

E [MeV]	ΔE [MeV]	σ [mb]	Δσ [mb]	Ref.

674 Reaction $^{90}_{40}Zr \rightarrow ^{86}_{39}Y$ (continued)

E [MeV]	ΔE [MeV]	σ [mb]	Δσ [mb]	Ref.
86.00		118.000	8.00	Kan 76
1000		23.600	1.60	Bel 78

675 Reaction $^{90}_{40}Zr \rightarrow ^{86m}_{39}Y$

E [MeV]	ΔE [MeV]	σ [mb]	Δσ [mb]	Ref.
24.00		5.000		Lev 91
24.80		6.700		Lev 91
25.70		7.700		Lev 91
26.60		11.500		Lev 91
27.60		14.700		Lev 91
28.50		16.500		Lev 91
29.50		19.500		Lev 91

676 Reaction $^{90}_{40}Zr \rightarrow ^{86m+g}_{39}Y$

E [MeV]	ΔE [MeV]	σ [mb]	Δσ [mb]	Ref.
22.20		1.000		Lev 91
23.10		4.100		Lev 91
24.00		8.200		Lev 91
24.80		12.500		Lev 91
25.70		18.000		Lev 91
26.60		30.000		Lev 91
27.60		34.000		Lev 91
28.50		41.000		Lev 91
29.50		45.000		Lev 91

677 Reaction $^{90}_{40}Zr \rightarrow ^{87}_{39}Y$ (Fig. 440, p. 272)

E [MeV]	ΔE [MeV]	σ [mb]	Δσ [mb]	Ref.
11.20		≤ 6.000		Del 63
12.00		1.000	0.10	Kan 76
16.00		5.100	0.30	Kan 76
18.30		7.000		Lev 91
19.00		12.100	0.70	Kan 76
19.30		7.000		Lev 91
20.30		9.000		Lev 91
21.40		10.200		Lev 91
22.20		11.200		Lev 91
23.10		10.000		Lev 91
24.00		26.000	2.00	Kan 76
24.00		10.500		Lev 91
24.80		8.200		Lev 91
25.70		7.200		Lev 91
26.60		6.500		Lev 91
27.60		5.000		Lev 91
28.50		5.200		Lev 91
29.50		5.200		Lev 91
30.00		18.000	1.00	Kan 76

E [MeV]	ΔE [MeV]	σ [mb]	Δσ [mb]	Ref.

677 Reaction $^{90}_{40}Zr \rightarrow ^{87}_{39}Y$ (continued)

E [MeV]	ΔE [MeV]	σ [mb]	Δσ [mb]	Ref.
38.00		9.000	3.00	Kan 76
47.00		40.000	10.00	Kan 76
56.00		108.000	25.00	Kan 76
68.00		114.000	16.00	Kan 76
86.00		102.000	22.00	Kan 76
1000		12.000	3.00	Bel 78

678 Reaction $^{90}_{40}Zr \rightarrow ^{87m}_{39}Y$ (Fig. 441, p. 272)

E [MeV]	ΔE [MeV]	σ [mb]	Δσ [mb]	Ref.
15.80		1.200		Lev 91
16.70		5.500		Lev 91
17.50		7.500		Lev 91
18.30		8.500		Lev 91
19.30		11.400		Lev 91
20.30		13.000		Lev 91
21.40		15.000		Lev 91
22.20		17.000		Lev 91
23.10		17.000		Lev 91
24.00		15.000		Lev 91
24.80		14.000		Lev 91
25.70		14.000		Lev 91
26.60		11.300		Lev 91
27.60		9.000		Lev 91
28.50		7.200		Lev 91
29.50		7.000		Lev 91

679 Reaction $^{90}_{40}Zr \rightarrow ^{87}_{39}Y$ (cumulative)

E [MeV]	ΔE [MeV]	σ [mb]	Δσ [mb]	Ref.
1000		27.500	1.90	Bel 78

680 Reaction $^{90}_{40}Zr \rightarrow ^{88}_{39}Y$ (Fig. 442, p. 272)

E [MeV]	ΔE [MeV]	σ [mb]	Δσ [mb]	Ref.
11.20		≤ 1.000		Del 63
30.00		3.600	0.10	Kan 76
38.00		27.600	1.10	Kan 76
44.00		40.000	2.00	Kan 76
50.00		46.000	3.00	Kan 76
56.00		45.000	3.00	Kan 76
62.00		38.000	4.00	Kan 76
68.00		42.000	3.00	Kan 76
74.00		41.000	3.00	Kan 76
80.00		41.000	3.00	Kan 76
86.00		44.000	3.00	Kan 76
1000		27.300	2.00	Bel 78

E [MeV]	ΔE [MeV]	σ [mb]	Δσ [mb]	Ref.
681 Reaction $^{90}_{40}Zr \rightarrow ^{86}_{40}Zr$ (cumulative)				
1000		6.300	0.40	Bel 78
682 Reaction $^{90}_{40}Zr \rightarrow ^{88}_{40}Zr$ (cumulative)				
1000		27.000	1.70	Bel 78
683 Reaction $^{90}_{40}Zr \rightarrow ^{89}_{40}Zr$ (Fig. 443, p. 272)				
11.20		≤ 2.000		Del 63
12.10		11.000		Lev 91
12.80		29.000		Lev 91
13.80		109.000		Lev 91
14.80		216.000		Lev 91
15.80		420.000		Lev 91
16.70		602.000		Lev 91
17.50		665.000		Lev 91
18.30		740.000		Lev 91
19.30		885.000		Lev 91
20.30		1007.000		Lev 91
21.40		1082.000		Lev 91
22.20		1147.000		Lev 91
23.10		1200.000		Lev 91
24.00		1182.000		Lev 91
24.80		1177.000		Lev 91
25.70		1227.000		Lev 91
26.60		1195.000		Lev 91
27.60		1172.000		Lev 91
28.50		1117.000		Lev 91
29.50		1067.000		Lev 91
684 Reaction $^{90}_{40}Zr \rightarrow ^{89}_{40}Zr$ (cumulative)				
1000		45.400	3.00	Bel 78
685 Reaction $^{90}_{40}Zr \rightarrow ^{89m}_{41}Nb$ (Fig. 444, p. 272)				
15.80		13.000		Lev 91
16.70		35.000		Lev 91
17.50		61.000		Lev 91
18.30		90.000		Lev 91
19.30		141.000		Lev 91
20.30		193.000		Lev 91
21.40		198.000		Lev 91
22.20		225.000		Lev 91
23.10		211.000		Lev 91
24.00		210.000		Lev 91

685 Reaction $^{90}_{40}Zr \rightarrow ^{89m}_{41}Nb$ (continued)

E [MeV]	ΔE [MeV]	σ [mb]	Δσ [mb]	Ref.
24.80		200.000		Lev 91
25.70		204.000		Lev 91
26.60		185.000		Lev 91
27.60		166.000		Lev 91
28.50		144.000		Lev 91
29.50		120.000		Lev 91

686 Reaction $^{90}_{40}Zr \rightarrow ^{90}_{41}Nb$ (Fig. 445, p. 273)

E [MeV]	ΔE [MeV]	σ [mb]	Δσ [mb]	Ref.
6.91		2.000		Del 63
7.70		520.000		Lev 91
8.00		61.000	7.00	Ska 87
8.15		57.000		Del 63
8.50		113.000	12.00	Ska 87
8.60		655.000		Lev 91
9.00		157.000	21.00	Ska 87
9.27		407.000		Del 63
9.50		743.000		Lev 91
10.26		582.000		Del 63
10.40		757.000		Lev 91
11.20		712.000	21.00	Del 63
11.30		855.000		Lev 91
12.10		880.000		Lev 91
12.80		900.000		Lev 91
13.80		1000.000		Lev 91
14.80		937.000		Lev 91
15.80		851.000		Lev 91
16.70		661.000		Lev 91
17.50		537.000		Lev 91
18.30		447.000		Lev 91
19.30		380.000		Lev 91
20.30		240.000		Lev 91
21.40		190.000		Lev 91
22.20		155.000		Lev 91
23.10		138.000		Lev 91
24.00		107.000		Lev 91
24.80		100.000		Lev 91
25.70		95.000		Lev 91
26.60		87.000		Lev 91
27.60		79.000		Lev 91
28.50		78.000		Lev 91
29.50		76.000		Lev 91

E [MeV]	ΔE [MeV]	σ [mb]	Δσ [mb]	Ref.
687 Reaction $^{90}_{40}Zr \rightarrow {}^{90m}_{41}Nb$				
8.00		122.000	10.00	Ska 87
8.50		193.000	15.00	Ska 87
9.00		293.000	23.00	Ska 87
400.0		2.180	0.29	Chu 69
688 Reaction $^{91}_{40}Zr \rightarrow {}^{71}_{33}As$ (cumulative)				
1000		12.800	0.90	Bel 78
689 Reaction $^{91}_{40}Zr \rightarrow {}^{72}_{33}As$				
1000		16.700	1.70	Bel 78
690 Reaction $^{91}_{40}Zr \rightarrow {}^{74}_{33}As$				
1000		4.700	0.40	Bel 78
691 Reaction $^{91}_{40}Zr \rightarrow {}^{72}_{34}Se$ (cumulative)				
1000		4.100	0.40	Bel 78
692 Reaction $^{91}_{40}Zr \rightarrow {}^{73}_{34}Se$ (cumulative)				
1000		11.500	0.70	Bel 78
693 Reaction $^{91}_{40}Zr \rightarrow {}^{75}_{34}Se$ (cumulative)				
1000		22.800	1.80	Bel 78
694 Reaction $^{91}_{40}Zr \rightarrow {}^{71}_{35}Br$				
1000		0.400E-1	0.400E-1	Bel 78
695 Reaction $^{91}_{40}Zr \rightarrow {}^{72}_{35}Br$				
1000		0.300	0.10	Bel 78
696 Reaction $^{91}_{40}Zr \rightarrow {}^{73}_{35}Br$				
1000		0.520	0.14	Bel 78
697 Reaction $^{91}_{40}Zr \rightarrow {}^{74}_{35}Br$				
1000		3.700	0.40	Bel 78

E [MeV]	ΔE [MeV]	σ [mb]	Δσ [mb]	Ref.
698 Reaction $^{91}_{40}Zr \rightarrow ^{75}_{35}Br$				
1000		10.700	0.30	Bel 78
699 Reaction $^{91}_{40}Zr \rightarrow ^{76}_{35}Br$				
1000		14.100	0.40	Bel 78
700 Reaction $^{91}_{40}Zr \rightarrow ^{77}_{35}Br$ (cumulative)				
1000		16.000	0.60	Bel 78
1000		20.000	1.40	Bel 78
701 Reaction $^{91}_{40}Zr \rightarrow ^{78}_{35}Br$				
1000		9.900	1.40	Bel 78
702 Reaction $^{91}_{40}Zr \rightarrow ^{80}_{35}Br$				
1000		2.800	1.40	Bel 78
703 Reaction $^{91}_{40}Zr \rightarrow ^{82}_{35}Br$				
1000		0.450	0.20	Bel 78
704 Reaction $^{91}_{40}Zr \rightarrow ^{83}_{35}Br$				
1000		0.270	0.17	Bel 78
705 Reaction $^{91}_{40}Zr \rightarrow ^{84}_{35}Br$				
1000		0.500E-1	0.500E-1	Bel 78
706 Reaction $^{91}_{40}Zr \rightarrow ^{76}_{37}Rb$				
1000		0.190	0.80E-1	Bel 78
707 Reaction $^{91}_{40}Zr \rightarrow ^{77}_{37}Rb$				
1000		0.400	0.12	Bel 78
708 Reaction $^{91}_{40}Zr \rightarrow ^{78}_{37}Rb$				
1000		2.520	0.24	Bel 78
709 Reaction $^{91}_{40}Zr \rightarrow ^{79}_{37}Rb$				
1000		8.900	0.40	Bel 78

E [MeV]	ΔE [MeV]	σ [mb]	Δσ [mb]	Ref.

710 Reaction $^{91}_{40}Zr \rightarrow {}^{80}_{37}Rb$

1000		12.700	0.60	Bel 78

711 Reaction $^{91}_{40}Zr \rightarrow {}^{81}_{37}Rb$ (cumulative)

1000		18.500	1.20	Bel 78
1000		21.300	1.20	Bel 78

712 Reaction $^{91}_{40}Zr \rightarrow {}^{82}_{37}Rb$

1000		13.300	1.20	Bel 78
1000		18.100	0.80	Bel 78

713 Reaction $^{91}_{40}Zr \rightarrow {}^{83}_{37}Rb$

1000		13.100	3.00	Bel 78
1000		12.200	0.60	Bel 78

714 Reaction $^{91}_{40}Zr \rightarrow {}^{84}_{37}Rb$

1000		5.600	0.50	Bel 78
1000		6.300	0.40	Bel 78

715 Reaction $^{91}_{40}Zr \rightarrow {}^{86}_{37}Rb$

1000		1.600	0.40	Bel 78

716 Reaction $^{91}_{40}Zr \rightarrow {}^{88}_{37}Rb$

1000		0.240	0.12	Bel 78

717 Reaction $^{91}_{40}Zr \rightarrow {}^{82}_{38}Sr$ (cumulative)

1000		12.300	1.40	Bel 78

718 Reaction $^{91}_{40}Zr \rightarrow {}^{83}_{38}Sr$

1000		18.500	1.40	Bel 78

719 Reaction $^{91}_{40}Zr \rightarrow {}^{86}_{39}Y$

1000		20.700	1.40	Bel 78

720 Reaction $^{91}_{40}Zr \rightarrow {}^{87}_{39}Y$

23.10		5.000		Lev 91
24.00		5.600		Lev 91

E [MeV]	ΔE [MeV]	σ [mb]	Δσ [mb]	Ref.

720 Reaction $^{91}_{40}Zr \rightarrow ^{87}_{39}Y$ (continued)

E [MeV]	ΔE [MeV]	σ [mb]	Δσ [mb]	Ref.
24.80		6.200		Lev 91
25.70		6.800		Lev 91
26.60		8.000		Lev 91
27.60		9.500		Lev 91
28.50		10.000		Lev 91
29.50		11.000		Lev 91
1000		9.300	2.40	Bel 78

721 Reaction $^{91}_{40}Zr \rightarrow ^{87}_{39}Y$ (cumulative)

E [MeV]	ΔE [MeV]	σ [mb]	Δσ [mb]	Ref.
1000		25.500	2.00	Bel 78

722 Reaction $^{91}_{40}Zr \rightarrow ^{87m}_{39}Y$

E [MeV]	ΔE [MeV]	σ [mb]	Δσ [mb]	Ref.
21.40		8.500		Lev 91
22.20		14.000		Lev 91
23.10		18.000		Lev 91
24.00		19.000		Lev 91
24.80		23.000		Lev 91
25.70		26.000		Lev 91
26.60		28.000		Lev 91
27.60		30.000		Lev 91
28.50		36.000		Lev 91
29.50		37.000		Lev 91

723 Reaction $^{91}_{40}Zr \rightarrow ^{88}_{39}Y$ (Fig. 446, p. 273)

E [MeV]	ΔE [MeV]	σ [mb]	Δσ [mb]	Ref.
9.50		2.200		Lev 91
10.40		3.000		Lev 91
11.20		≤ 4.000		Del 63
11.30		4.500		Lev 91
12.10		4.800		Lev 91
12.80		5.000		Lev 91
13.80		7.400		Lev 91
14.80		9.500		Lev 91
15.80		12.000		Lev 91
16.70		13.000		Lev 91
17.50		15.000		Lev 91
18.30		16.000		Lev 91
19.30		19.000		Lev 91
20.30		19.000		Lev 91
21.40		17.000		Lev 91
22.20		16.000		Lev 91
23.10		14.000		Lev 91
24.00		13.000		Lev 91
24.80		12.000		Lev 91

E [MeV]	ΔE [MeV]	σ [mb]	Δσ [mb]	Ref.
723 Reaction $^{91}_{40}Zr \rightarrow ^{88}_{39}Y$ (continued)				
25.70		10.000		Lev 91
26.60		7.800		Lev 91
27.60		9.800		Lev 91
1000		25.200	1.80	Bel 78
724 Reaction $^{91}_{40}Zr \rightarrow ^{86}_{40}Zr$ (cumulative)				
1000		4.600	0.40	Bel 78
725 Reaction $^{91}_{40}Zr \rightarrow ^{88}_{40}Zr$ (cumulative)				
1000		20.300	1.50	Bel 78
726 Reaction $^{91}_{40}Zr \rightarrow ^{89}_{40}Zr$				
23.10		12.000		Lev 91
24.00		57.000		Lev 91
24.80		100.000		Lev 91
25.70		141.000		Lev 91
26.60		229.000		Lev 91
27.60		323.000		Lev 91
28.50		380.000		Lev 91
29.50		549.000		Lev 91
727 Reaction $^{91}_{40}Zr \rightarrow ^{89}_{40}Zr$ (cumulative)				
1000		29.500	2.00	Bel 78
728 Reaction $^{91}_{40}Zr \rightarrow ^{90}_{41}Nb$ (Fig. 447, p. 273)				
14.80		178.000		Lev 91
15.80		479.000		Lev 91
16.70		575.000		Lev 91
17.50		741.000		Lev 91
18.30		794.000		Lev 91
19.30		891.000		Lev 91
20.30		940.000		Lev 91
21.40		933.000		Lev 91
22.20		977.000		Lev 91
23.10		960.000		Lev 91
24.00		964.000		Lev 91
24.80		871.000		Lev 91
25.70		776.000		Lev 91
26.60		724.000		Lev 91
27.60		602.000		Lev 91
28.50		461.000		Lev 91
29.50		434.000		Lev 91

E [MeV]	ΔE [MeV]	σ [mb]	Δσ [mb]	Ref.

729 Reaction $^{91}_{40}Zr \rightarrow ^{91}_{41}Nb$ (Fig. 448, p. 273)

E [MeV]	ΔE [MeV]	σ [mb]	Δσ [mb]	Ref.
3.50		2.000		Bla 51*
3.80		2.900		Bla 51*
3.90		4.300		Bla 51*
4.20		5.700		Bla 51*
4.40		7.000		Bla 51*
5.10		19.000		Bla 51*
5.40		26.000		Bla 51*
5.80		30.000		Bla 51*
6.10		39.000		Bla 51*
6.40		44.000		Bla 51*
6.70		50.000		Bla 51*
6.80		51.000		Bla 51*

730 Reaction $^{92}_{40}Zr \rightarrow ^{88}_{39}Y$ (Fig. 449, p. 273)

E [MeV]	ΔE [MeV]	σ [mb]	Δσ [mb]	Ref.
18.30		3.700		Lev 91
19.30		5.700		Lev 91
20.30		13.000		Lev 91
21.40		17.000		Lev 91
22.20		20.000		Lev 91
23.10		22.000		Lev 91
24.00		27.000		Lev 91
24.80		32.000		Lev 91
25.70		31.000		Lev 91
26.60		41.000		Lev 91
27.60		44.000		Lev 91
28.50		45.000		Lev 91
29.50		48.000		Lev 91

731 Reaction $^{92}_{40}Zr \rightarrow ^{90}_{41}Nb$

E [MeV]	ΔE [MeV]	σ [mb]	Δσ [mb]	Ref.
23.10		7.400		Lev 91
24.00		25.000		Lev 91
24.80		83.000		Lev 91
25.70		186.000		Lev 91
26.60		302.000		Lev 91
27.60		417.000		Lev 91
28.50		525.000		Lev 91
29.50		617.000		Lev 91

732 Reaction $^{92}_{40}Zr \rightarrow ^{92}_{41}Nb$ (Fig. 450, p. 274)

E [MeV]	ΔE [MeV]	σ [mb]	Δσ [mb]	Ref.
3.50		2.900		Bla 51*
3.90		7.500		Bla 51*
4.40		20.000		Bla 51*
4.80		30.000		Bla 51*

E [MeV]	ΔE [MeV]	σ [mb]	Δσ [mb]	Ref.

732 Reaction $^{92}_{40}Zr \rightarrow ^{92}_{41}Nb$ (continued)

E [MeV]	ΔE [MeV]	σ [mb]	Δσ [mb]	Ref.
5.10		50.000		Bla 51*
5.40		70.000		Bla 51*
5.80		90.000		Bla 51*
6.10		120.000		Bla 51*
6.40		160.000		Bla 51*
6.70		180.000		Bla 51*
6.80		300.000		Bla 51*

733 Reaction $^{92}_{40}Zr \rightarrow ^{92m}_{41}Nb$ (Fig. 451, p. 274)

E [MeV]	ΔE [MeV]	σ [mb]	Δσ [mb]	Ref.
4.50		10.000	0.90	Ska 87
5.00		29.500	2.20	Ska 87
5.50		73.000	5.00	Ska 87
6.00		116.000	9.00	Ska 87
6.50		181.000	13.00	Ska 87
7.00		272.000	20.00	Ska 87
7.50		379.000	28.00	Ska 87
8.00		438.000	33.00	Ska 87
8.50		503.000	38.00	Ska 87
9.00		662.000	49.00	Ska 87

734 Reaction $^{94}_{40}Zr \rightarrow ^{71}_{33}As$ (cumulative)

E [MeV]	ΔE [MeV]	σ [mb]	Δσ [mb]	Ref.
1000		9.300	0.70	Bel 78

735 Reaction $^{94}_{40}Zr \rightarrow ^{72}_{33}As$

E [MeV]	ΔE [MeV]	σ [mb]	Δσ [mb]	Ref.
1000		15.000	1.40	Bel 78

736 Reaction $^{94}_{40}Zr \rightarrow ^{74}_{33}As$

E [MeV]	ΔE [MeV]	σ [mb]	Δσ [mb]	Ref.
1000		7.300	0.50	Bel 78

737 Reaction $^{94}_{40}Zr \rightarrow ^{72}_{34}Se$ (cumulative)

E [MeV]	ΔE [MeV]	σ [mb]	Δσ [mb]	Ref.
1000		3.100	0.30	Bel 78

738 Reaction $^{94}_{40}Zr \rightarrow ^{73}_{34}Se$ (cumulative)

E [MeV]	ΔE [MeV]	σ [mb]	Δσ [mb]	Ref.
1000		8.500	0.60	Bel 78

739 Reaction $^{94}_{40}Zr \rightarrow ^{75}_{34}Se$ (cumulative)

E [MeV]	ΔE [MeV]	σ [mb]	Δσ [mb]	Ref.
1000		20.500	1.60	Bel 78

E [MeV]	ΔE [MeV]	σ [mb]	Δσ [mb]	Ref.
740 Reaction $^{94}_{40}Zr \rightarrow {}^{77}_{35}Br$ (cumulative)				
1000		17.900	1.20	Bel 78
741 Reaction $^{94}_{40}Zr \rightarrow {}^{76}_{37}Rb$				
1000		0.120	0.11	Bel 78
742 Reaction $^{94}_{40}Zr \rightarrow {}^{77}_{37}Rb$				
1000		0.320	0.11	Bel 78
743 Reaction $^{94}_{40}Zr \rightarrow {}^{78}_{37}Rb$				
1000		2.030	0.22	Bel 78
744 Reaction $^{94}_{40}Zr \rightarrow {}^{79}_{37}Rb$				
1000		6.410	1.10	Bel 78
745 Reaction $^{94}_{40}Zr \rightarrow {}^{80}_{37}Rb$				
1000		12.500	0.80	Bel 78
746 Reaction $^{94}_{40}Zr \rightarrow {}^{81}_{37}Rb$ (cumulative)				
1000		22.500	0.90	Bel 78
1000		14.300	0.90	Bel 78
747 Reaction $^{94}_{40}Zr \rightarrow {}^{82}_{37}Rb$				
1000		17.000	1.10	Bel 78
1000		19.000	1.40	Bel 78
748 Reaction $^{94}_{40}Zr \rightarrow {}^{83}_{37}Rb$				
1000		18.200	1.50	Bel 78
1000		18.200	2.40	Bel 78
749 Reaction $^{94}_{40}Zr \rightarrow {}^{84}_{37}Rb$				
1000		11.100	1.10	Bel 78
1000		11.300	0.80	Bel 78
750 Reaction $^{94}_{40}Zr \rightarrow {}^{86}_{37}Rb$				
1000		2.260	0.30	Bel 78

E [MeV]	ΔE [MeV]	σ [mb]	Δσ [mb]	Ref.
751 Reaction $^{94}_{40}Zr \rightarrow {}^{88}_{37}Rb$				
1000		0.420	0.17	Bel 78
752 Reaction $^{94}_{40}Zr \rightarrow {}^{89}_{37}Rb$				
1000		0.190	0.10	Bel 78
753 Reaction $^{94}_{40}Zr \rightarrow {}^{90}_{37}Rb$				
1000		0.700E-1	0.700E-1	Bel 78
754 Reaction $^{94}_{40}Zr \rightarrow {}^{82}_{38}Sr$ (cumulative)				
1000		6.700	0.70	Bel 78
755 Reaction $^{94}_{40}Zr \rightarrow {}^{83}_{38}Sr$ (cumulative)				
1000		12.100	0.90	Bel 78
756 Reaction $^{94}_{40}Zr \rightarrow {}^{86}_{39}Y$				
1000		15.200	1.00	Bel 78
757 Reaction $^{94}_{40}Zr \rightarrow {}^{87}_{39}Y$				
1000		5.000	2.00	Bel 78
758 Reaction $^{94}_{40}Zr \rightarrow {}^{87}_{39}Y$ (cumulative)				
1000		19.600	1.80	Bel 78
759 Reaction $^{94}_{40}Zr \rightarrow {}^{88}_{39}Y$				
1000		24.500	1.80	Bel 78
760 Reaction $^{94}_{40}Zr \rightarrow {}^{86}_{40}Zr$ (cumulative)				
1000		1.600	0.20	Bel 78
761 Reaction $^{94}_{40}Zr \rightarrow {}^{88}_{40}Zr$				
1000		10.800	0.80	Bel 78
762 Reaction $^{94}_{40}Zr \rightarrow {}^{89}_{40}Zr$				
11.20		≤ 2.000		Del 63

E [MeV]	ΔE [MeV]	σ [mb]	Δσ [mb]	Ref.
763 Reaction $^{94}_{40}Zr \rightarrow ^{89}_{40}Zr$ (cumulative)				
1000		17.500	1.10	Bel 78
764 Reaction $^{96}_{40}Zr \rightarrow ^{72}_{30}Zn$ (cumulative)				
1800		0.110	0.20E-1	Por 64a
765 Reaction $^{96}_{40}Zr \rightarrow ^{66}_{31}Ga$				
1800		2.000	0.40	Por 64a
766 Reaction $^{96}_{40}Zr \rightarrow ^{67}_{31}Ga$ (cumulative)				
1800		6.400	0.80	Por 64a
767 Reaction $^{96}_{40}Zr \rightarrow ^{72}_{31}Ga$				
1800		0.900	0.10	Por 64a
768 Reaction $^{96}_{40}Zr \rightarrow ^{73}_{31}Ga$ (cumulative)				
1800		0.340	0.50E-1	Por 64a
769 Reaction $^{96}_{40}Zr \rightarrow ^{71}_{33}As$ (cumulative)				
1800		4.400	1.00	Por 64a
770 Reaction $^{96}_{40}Zr \rightarrow ^{72}_{33}As$				
1800		9.000	1.00	Por 64a
771 Reaction $^{96}_{40}Zr \rightarrow ^{74}_{33}As$				
1800		5.100	1.00	Por 64a
772 Reaction $^{96}_{40}Zr \rightarrow ^{72}_{34}Se$ (cumulative)				
1800		0.700	0.15	Por 64a
773 Reaction $^{96}_{40}Zr \rightarrow ^{95}_{39}Y$				
250.0	12.50	35.800	3.60	Str 61
300.0		35.000	4.20	Str 61
350.0		37.400	4.60	Str 61
400.0		40.400	3.20	Str 61
440.0	8.80	37.800	2.40	Str 61

E [MeV]	ΔE [MeV]	σ [mb]	Δσ [mb]	Ref.

774 Reaction $^{96}_{40}Zr \rightarrow ^{95}_{40}Zr$ (Fig. 452, p. 274)

E [MeV]	ΔE [MeV]	σ [mb]	Δσ [mb]	Ref.
12.10		32.000		Lev 91
12.80		37.000		Lev 91
13.80		57.000		Lev 91
14.80		71.000		Lev 91
15.80		86.000		Lev 91
16.70		100.000		Lev 91
17.50		110.000		Lev 91
18.30		120.000		Lev 91
19.30		135.000		Lev 91
20.30		148.000		Lev 91
21.40		160.000		Lev 91
22.20		178.000		Lev 91
23.10		195.000		Lev 91
24.00		209.000		Lev 91
24.80		186.000		Lev 91
25.70		182.000		Lev 91
26.60		186.000		Lev 91
27.60		174.000		Lev 91
28.50		190.000		Lev 91
29.50		200.000		Lev 91
250.0	12.50	69.800	5.40	Str 61
300.0		65.700	4.40	Str 61
350.0		63.300	2.00	Str 61
400.0		61.400	3.60	Str 61
440.0	8.80	61.300	2.40	Str 61

775 Reaction $^{96}_{40}Zr \rightarrow ^{95}_{41}Nb$ (Fig. 453, p. 274)

E [MeV]	ΔE [MeV]	σ [mb]	Δσ [mb]	Ref.
7.70		84.000		Lev 91
8.60		164.000		Lev 91
9.50		277.000		Lev 91
10.40		442.000		Lev 91
11.30		460.000		Lev 91
12.10		532.000		Lev 91
12.80		722.000		Lev 91
13.80		853.000		Lev 91
14.80		842.000		Lev 91
15.80		880.000		Lev 91
16.70		942.000		Lev 91
17.50		880.000		Lev 91
18.30		761.000		Lev 91
19.30		755.000		Lev 91
20.30		535.000		Lev 91
21.40		422.000		Lev 91
22.20		300.000		Lev 91
23.10		275.000		Lev 91

E [MeV]	ΔE [MeV]	σ [mb]	Δσ [mb]	Ref.

775 Reaction $^{96}_{40}Zr \rightarrow ^{95}_{41}Nb$ (continued)

E [MeV]	ΔE [MeV]	σ [mb]	Δσ [mb]	Ref.
24.00		229.000		Lev 91
24.80		188.000		Lev 91
25.70		147.000		Lev 91
26.60		136.000		Lev 91
27.60		106.000		Lev 91
28.50		104.000		Lev 91
29.50		100.000		Lev 91
250.0	12.50	20.600	1.60	Str 61
300.0		19.000	1.00	Str 61
350.0		15.200	3.20	Str 61
400.0		14.400	4.20	Str 61
440.0	8.80	13.900	3.70	Str 61
400.0		4.450	0.61	Chu 71
400.0		3.200	0.80	Chu 71

776 Reaction $^{96}_{40}Zr \rightarrow ^{95m}_{41}Nb$ (Fig. 454, p. 274)

E [MeV]	ΔE [MeV]	σ [mb]	Δσ [mb]	Ref.
7.70		48.000		Lev 91
8.60		103.000		Lev 91
9.50		159.000		Lev 91
10.40		255.000		Lev 91
11.30		277.000		Lev 91
12.10		262.000		Lev 91
12.80		332.000		Lev 91
13.80		325.000		Lev 91
14.80		305.000		Lev 91
15.80		290.000		Lev 91
16.70		287.000		Lev 91
17.50		270.000		Lev 91
18.30		208.000		Lev 91
19.30		202.000		Lev 91
20.30		140.000		Lev 91
21.40		104.000		Lev 91
22.20		77.000		Lev 91
23.10		73.000		Lev 91
24.00		64.000		Lev 91
24.80		53.000		Lev 91
25.70		44.000		Lev 91
26.60		39.000		Lev 91
27.60		35.000		Lev 91
28.50		34.000		Lev 91
29.50		33.000		Lev 91
400.0		1.060	0.18	Chu 71

E [MeV]	ΔE [MeV]	σ [mb]	Δσ [mb]	Ref.

777 Reaction $^{96}_{40}Zr \rightarrow ^{96}_{41}Nb$ (Fig. 455, p. 275)

E [MeV]	ΔE [MeV]	σ [mb]	Δσ [mb]	Ref.
2.80		0.100		Bla 51*
2.90		0.600		Bla 51*
3.00		1.000		Bla 51*
3.40		4.200		Bla 51*
3.50		5.800		Bla 51*
3.80		12.000		Bla 51*
3.90		20.000		Bla 51*
4.00		1.600	0.30	Ska 87
4.40		30.000		Bla 51*
4.50		13.600	1.30	Ska 87
4.80		50.000		Bla 51*
5.00		40.000	3.00	Ska 87
5.10		80.000		Bla 51*
5.40		140.000		Bla 51*
5.50		86.000	7.00	Ska 87
6.00		179.000	15.00	Ska 87
6.00		210.000		Bla 51*
6.40		280.000		Bla 51*
6.50		300.000	30.00	Ska 87
6.80		300.000		Bla 51*
7.00		430.000	40.00	Ska 87
7.50		580.000	50.00	Ska 87
7.70		537.000		Lev 91
8.00		630.000	60.00	Ska 87
8.50		620.000	60.00	Ska 87
8.60		468.000		Lev 91
9.00		550.000	50.00	Ska 87
9.50		309.000		Lev 91
10.40		214.000		Lev 91
11.30		155.000		Lev 91
12.10		105.000		Lev 91
12.80		102.000		Lev 91
13.80		69.000		Lev 91
14.80		55.000		Lev 91
15.80		49.000		Lev 91
16.70		44.000		Lev 91
17.50		37.000		Lev 91
18.30		35.000		Lev 91
400.0		0.760	0.12	Chu 69

778 Reaction $_{41}Nb \rightarrow ^{3}_{1}H$

E [MeV]	ΔE [MeV]	σ [mb]	Δσ [mb]	Ref.
150.0		20.000	3.0	Bru 62b
12000		229.000	34.00	Nog 91

E [MeV]	ΔE [MeV]	σ [mb]	Δσ [mb]	Ref.
779 Reaction $_{41}$Nb → $^{7}_{4}$Be				
800.0		2.300	0.24	Dit 90
1200.0		3.880	0.30	Dit 90
2600.0		7.280	0.56	Dit 90
12000		9.720	2.56	Tom 84
780 Reaction $_{41}$Nb → $^{22}_{11}$Na				
2600.0		0.820	0.14	Dit 90
781 Reaction $_{41}$Nb → $^{24}_{11}$Na (cumulative)				
12000		3.640	0.23	Tom 84
782 Reaction $_{41}$Nb → $^{47}_{20}$Ca (cumulative)				
12000		1.020	0.12	Tom 84
783 Reaction $_{41}$Nb → $^{44m}_{21}$Sc				
12000		2.360	0.21	Tom 84
784 Reaction $_{41}$Nb → $^{46}_{21}$Sc				
12000		2.360	0.40	Tom 84
785 Reaction ^{41}Nb → $^{47}_{21}$Sc				
12000		1.320	0.46	Tom 84
786 Reaction $_{41}$Nb → $^{48}_{21}$Sc				
12000		0.310	0.12	Tom 84
787 Reaction $_{41}$Nb → $^{48}_{23}$V				
12000		2.630	0.12	Tom 84
788 Reaction $_{41}$Nb → $^{52}_{25}$Mn (cumulative)				
12000		1.610	0.10	Tom 84
789 Reaction $_{41}$Nb → $^{54}_{25}$Mn				
12000		5.260	0.56	Tom 84

E [MeV]	ΔE [MeV]	σ [mb]	Δσ [mb]	Ref.
790 Reaction $_{41}Nb \rightarrow {}^{56}_{25}Mn$ (cumulative)				
12000		1.180	0.27	Tom 84
791 Reaction $_{41}Nb \rightarrow {}^{56}_{27}Co$				
12000		1.560	0.33	Tom 84
792 Reaction $_{41}Nb \rightarrow {}^{58}_{27}Co$				
12000		6.370	1.08	Tom 84
793 Reaction $_{41}Nb \rightarrow {}^{65}_{30}Zn$ (cumulative)				
12000		8.080	0.89	Tom 84
794 Reaction $_{41}Nb \rightarrow {}^{66}_{31}Ga$				
12000		4.010	0.48	Tom 84
795 Reaction $_{41}Nb \rightarrow {}^{67}_{31}Ga$ (cumulative)				
12000		8.400	0.53	Tom 84
796 Reaction $_{41}Nb \rightarrow {}^{69}_{32}Ge$ (cumulative)				
12000		8.570	0.74	Tom 84
797 Reaction $_{41}Nb \rightarrow {}^{71}_{33}As$ (cumulative)				
12000		6.810	1.94	Tom 84
798 Reaction $_{41}Nb \rightarrow {}^{72}_{33}As$				
12000		7.760	0.60	Tom 84
799 Reaction $_{41}Nb \rightarrow {}^{74}_{33}As$				
1000		4.600	0.30	Ale 84
12000		2.360	0.31	Tom 84
800 Reaction $_{41}Nb \rightarrow {}^{73}_{34}Se$ (cumulative)				
12000		4.820	0.43	Tom 84
801 Reaction $_{41}Nb \rightarrow {}^{75}_{34}Se$				
12000		4.170	0.86	Tom 84

E [MeV]	ΔE [MeV]	σ [mb]	Δσ [mb]	Ref.
802 Reaction $_{41}Nb \rightarrow {}^{75}_{35}Br$ (cumulative)				
12000		4.640	0.60	Tom 84
803 Reaction $_{41}Nb \rightarrow {}^{76}_{35}Br$				
12000		5.980	1.57	Tom 84
804 Reaction $_{41}Nb \rightarrow {}^{77}_{35}Br$ (cumulative)				
1000		19.400	1.00	Ale 84
12000		10.500	1.40	Tom 84
805 Reaction $_{41}Nb \rightarrow {}^{76}_{36}Kr$ (cumulative)				
12000		1.550	0.62	Tom 84
806 Reaction $_{41}Nb \rightarrow {}^{77}_{36}Kr$ (cumulative)				
12000		2.240	0.17	Tom 84
807 Reaction $_{41}Nb \rightarrow {}^{79}_{36}Kr$ (cumulative)				
1000		20.100	3.40	Ale 84
12000		11.900	0.91	Tom 84
808 Reaction $_{41}Nb \rightarrow {}^{81}_{37}Rb$ (cumulative)				
12000		16.400	2.50	Tom 84
809 Reaction $_{41}Nb \rightarrow {}^{82m}_{37}Rb$				
12000		4.700	0.75	Tom 84
810 Reaction $_{41}Nb \rightarrow {}^{83}_{37}Rb$				
12000		5.200	2.80	Tom 84
811 Reaction $_{41}Nb \rightarrow {}^{84}_{37}Rb$				
1000		4.100	0.40	Ale 84
12000		2.740	0.22	Tom 84
812 Reaction $_{41}Nb \rightarrow {}^{82}_{38}Sr$ (cumulative)				
12000		6.630	0.88	Tom 84

E [MeV]	ΔE [MeV]	σ [mb]	Δσ [mb]	Ref.
813 Reaction $_{41}Nb \rightarrow {}^{83}_{38}Sr$ (cumulative)				
12000		11.800	1.30	Tom 84
814 Reaction $_{41}Nb \rightarrow {}^{84}_{39}Y$ (cumulative)				
12000		2.230	0.29	Tom 84
815 Reaction $_{41}Nb \rightarrow {}^{86}_{39}Y$				
12000		5.740	0.41	Tom 84
816 Reaction $_{41}Nb \rightarrow {}^{86m}_{39}Y$				
12000		6.550	0.49	Tom 84
817 Reaction $_{41}Nb \rightarrow {}^{87}_{39}Y$				
12000		2.880	0.92	Tom 84
818 Reaction $_{41}Nb \rightarrow {}^{87m}_{39}Y$ (cumulative)				
12000		22.800	4.00	Tom 84
819 Reaction $_{41}Nb \rightarrow {}^{88}_{39}Y$				
12000		10.900	2.30	Tom 84
820 Reaction $_{41}Nb \rightarrow {}^{90m}_{39}Y$				
12000		1.050	0.50E-1	Tom 84
821 Reaction $_{41}Nb \rightarrow {}^{86}_{40}Zr$ (cumulative)				
12000		3.560	0.24	Tom 84
822 Reaction $_{41}Nb \rightarrow {}^{88}_{40}Zr$ (cumulative)				
12000		13.600	3.80	Tom 84
823 Reaction $_{41}Nb \rightarrow {}^{89}_{40}Zr$ (Fig. 456, p. 275)				
16.70		4.000		Lev 91
17.50		7.000		Lev 91
18.30		10.000		Lev 91
19.30		17.000		Lev 91
20.30		26.000		Lev 91
21.40		33.000		Lev 91

E [MeV]	ΔE [MeV]	σ [mb]	Δσ [mb]	Ref.

823 Reaction $_{41}$Nb → $^{89}_{40}$Zr (continued)

E [MeV]	ΔE [MeV]	σ [mb]	Δσ [mb]	Ref.
22.20		34.000		Lev 91
23.10		37.000		Lev 91
24.00		42.000		Lev 91
24.80		47.000		Lev 91
25.70		56.000		Lev 91
26.60		56.000		Lev 91
27.60		59.000		Lev 91
28.50		56.000		Lev 91
29.50		57.000		Lev 91

824 Reaction $_{41}$Nb → $^{89}_{40}$Zr (cumulative)

E [MeV]	ΔE [MeV]	σ [mb]	Δσ [mb]	Ref.
12000		31.800	2.30	Tom 84

825 Reaction $_{41}$Nb → $^{89}_{41}$Nb

E [MeV]	ΔE [MeV]	σ [mb]	Δσ [mb]	Ref.
12000		1.180	0.18	Tom 84

826 Reaction $_{41}$Nb → $^{90}_{41}$Nb

E [MeV]	ΔE [MeV]	σ [mb]	Δσ [mb]	Ref.
12000		12.800	4.20	Tom 84

827 Reaction $_{41}$Nb → $^{92m}_{41}$Nb (Fig. 457, p. 275)

E [MeV]	ΔE [MeV]	σ [mb]	Δσ [mb]	Ref.
12.10		5.500		Lev 91
12.80		9.000		Lev 91
13.80		22.000		Lev 91
14.80		36.000		Lev 91
15.80		56.000		Lev 91
16.70		71.000		Lev 91
17.50		79.000		Lev 91
18.30		89.000		Lev 91
19.30		98.000		Lev 91
20.30		115.000		Lev 91
21.40		116.000		Lev 91
22.20		117.000		Lev 91
23.10		123.000		Lev 91
24.00		110.000		Lev 91
24.80		112.000		Lev 91
25.70		102.000		Lev 91
26.60		100.000		Lev 91
27.60		91.000		Lev 91
28.50		85.000		Lev 91
29.50		83.000		Lev 91
12000		15.600		Tom 84

E [MeV]	ΔE [MeV]	σ [mb]	Δσ [mb]	Ref.

828 Reaction $_{41}$Nb → $^{93}{}_{42}$Mo (Fig. 458, p. 275)

E [MeV]	ΔE [MeV]	σ [mb]	Δσ [mb]	Ref.
3.90		0.400E-2		Bla 51*
4.30		0.230E-1		Bla 51*
4.70		0.500E-1		Bla 51*
4.80		0.550E-1		Bla 51*
4.85		0.850E-1		Bla 51*
4.90		0.120		Bla 51*
5.10		0.140		Bla 51*
5.30		0.170		Bla 51*
5.40		0.200		Bla 51*
5.50		0.300		Bla 51*
5.70		0.400		Bla 51*
5.80		0.450		Bla 51*
5.90		0.500		Bla 51*
6.00		0.600		Bla 51*
6.20		1.000		Bla 51*
6.40		1.200		Bla 51*
6.50		1.500		Bla 51*
6.70		1.800		Bla 51
6.80		1.900		Bla 51*

829 Reaction $_{41}$Nb → $^{93m}{}_{42}$Mo (Fig. 459, p. 276)

E [MeV]	ΔE [MeV]	σ [mb]	Δσ [mb]	Ref.
5.00		0.700		For 53*
6.50		2.000		For 53*
7.00		13.000		For 53*
7.50		5.000		Jam 54*
8.00		50.000		For 53*
8.00		10.000		Jam 54*
9.00		130.000		For 53*
10.50		130.000		Jam 54*
11.00		105.000		Jam 54*
12.50		190.000		Jam 54*
12.50		270.000		For 53*
13.00		250.000		Jam 54*
13.00		350.000		For 53*
13.50		225.000		Jam 54*
13.50		350.000		For 53*
14.00		270.000		Jam 54*
14.50		265.000		Jam 54*
15.00		290.000		For 53*
15.50		215.000		Jam 54*
16.00		230.000		For 53*
17.00		150.000		Jam 54*
18.00		150.000		For 53*
18.00		135.000		Jam 54*

E [MeV]	ΔE [MeV]	σ [mb]	Δσ [mb]	Ref.

829 Reaction $_{41}Nb \rightarrow {}^{93m}{}_{42}Mo$ (continued)

E [MeV]	ΔE [MeV]	σ [mb]	Δσ [mb]	Ref.
19.00		110.000		Jam 54*
19.50		95.000		Jam 54*
20.50		90.000		Jam 54*

830 Reaction $_{42}Mo \rightarrow {}^{46}{}_{21}Sc$ (cumulative)

E [MeV]	ΔE [MeV]	σ [mb]	Δσ [mb]	Ref.
800.0		0.210	0.10E-1	Gra 83

831 Reaction $_{42}Mo \rightarrow {}^{48}{}_{23}V$ (cumulative)

E [MeV]	ΔE [MeV]	σ [mb]	Δσ [mb]	Ref.
800.0		0.260	0.10E-1	Gra 83

832 Reaction $_{42}Mo \rightarrow {}^{51}{}_{24}Cr$ (cumulative)

E [MeV]	ΔE [MeV]	σ [mb]	Δσ [mb]	Ref.
800.0		0.750	0.80E-1	Gra 83

833 Reaction $_{42}Mo \rightarrow {}^{52}{}_{25}Mn$ (cumulative)

E [MeV]	ΔE [MeV]	σ [mb]	Δσ [mb]	Ref.
800.0		0.310	0.10E-1	Gra 83

834 Reaction $_{42}Mo \rightarrow {}^{59}{}_{26}Fe$ (cumulative)

E [MeV]	ΔE [MeV]	σ [mb]	Δσ [mb]	Ref.
800.0		0.180	0.10E-1	Gra 83

835 Reaction $_{42}Mo \rightarrow {}^{56}{}_{27}Co$ (cumulative)

E [MeV]	ΔE [MeV]	σ [mb]	Δσ [mb]	Ref.
800.0		0.440	0.20E-1	Gra 83

836 Reaction $_{42}Mo \rightarrow {}^{57}{}_{27}Co$ (cumulative)

E [MeV]	ΔE [MeV]	σ [mb]	Δσ [mb]	Ref.
800.0		1.200	0.10	Gra 83

837 Reaction $_{42}Mo \rightarrow {}^{58}{}_{27}Co$ (cumulative)

E [MeV]	ΔE [MeV]	σ [mb]	Δσ [mb]	Ref.
800.0		2.200	0.10	Gra 83

838 Reaction $_{42}Mo \rightarrow {}^{65}{}_{30}Zn$ (cumulative)

E [MeV]	ΔE [MeV]	σ [mb]	Δσ [mb]	Ref.
800.0		7.600	0.20	Gra 83

839 Reaction $_{42}Mo \rightarrow {}^{71}{}_{33}As$ (cumulative)

E [MeV]	ΔE [MeV]	σ [mb]	Δσ [mb]	Ref.
800.0		12.000	0.70	Gra 83

E [MeV]	ΔE [MeV]	σ [mb]	Δσ [mb]	Ref.
840 Reaction $_{42}Mo \rightarrow {}^{74}_{33}As$				
800.0		4.600	0.40E-1	Gra 83
841 Reaction $_{42}Mo \rightarrow {}^{75}_{34}Se$ (cumulative)				
800.0		26.000	0.30	Gra 83
842 Reaction $_{42}Mo \rightarrow {}^{77}_{35}Br$ (cumulative)				
800.0		27.000	0.90	Gra 83
843 Reaction $_{42}Mo \rightarrow {}^{76}_{36}Kr$ (cumulative)				
1000		2.400	0.50	Ale 84
844 Reaction $_{42}Mo \rightarrow {}^{77}_{36}Kr$ (cumulative)				
1000		8.400	1.60	Ale 84
845 Reaction $_{42}Mo \rightarrow {}^{79}_{36}Kr$ (cumulative)				
1000		18.600	2.60	Ale 84
846 Reaction $_{42}Mo \rightarrow {}^{83}_{37}Rb$ (cumulative)				
800.0		46.000	0.40	Gra 83
847 Reaction $_{42}Mo \rightarrow {}^{84}_{37}Rb$ (cumulative)				
800.0		5.700	0.50E-1	Gra 83
848 Reaction $_{42}Mo \rightarrow {}^{86}_{37}Rb$ (cumulative)				
800.0		1.400	0.10E-1	Gra 83
849 Reaction $_{42}Mo \rightarrow {}^{82}_{38}Sr$ (cumulative)				
800.0		23.000	1.00	Gra 83
850 Reaction $_{42}Mo \rightarrow {}^{85}_{38}Sr$ (cumulative)				
800.0		48.000	0.70	Gra 83
851 Reaction $_{42}Mo \rightarrow {}^{87}_{39}Y$ (cumulative)				
800.0		59.000	1.00	Gra 83

E [MeV]	ΔE [MeV]	σ [mb]	Δσ [mb]	Ref.

852 Reaction $_{42}Mo \rightarrow {}^{88}_{40}Zr$ (cumulative)

E [MeV]	ΔE [MeV]	σ [mb]	Δσ [mb]	Ref.
800.0		43.000	0.20	Gra 83

853 Reaction $_{42}Mo \rightarrow {}^{89}_{40}Zr$ (cumulative)

E [MeV]	ΔE [MeV]	σ [mb]	Δσ [mb]	Ref.
800.0		51.000	8.0	Gra 83

854 Reaction $_{42}Mo \rightarrow {}^{91m}_{41}Nb$ (cumulative)

E [MeV]	ΔE [MeV]	σ [mb]	Δσ [mb]	Ref.
800.0		7.100	0.60	Gra 83

855 Reaction $_{42}Mo \rightarrow {}^{92m}_{41}Nb$

E [MeV]	ΔE [MeV]	σ [mb]	Δσ [mb]	Ref.
800.0		9.700	0.10	Gra 83

856 Reaction $_{42}Mo \rightarrow {}^{90}_{42}Mo$ (Fig. 460, p. 276)

E [MeV]	ΔE [MeV]	σ [mb]	Δσ [mb]	Ref.
30.00		5.000		Lag 91*
31.50		45.000		Lag 91*
33.00		70.000		Lag 91*
34.00		135.000		Lag 91*
35.00		95.000		Lag 91*
36.00		140.000		Lag 91*
37.50		190.000		Lag 91*
38.50		235.000		Lag 91*
39.00		180.000		Lag 91*
40.50		190.000		Lag 91*
42.00		215.000		Lag 91*
43.00		220.000		Lag 91*
44.00		220.000		Lag 91*
45.00		220.000		Lag 91*
46.00		210.000		Lag 91*
47.00		190.000		Lag 91*
48.00		185.000		Lag 91*
49.00		170.000		Lag 91*
50.00		155.000		Lag 91*
51.00		145.000		Lag 91*
52.00		130.000		Lag 91*
53.00		80.000		Lag 91*
54.00		75.000		Lag 91*
55.00		70.000		Lag 91*
55.50		65.000		Lag 91*
56.50		70.000		Lag 91*
57.50		70.000		Lag 91*
58.00		60.000		Lag 91*
59.00		65.000		Lag 91*
60.00		70.000		Lag 91*

E [MeV]	ΔE [MeV]	σ [mb]	Δσ [mb]	Ref.

856 Reaction $_{42}$Mo → $^{90}{}_{42}$Mo (continued)

E [MeV]	ΔE [MeV]	σ [mb]	Δσ [mb]	Ref.
61.00		65.000		Lag 91*
61.50		65.000		Lag 91*
62.50		70.000		Lag 91*
63.00		50.000		Lag 91*
64.00		50.000		Lag 91*
65.00		50.000		Lag 91*
65.50		50.000		Lag 91*
66.50		50.000		Lag 91*
67.00		55.000		Lag 91*

857 Reaction $_{42}$Mo → $^{93m}{}_{42}$Mo **(Fig. 461, p. 276)**

E [MeV]	ΔE [MeV]	σ [mb]	Δσ [mb]	Ref.
26.00		6.000		Lag 91*
27.50		7.000		Lag 91*
29.00		9.000		Lag 91*
30.00		15.000		Lag 91*
31.50		10.000		Lag 91*
33.00		10.000		Lag 91*
34.00		12.000		Lag 91*
35.50		15.000		Lag 91*
36.50		12.000		Lag 91*
37.50		20.000		Lag 91*
38.50		20.000		Lag 91*
39.50		20.000		Lag 91*
41.00		20.000		Lag 91*
42.00		25.000		Lag 91*
43.00		30.000		Lag 91*
44.00		15.000		Lag 91*
45.00		20.000		Lag 91*
46.00		20.000		Lag 91*
47.00		20.000		Lag 91*
48.00		20.000		Lag 91*
49.00		30.000		Lag 91*
50.00		11.000		Lag 91*
51.00		20.000		Lag 91*
52.00		15.000		Lag 91*
53.00		20.000		Lag 91*
54.00		18.000		Lag 91*
54.50		15.000		Lag 91*
55.50		18.000		Lag 91*
56.50		10.000		Lag 91*
57.50		14.000		Lag 91*
58.00		18.000		Lag 91*
59.00		18.000		Lag 91*
60.00		20.000		Lag 91*
60.50		15.000		Lag 91*

E [MeV]	ΔE [MeV]	σ [mb]	Δσ [mb]	Ref.

857 Reaction $_{42}$Mo → $^{93m}{}_{42}$Mo (continued)

E [MeV]	ΔE [MeV]	σ [mb]	Δσ [mb]	Ref.
61.50		17.000		Lag 91*
62.50		10.000		Lag 91*
63.00		12.000		Lag 91*
64.00		24.000		Lag 91*
65.00		15.000		Lag 91*
65.50		15.000		Lag 91*
66.50		12.000		Lag 91*
67.00		14.000		Lag 91*

858 Reaction $_{42}$Mo → $^{99}{}_{42}$Mo **(Fig. 462, p. 277)**

E [MeV]	ΔE [MeV]	σ [mb]	Δσ [mb]	Ref.
9.10	1.50	1.000		Lag 91
12.00	1.30	2.000		Lag 91
14.50	0.80	3.000		Lag 91
16.80	1.00	7.000		Lag 91
18.80	1.00	18.000		Lag 91
20.70	0.90	35.000		Lag 91
22.50	0.90	46.000		Lag 91
24.20	0.80	59.000		Lag 91
25.80	0.80	63.000		Lag 91
27.30	0.80	65.000		Lag 91
28.80	0.80	68.000		Lag 91
30.20	0.70	68.000		Lag 91
31.60	0.60	68.000		Lag 91
32.90	0.70	69.000		Lag 91
34.20	0.70	72.000		Lag 91
34.80	0.60	77.000		Lag 91
36.10	0.60	80.000		Lag 91
37.30	0.60	82.000		Lag 91
38.50	0.50	82.000		Lag 91
39.60	0.60	83.000		Lag 91
40.70	0.60	83.000		Lag 91
41.80	0.60	79.000		Lag 91
42.90	0.60	81.000		Lag 91
44.00	0.50	81.000		Lag 91
45.00	0.50	79.000		Lag 91
46.10	0.50	83.000		Lag 91
47.10	0.50	79.000		Lag 91
48.10	0.50	83.000		Lag 91
49.00	0.50	78.000		Lag 91
50.00	0.50	81.000		Lag 91
50.90	0.50	78.000		Lag 91
51.90	0.50	77.000		Lag 91
52.80	0.50	80.000		Lag 91
53.80	0.50	79.000		Lag 91
54.70	0.50	79.000		Lag 91

E [MeV]	ΔE [MeV]	σ [mb]	Δσ [mb]	Ref.

858 Reaction $_{42}$Mo → $^{99}{}_{42}$Mo (continued)

E [MeV]	ΔE [MeV]	σ [mb]	Δσ [mb]	Ref.
55.60	0.50	81.000		Lag 91
56.40	0.50	75.000		Lag 91
57.30	0.40	74.000		Lag 91
58.10	0.50	75.000		Lag 91
59.00	0.40	71.000		Lag 91
59.90	0.50	75.000		Lag 91
60.70	0.40	76.000		Lag 91
61.50	0.40	74.000		Lag 91
62.40	0.50	73.000		Lag 91
63.20	0.40	74.000		Lag 91
64.00	0.40	75.000		Lag 91
64.80	0.40	74.000		Lag 91
65.50	0.40	68.000		Lag 91
66.30	0.40	71.000		Lag 91
67.10	0.40	71.000		Lag 91

859 Reaction $_{42}$Mo → $^{99}{}_{42}$Mo **(cumulative)**

E [MeV]	ΔE [MeV]	σ [mb]	Δσ [mb]	Ref.
800.0		8.800	0.20	Gra 83

860 Reaction $_{42}$Mo → $^{93}{}_{43}$Tc **(Fig. 463, p. 277)**

E [MeV]	ΔE [MeV]	σ [mb]	Δσ [mb]	Ref.
14.50		1.000		Lag 91*
17.00		5.000		Lag 91*
19.00		260.000		Lag 91*
21.00		400.000		Lag 91*
24.00		240.000		Lag 91*
26.00		235.000		Lag 91*
27.50		285.000		Lag 91*
29.00		315.000		Lag 91*
30.00		300.000		Lag 91*
32.00		325.000		Lag 91*
33.00		235.000		Lag 91*
34.00		245.000		Lag 91*
35.00		230.000		Lag 91*
36.50		135.000		Lag 91*
37.50		145.000		Lag 91*
38.50		130.000		Lag 91*
39.50		180.000		Lag 91*
41.00		140.000		Lag 91*
42.00		210.000		Lag 91*
43.00		135.000		Lag 91*
44.00		115.000		Lag 91*
45.00		120.000		Lag 91*
46.00		160.000		Lag 91*
47.00		125.000		Lag 91*

860 Reaction $_{42}Mo \rightarrow {}^{93}_{43}Tc$ (continued)

E [MeV]	ΔE [MeV]	σ [mb]	Δσ [mb]	Ref.
48.00		105.000		Lag 91*
49.00		75.000		Lag 91*
50.00		80.000		Lag 91*
51.00		60.000		Lag 91*
52.00		80.000		Lag 91*
53.00		80.000		Lag 91*
54.00		75.000		Lag 91*
55.00		60.000		Lag 91*
55.50		50.000		Lag 91*
56.50		40.000		Lag 91*
57.50		45.000		Lag 91*
58.00		85.000		Lag 91*
60.00		85.000		Lag 91*
61.00		95.000		Lag 91*
61.50		65.000		Lag 91*
62.50		45.000		Lag 91*
63.00		75.000		Lag 91*
64.00		40.000		Lag 91*
65.00		50.000		Lag 91*
65.50		45.000		Lag 91*
66.50		40.000		Lag 91*
67.00		45.000		Lag 91*

861 Reaction $_{42}Mo \rightarrow {}^{93}_{43}Tc$ (Fig. 464, p. 277)

E [MeV]	ΔE [MeV]	σ [mb]	Δσ [mb]	Ref.
9.00		2.000		Lag 91*
12.00		10.000		Lag 91*
14.50		30.000		Lag 91*
17.00		55.000		Lag 91*
19.00		120.000		Lag 91*
21.00		155.000		Lag 91*
22.50		175.000		Lag 91*
24.00		180.000		Lag 91*
26.00		125.000		Lag 91*
27.50		135.000		Lag 91*
29.00		160.000		Lag 91*
30.00		155.000		Lag 91*
31.50		145.000		Lag 91*
33.00		145.000		Lag 91*
34.00		135.000		Lag 91*
35.00		105.000		Lag 91*
35.50		110.000		Lag 91*
37.50		95.000		Lag 91*
39.00		95.000		Lag 91*
39.50		90.000		Lag 91*
41.00		85.000		Lag 91*

E [MeV]	ΔE [MeV]	σ [mb]	Δσ [mb]	Ref.

861 Reaction $_{42}$Mo → $^{93}_{43}$Tc (continued)

E [MeV]	ΔE [MeV]	σ [mb]	Δσ [mb]	Ref.
42.00		85.000		Lag 91*
43.00		70.000		Lag 91*
44.00		80.000		Lag 91*
45.00		70.000		Lag 91*
46.00		70.000		Lag 91*
47.00		75.000		Lag 91*
48.00		75.000		Lag 91*
49.00		65.000		Lag 91*
50.00		70.000		Lag 91*
51.00		75.000		Lag 91*
52.00		70.000		Lag 91*
53.00		65.000		Lag 91*
54.00		55.000		Lag 91*
54.50		65.000		Lag 91*
55.50		75.000		Lag 91*
56.50		75.000		Lag 91*
57.50		80.000		Lag 91*
58.00		60.000		Lag 91*
59.00		55.000		Lag 91*
60.00		50.000		Lag 91*
61.00		60.000		Lag 91*
61.50		55.000		Lag 91*
62.50		55.000		Lag 91*
63.00		55.000		Lag 91*
64.00		45.000		Lag 91*
65.00		40.000		Lag 91*
65.50		45.000		Lag 91*
66.50		40.000		Lag 91*
67.00		35.000		Lag 91*

862 Reaction $_{42}$Mo → $^{95}_{43}$Tc (Fig. 465, p. 277)

E [MeV]	ΔE [MeV]	σ [mb]	Δσ [mb]	Ref.
5.00		1.000		Lag 91*
12.00		105.000		Lag 91*
14.50		415.000		Lag 91*
17.00		305.000		Lag 91*
19.00		360.000		Lag 91*
21.00		350.000		Lag 91*
22.50		295.000		Lag 91*
24.00		325.000		Lag 91*
26.00		330.000		Lag 91*
27.50		320.000		Lag 91*
29.00		280.000		Lag 91*
30.00		160.000		Lag 91*
31.50		150.000		Lag 91*
33.00		140.000		Lag 91*

E [MeV]	ΔE [MeV]	σ [mb]	Δσ [mb]	Ref.

862 Reaction $_{42}Mo \rightarrow {}^{95}_{43}Tc$ (continued)

E [MeV]	ΔE [MeV]	σ [mb]	Δσ [mb]	Ref.
34.00		145.000		Lag 91*
35.00		155.000		Lag 91*
36.00		160.000		Lag 91*
37.50		180.000		Lag 91*
38.50		180.000		Lag 91*
39.00		185.000		Lag 91*
40.00		180.000		Lag 91*
41.50		205.000		Lag 91*
42.50		180.000		Lag 91*
43.50		175.000		Lag 91*
44.50		155.000		Lag 91*
45.00		145.000		Lag 91*
46.00		140.000		Lag 91*
47.00		130.000		Lag 91*
48.00		130.000		Lag 91*
49.00		110.000		Lag 91*
50.00		100.000		Lag 91*
52.00		95.000		Lag 91*
53.00		90.000		Lag 91*
54.00		80.000		Lag 91*
54.50		85.000		Lag 91*
55.50		70.000		Lag 91*
56.50		70.000		Lag 91*
57.50		85.000		Lag 91*
58.00		70.000		Lag 91*
59.00		75.000		Lag 91*
60.00		70.000		Lag 91*
60.50		65.000		Lag 91*
61.50		60.000		Lag 91*
62.50		60.000		Lag 91*
63.00		70.000		Lag 91*
64.00		65.000		Lag 91*
65.00		55.000		Lag 91*
65.50		55.000		Lag 91*
66.50		65.000		Lag 91*
67.00		55.000		Lag 91*

863 Reaction $_{42}Mo \rightarrow {}^{95m}_{43}Tc$ (Fig. 466, p. 278)

E [MeV]	ΔE [MeV]	σ [mb]	Δσ [mb]	Ref.
9.00		15.000		Lag 91*
12.00		75.000		Lag 91*
14.50		165.000		Lag 91*
17.00		115.000		Lag 91*
19.00		145.000		Lag 91*
21.00		110.000		Lag 91*
22.50		105.000		Lag 91*

863 Reaction $_{42}Mo \rightarrow {}^{95m}_{43}Tc$ (continued)

E [MeV]	ΔE [MeV]	σ [mb]	Δσ [mb]	Ref.
24.00		110.000		Lag 91*
26.00		95.000		Lag 91*
27.50		95.000		Lag 91*
29.00		70.000		Lag 91*
30.00		45.000		Lag 91*
31.50		35.000		Lag 91*
33.00		30.000		Lag 91*
34.00		45.000		Lag 91*
35.50		45.000		Lag 91*
36.00		40.000		Lag 91*
37.50		45.000		Lag 91*
38.50		40.000		Lag 91*
39.00		50.000		Lag 91*
40.00		45.000		Lag 91*
41.50		50.000		Lag 91*
42.50		50.000		Lag 91*
43.50		40.000		Lag 91*
44.50		35.000		Lag 91*
46.00		40.000		Lag 91*
47.00		30.000		Lag 91*
48.00		20.000		Lag 91*
49.00		20.000		Lag 91*
50.00		20.000		Lag 91*
51.00		20.000		Lag 91*
52.00		20.000		Lag 91*
53.00		20.000		Lag 91*
54.00		20.000		Lag 91*
54.50		20.000		Lag 91*
55.50		20.000		Lag 91*
56.50		20.000		Lag 91*
57.50		18.000		Lag 91*
58.00		18.000		Lag 91*
59.00		15.000		Lag 91*
60.00		20.000		Lag 91*
61.00		20.000		Lag 91*
61.50		18.000		Lag 91*
62.50		15.000		Lag 91*
63.00		10.000		Lag 91*
64.00		12.000		Lag 91*
65.00		18.000		Lag 91*
65.50		20.000		Lag 91*
66.50		12.000		Lag 91*
67.00		15.000		Lag 91*

E [MeV]	ΔE [MeV]	σ [mb]	Δσ [mb]	Ref.

864 Reaction $_{42}Mo \rightarrow {}^{95m}{}_{43}Tc$ (cumulative)

E [MeV]	ΔE [MeV]	σ [mb]	Δσ [mb]	Ref.
800.0		0.560	0.40E-1	Gra 83

865 Reaction $_{42}Mo \rightarrow {}^{96}{}_{43}Tc$ (Fig. 467, p. 278)

E [MeV]	ΔE [MeV]	σ [mb]	Δσ [mb]	Ref.
5.00		2.000		Lag 91*
9.00		35.000		Lag 91*
12.00		130.000		Lag 91*
14.50		315.000		Lag 91*
17.00		450.000		Lag 91*
19.00		375.000		Lag 91*
21.00		315.000		Lag 91*
22.50		175.000		Lag 91*
24.00		190.000		Lag 91*
26.00		245.000		Lag 91*
27.50		305.000		Lag 91*
29.00		335.000		Lag 91*
30.00		340.000		Lag 91*
31.50		340.000		Lag 91*
32.00		330.000		Lag 91*
34.00		295.000		Lag 91*
35.00		335.000		Lag 91*
35.50		245.000		Lag 91*
37.50		210.000		Lag 91*
38.50		180.000		Lag 91*
39.00		155.000		Lag 91*
40.00		130.000		Lag 91*
42.00		115.000		Lag 91*
43.00		105.000		Lag 91*
44.00		95.000		Lag 91*
45.00		100.000		Lag 91*
46.00		105.000		Lag 91*
47.00		105.000		Lag 91*
48.00		105.000		Lag 91*
49.00		100.000		Lag 91*
50.00		100.000		Lag 91*
51.00		100.000		Lag 91*
52.00		100.000		Lag 91*
53.00		95.000		Lag 91*
54.00		100.000		Lag 91*
55.00		95.000		Lag 91*
55.50		90.000		Lag 91*
56.50		85.000		Lag 91*
57.50		90.000		Lag 91*
58.00		85.000		Lag 91*
59.00		75.000		Lag 91*
60.00		75.000		Lag 91*

E [MeV]	ΔE [MeV]	σ [mb]	Δσ [mb]	Ref.

865 Reaction $_{42}Mo \rightarrow {}^{96}{}_{43}Tc$ (continued)

E [MeV]	ΔE [MeV]	σ [mb]	Δσ [mb]	Ref.
61.00		70.000		Lag 91*
61.50		70.000		Lag 91*
62.50		65.000		Lag 91*
63.00		65.000		Lag 91*
64.00		65.000		Lag 91*
65.00		65.000		Lag 91*
65.50		60.000		Lag 91*
66.50		55.000		Lag 91*
67.00		55.000		Lag 91*

866 Reaction $_{42}Mo \rightarrow {}^{96}{}_{43}Tc$ (cumulative)

E [MeV]	ΔE [MeV]	σ [mb]	Δσ [mb]	Ref.
800.0		1.800	0.20	Gra 83

867 Reaction $_{42}Mo \rightarrow {}^{99m}{}_{43}Tc$ (Fig. 468, p. 278)

E [MeV]	ΔE [MeV]	σ [mb]	Δσ [mb]	Ref.
5.10	2.50	7.000		Lag 91
9.10	1.50	39.000		Lag 91
12.00	1.30	83.000		Lag 91
14.50	0.80	190.000		Lag 91
16.80	1.00	251.000		Lag 91
18.80	1.00	286.000		Lag 91
20.70	0.90	265.000		Lag 91
22.50	0.90	219.000		Lag 91
24.20	0.80	169.000		Lag 91
25.80	0.80	142.000		Lag 91
27.30	0.80	89.000		Lag 91
28.80	0.80	81.000		Lag 91
30.20	0.70	82.000		Lag 91
31.60	0.60	108.000		Lag 91
32.90	0.70	204.000		Lag 91
34.20	0.70	257.000		Lag 91
35.40	0.60	260.000		Lag 91
36.60	0.60	286.000		Lag 91
37.80	0.60	405.000		Lag 91
39.00	0.60	430.000		Lag 91
40.20	0.60	451.000		Lag 91
41.30	0.60	481.000		Lag 91
42.50	0.50	494.000		Lag 91
43.50	0.60	461.000		Lag 91
44.50	0.50	403.000		Lag 91
46.10	0.60	396.000		Lag 91
47.10	0.50	381.000		Lag 91
48.10	0.50	358.000		Lag 91
49.00	0.50	320.000		Lag 91
50.00	0.50	311.000		Lag 91

E [MeV]	ΔE [MeV]	σ [mb]	Δσ [mb]	Ref.

867 Reaction $_{42}$Mo → $^{99m}{}_{43}$Tc (continued)

E [MeV]	ΔE [MeV]	σ [mb]	Δσ [mb]	Ref.
50.90	0.50	290.000		Lag 91
51.90	0.50	256.000		Lag 91
52.80	0.50	241.000		Lag 91
53.80	0.50	248.000		Lag 91
54.70	0.50	250.000		Lag 91
55.60	0.50	237.000		Lag 91
56.40	0.50	219.000		Lag 91
57.30	0.40	237.000		Lag 91
58.10	0.50	259.000		Lag 91
59.00	0.40	233.000		Lag 91
59.90	0.50	229.000		Lag 91
60.70	0.40	224.000		Lag 91
61.50	0.40	200.000		Lag 91
62.40	0.50	210.000		Lag 91
63.20	0.40	218.000		Lag 91
64.00	0.40	211.000		Lag 91
64.80	0.40	209.000		Lag 91
65.50	0.40	234.000		Lag 91
66.30	0.40	240.000		Lag 91
67.10	0.40	233.000		Lag 91

868 Reaction $^{92}{}_{42}$Mo → $^{88}{}_{40}$Zr

E [MeV]	ΔE [MeV]	σ [mb]	Δσ [mb]	Ref.
21.40		2.700		Lev 91
22.20		6.200		Lev 91
23.10		12.000		Lev 91
24.00		19.600		Lev 91
24.80		28.000		Lev 91
25.70		34.000		Lev 91
26.60		41.000		Lev 91
27.60		57.000		Lev 91
28.50		70.000		Lev 91
29.50		80.000		Lev 91

869 Reaction $^{92}{}_{42}$Mo → $^{88}{}_{41}$Nb

E [MeV]	ΔE [MeV]	σ [mb]	Δσ [mb]	Ref.
26.60		3.700		Lev 91
27.60		6.200		Lev 91
28.50		9.900		Lev 91
29.50		10.900		Lev 91

870 Reaction $^{92}{}_{42}$Mo → $^{88m}{}_{41}$Nb

E [MeV]	ΔE [MeV]	σ [mb]	Δσ [mb]	Ref.
26.60		4.500		Lev 91
27.60		8.000		Lev 91
28.50		12.100		Lev 91
29.50		12.900		Lev 91

E [MeV]	ΔE [MeV]	σ [mb]	Δσ [mb]	Ref.

871 Reaction $^{92}_{42}Mo \rightarrow {}^{89}_{41}Nb$ (Fig. 469, p. 278)

E [MeV]	ΔE [MeV]	σ [mb]	Δσ [mb]	Ref.
15.00		8.000		Lev 84*
15.80		13.500		Lev 84*
16.80		16.500		Lev 84*
17.80		23.000		Lev 84*
18.40		26.000		Lev 84*
19.20		27.800		Lev 84*
20.00		32.000		Lev 84*
20.70		38.000		Lev 84*
21.60		48.500		Lev 84*
22.20		48.500		Lev 84*
23.40		46.000		Lev 84*
24.40		46.000		Lev 84*
25.50		42.500		Lev 84*
26.30		37.800		Lev 84*
27.40		30.000		Lev 84*
28.40		26.200		Lev 84*
29.20		21.000		Lev 84*

872 Reaction $^{92}_{42}Mo \rightarrow {}^{89m}_{41}Nb$ (Fig. 470, p. 278)

E [MeV]	ΔE [MeV]	σ [mb]	Δσ [mb]	Ref.
14.80		2.300		Lev 91
15.80		4.800		Lev 91
16.70		5.300		Lev 91
17.50		7.300		Lev 91
18.30		7.800		Lev 91
19.30		9.500		Lev 91
20.30		11.700		Lev 91
21.40		12.200		Lev 91
22.20		12.800		Lev 91
23.10		11.800		Lev 91
24.00		10.500		Lev 91
24.80		9.900		Lev 91
25.70		6.900		Lev 91
26.60		5.400		Lev 91
27.60		5.500		Lev 91
28.50		4.300		Lev 91
29.50		3.200		Lev 91

873 Reaction $^{92}_{42}Mo \rightarrow {}^{89m+g}_{41}Nb$ (Fig. 471, p. 279)

E [MeV]	ΔE [MeV]	σ [mb]	Δσ [mb]	Ref.
14.80		6.600		Lev 91
15.80		12.700		Lev 91
16.70		13.600		Lev 91
17.50		20.000		Lev 91
18.30		24.000		Lev 91
19.30		29.000		Lev 91

E [MeV]	ΔE [MeV]	σ [mb]	Δσ [mb]	Ref.

873 Reaction $^{92}_{42}Mo \rightarrow ^{89m+g}_{41}Nb$ (continued)

E [MeV]	ΔE [MeV]	σ [mb]	Δσ [mb]	Ref.
20.30		34.000		Lev 91
21.40		42.000		Lev 91
22.20		43.000		Lev 91
23.10		42.000		Lev 91
24.00		41.000		Lev 91
24.80		39.000		Lev 91
25.70		32.000		Lev 91
26.60		25.000		Lev 91
27.60		23.000		Lev 91
28.50		20.000		Lev 91
29.50		17.600		Lev 91

874 Reaction $^{92}_{42}Mo \rightarrow ^{90}_{41}Nb$

E [MeV]	ΔE [MeV]	σ [mb]	Δσ [mb]	Ref.
26.60		3.600		Lev 91
27.60		8.200		Lev 91
28.50		19.400		Lev 91
29.50		46.000		Lev 91
27.30		1.700		Lev 84
28.20		8.500		Lev 84
29.50		33.000		Lev 84

875 $^{92}_{42}Mo \rightarrow ^{90}_{42}Mo$

E [MeV]	ΔE [MeV]	σ [mb]	Δσ [mb]	Ref.
26.60		2.200		Lev 91
27.60		4.600		Lev 91
28.50		9.300		Lev 91
29.50		21.000		Lev 91

876 Reaction $^{94}_{42}Mo \rightarrow ^{90}_{41}Nb$ (Fig. 472, p. 279)

E [MeV]	ΔE [MeV]	σ [mb]	Δσ [mb]	Ref.
19.30		5.000		Lev 91
20.30		10.500		Lev 91
20.70		2.000		Lev 84*
21.40		14.800		Lev 91
21.50		10.000		Lev 84*
22.10		16.000		Lev 84*
22.20		22.000		Lev 91
23.10		27.000		Lev 91
23.50		30.000		Lev 84*
24.00		38.000		Lev 91
24.40		40.000		Lev 84*
24.80		40.000		Lev 91
25.40		50.000		Lev 84*
25.70		49.000		Lev 91
26.20		60.000		Lev 84*

E [MeV]	ΔE [MeV]	σ [mb]	Δσ [mb]	Ref.

876 Reaction $^{94}_{42}Mo \rightarrow {}^{90}_{41}Nb$ (continued)

E [MeV]	ΔE [MeV]	σ [mb]	Δσ [mb]	Ref.
26.60		57.000		Lev 91
27.40		70.000		Lev 84*
27.60		65.000		Lev 91
28.40		80.000		Lev 84*
28.50		69.000		Lev 91
29.10		90.000		Lev 84*
29.50		69.000		Lev 91

877 Reaction $^{94}_{42}Mo \rightarrow {}^{91}_{41}Nb$

E [MeV]	ΔE [MeV]	σ [mb]	Δσ [mb]	Ref.
11.20		≤ 5.000		Del 63

878 Reaction $^{94}_{42}Mo \rightarrow {}^{93}_{42}Mo$

E [MeV]	ΔE [MeV]	σ [mb]	Δσ [mb]	Ref.
11.20		53.000		Del 63

879 Reaction $^{94}_{42}Mo \rightarrow {}^{93m}_{42}Mo$

E [MeV]	ΔE [MeV]	σ [mb]	Δσ [mb]	Ref.
23.10		6.200		Lev 91
24.00		8.900		Lev 91
24.80		11.300		Lev 91
25.70		12.100		Lev 91
26.60		15.100		Lev 91
27.60		18.000		Lev 91
28.50		18.100		Lev 91
29.50		17.500		Lev 91

880 Reaction $^{94}_{42}Mo \rightarrow {}^{93}_{43}Tc$ (Fig. 473, p. 279)

E [MeV]	ΔE [MeV]	σ [mb]	Δσ [mb]	Ref.
14.80		189.000		Lev 91
15.80		315.000		Lev 91
16.70		380.000		Lev 91
17.50		387.000		Lev 91
18.30		500.000		Lev 91
19.30		585.000		Lev 91
20.30		572.000		Lev 91
21.40		630.000		Lev 91
22.20		600.000		Lev 91
23.10		602.000		Lev 91
24.00		537.000		Lev 91
24.80		500.000		Lev 91
25.70		484.000		Lev 91
26.60		426.000		Lev 91
27.60		343.000		Lev 91
28.50		307.000		Lev 91
29.50		227.000		Lev 91

E [MeV]	ΔE [MeV]	σ [mb]	Δσ [mb]	Ref.

881 Reaction $^{94}_{42}Mo \rightarrow ^{93m}_{43}Tc$ (Fig. 474, p. 279)

E [MeV]	ΔE [MeV]	σ [mb]	Δσ [mb]	Ref.
13.80		14.000		Lev 91
14.80		25.000		Lev 91
15.80		88.000		Lev 91
16.70		101.000		Lev 91
17.50		105.000		Lev 91
18.30		139.000		Lev 91
19.30		159.000		Lev 91
20.30		146.000		Lev 91
21.40		153.000		Lev 91
22.20		145.000		Lev 91
23.10		128.000		Lev 91
24.00		109.000		Lev 91
24.80		90.000		Lev 91
25.70		87.000		Lev 91
26.60		64.000		Lev 91
27.60		56.000		Lev 91
28.50		45.000		Lev 91
29.50		36.000		Lev 91

882 Reaction $^{94}_{42}Mo \rightarrow ^{94}_{43}Tc$ (Fig. 475, p. 279)

E [MeV]	ΔE [MeV]	σ [mb]	Δσ [mb]	Ref.
6.00		1.250	0.14	Ska 87
6.50		5.000	0.40	Ska 87
7.00		13.100	1.00	Ska 87
7.50		29.800	2.10	Ska 87
7.70		31.000		Lev 91
8.00		36.500	2.40	Ska 87
8.50		49.000	4.00	Ska 87
8.60		98.000		Lev 91
9.00		65.000	5.00	Ska 87
9.50		131.000		Lev 91
10.40		143.000		Lev 91
11.20		203.000	6.00	Del 63
11.30		167.000		Lev 91
12.10		176.000		Lev 91
12.80		206.000		Lev 91
13.80		175.000		Lev 91
14.80		161.000		Lev 91
15.80		126.000		Lev 91
16.70		73.000		Lev 91
17.50		52.000		Lev 91
18.30		48.000		Lev 91

E [MeV]	ΔE [MeV]	σ [mb]	Δσ [mb]	Ref.
883 Reaction $^{94}_{42}Mo \rightarrow ^{94m}_{43}Tc$ (Fig. 476, p. 280)				
5.50		1.000	0.10	Ska 87
6.00		52.000	3.50	Ska 87
6.50		125.000	9.00	Ska 87
7.00		233.000	16.00	Ska 87
7.50		262.000	19.00	Ska 87
7.70		280.000		Lev 91
8.00		299.000	21.00	Ska 87
8.50		308.000	21.00	Ska 87
8.60		477.000		Lev 91
9.00		331.000	23.00	Ska 87
9.50		492.000		Lev 91
10.40		542.000		Lev 91
11.30		487.000		Lev 91
12.10		505.000		Lev 91
12.80		495.000		Lev 91
13.80		325.000		Lev 91
14.80		257.000		Lev 91
15.80		174.000		Lev 91
16.70		108.000		Lev 91
17.50		64.000		Lev 91
18.30		43.000		Lev 91
884 Reaction $^{95}_{42}Mo \rightarrow ^{92m}_{41}Nb$ (Fig. 477, p. 280)				
12.80		6.100		Lev 91
14.80		8.000		Lev 91
16.70		7.000		Lev 91
17.50		8.300		Lev 91
18.30		7.200		Lev 91
19.30		6.700		Lev 91
20.30		5.800		Lev 91
21.40		5.100		Lev 91
22.20		4.400		Lev 91
23.10		3.300		Lev 91
24.00		3.100		Lev 91
24.80		2.900		Lev 91
25.70		2.400		Lev 91
26.60		2.800		Lev 91
27.60		2.600		Lev 91
28.50		2.400		Lev 91
29.50		2.100		Lev 91
885 Reaction $^{95}_{42}Mo \rightarrow ^{93}_{43}Tc$				
22.20		41.000		Lev 91
23.10		88.000		Lev 91

E [MeV]	ΔE [MeV]	σ [mb]	Δσ [mb]	Ref.

885 Reaction $^{95}_{42}Mo \rightarrow ^{93}_{43}Tc$ (continued)

E [MeV]	ΔE [MeV]	σ [mb]	Δσ [mb]	Ref.
24.00		125.000		Lev 91
24.80		208.000		Lev 91
25.70		285.000		Lev 91
26.60		370.000		Lev 91
27.60		412.000		Lev 91
28.50		432.000		Lev 91
29.50		505.000		Lev 91

886 Reaction $^{95}_{42}Mo \rightarrow ^{93m}_{43}Tc$

E [MeV]	ΔE [MeV]	σ [mb]	Δσ [mb]	Ref.
24.00		13.000		Lev 91
24.80		25.000		Lev 91
25.70		36.000		Lev 91
26.60		47.000		Lev 91
27.60		58.000		Lev 91
28.50		60.000		Lev 91
29.50		68.000		Lev 91

887 Reaction $^{95}_{42}Mo \rightarrow ^{94}_{43}Tc$ (Fig. 478, p. 280)

E [MeV]	ΔE [MeV]	σ [mb]	Δσ [mb]	Ref.
12.80		32.000		Lev 91
13.80		175.000		Lev 91
14.80		247.000		Lev 91
15.80		325.000		Lev 91
16.70		430.000		Lev 91
17.50		482.000		Lev 91
18.30		487.000		Lev 91
19.30		518.000		Lev 91
20.30		522.000		Lev 91
21.40		557.000		Lev 91
22.20		535.000		Lev 91
23.10		527.000		Lev 91
24.00		450.000		Lev 91
24.80		415.000		Lev 91
25.70		330.000		Lev 91
26.60		282.000		Lev 91
27.60		194.000		Lev 91
28.50		166.000		Lev 91
29.50		152.000		Lev 91

888 Reaction $^{95}_{42}Mo \rightarrow ^{94m}_{43}Tc$ (Fig. 479, p. 280)

E [MeV]	ΔE [MeV]	σ [mb]	Δσ [mb]	Ref.
12.80		42.000		Lev 91
13.80		117.000		Lev 91
14.80		286.000		Lev 91
15.80		382.000		Lev 91

E [MeV]	ΔE [MeV]	σ [mb]	Δσ [mb]	Ref.

888 Reaction $^{95}_{42}$Mo → $^{94m}_{43}$Tc (continued)

E [MeV]	ΔE [MeV]	σ [mb]	Δσ [mb]	Ref.
16.70		384.000		Lev 91
17.50		456.000		Lev 91
18.30		442.000		Lev 91
19.30		472.000		Lev 91
20.30		425.000		Lev 91
21.40		447.000		Lev 91
22.20		375.000		Lev 91
23.10		337.000		Lev 91
24.00		270.000		Lev 91
24.80		241.000		Lev 91
25.70		180.000		Lev 91
26.60		141.000		Lev 91
27.60		99.000		Lev 91
28.50		84.000		Lev 91
29.50		76.000		Lev 91

889 Reaction $^{95}_{42}$Mo → $^{95}_{43}$Tc (Fig. 480, p. 281)

E [MeV]	ΔE [MeV]	σ [mb]	Δσ [mb]	Ref.
4.00		1.100	0.10	Ska 87
4.50		4.100	0.30	Ska 87
5.00		14.300	1.00	Ska 87
5.00		11.000		Izu 91*
5.50		35.000		Izu 91*
5.50		28.100	1.90	Ska 87
6.00		75.000	5.00	Ska 87
6.50		128.000	9.00	Ska 87
6.50		60.000		Izu 91*
7.00		140.000		Izu 91*
7.00		184.000	13.00	Ska 87
7.50		251.000	17.00	Ska 87
7.50		160.000		Izu 91*
7.70		370.000		Lev 91
8.00		190.000		Izu 91*
8.00		303.000	21.00	Ska 87
8.50		317.000	22.00	Ska 87
8.50		195.000		Izu 91*
8.60		459.000		Lev 91
9.00		230.000		Izu 91*
9.00		345.000	24.00	Ska 87
9.50		260.000		Izu 91*
9.50		518.000		Lev 91
10.40		605.000		Lev 91
10.50		370.000		Izu 91*
11.30		690.000		Lev 91
12.00		380.000		Izu 91*
12.10		668.000		Lev 91

889 Reaction $^{95}_{42}Mo \rightarrow ^{95}_{43}Tc$ (continued)

E [MeV]	ΔE [MeV]	σ [mb]	Δσ [mb]	Ref.
12.80		650.000		Lev 91
13.00		390.000		Izu 91*
13.80		409.000		Lev 91
14.00		280.000		Izu 91*
14.50		210.000		Izu 91*
14.80		285.000		Lev 91
15.00		200.000		Izu 91*
15.80		185.000		Lev 91
16.00		130.000		Izu 91*
16.50		85.000		Izu 91*
16.70		112.000		Lev 91
17.50		83.000		Lev 91
18.00		40.000		Izu 91*
18.30		75.000		Lev 91
20.50		19.000		Izu 91*
23.00		15.000		Izu 91*
26.00		12.000		Izu 91*
28.00		12.000		Izu 91*

890 Reaction $^{95}_{42}Mo \rightarrow ^{95m}_{43}Tc$ (Fig. 481, p. 281)

E [MeV]	ΔE [MeV]	σ [mb]	Δσ [mb]	Ref.
4.00		0.800	0.70	Ska 87
4.50		2.500	0.80	Ska 87
5.00		7.500	1.00	Ska 87
5.00		7.800		Izu 91*
5.50		18.000		Izu 91*
5.50		13.300	1.20	Ska 87
6.00		38.000	3.00	Ska 87
6.50		47.000	4.00	Ska 87
6.50		30.000		Izu 91*
7.00		50.000		Izu 91*
7.00		72.000	6.00	Ska 87
7.50		100.000	8.00	Ska 87
7.50		65.000		Izu 91*
7.70		161.000		Lev 91
8.00		85.000		Izu 91*
8.00		111.000	9.00	Ska 87
8.50		122.000	10.00	Ska 87
8.50		100.000		Izu 91*
8.60		232.000		Lev 91
9.00		110.000		Izu 91*
9.00		127.000	10.00	Ska 87
9.50		120.000		Izu 91*
9.50		220.000		Lev 91
10.40		257.000		Lev 91
10.50		130.000		Izu 91*

E [MeV]	ΔE [MeV]	σ [mb]	Δσ [mb]	Ref.

890 Reaction $^{95}_{42}Mo \rightarrow {}^{95m}_{43}Tc$ (continued)

E [MeV]	ΔE [MeV]	σ [mb]	Δσ [mb]	Ref.
11.30		260.000		Lev 91
12.00		130.000		Izu 91*
12.10		249.000		Lev 91
12.80		224.000		Lev 91
13.00		120.000		Izu 91*
13.80		143.000		Lev 91
14.00		85.000		Izu 91*
14.50		70.000		Izu 91*
14.80		86.000		Lev 91
15.00		60.000		Izu 91*
15.80		52.000		Lev 91
16.00		40.000		Izu 91*
16.50		25.000		Izu 91*
16.70		34.000		Lev 91
17.50		26.000		Lev 91
18.00		13.000		Izu 91*
18.30		21.000		Lev 91
20.50		6.000		Izu 91*
23.00		4.200		Izu 91*
26.00		4.300		Izu 91*
28.00		3.900		Izu 91*

891 Reaction $^{95}_{42}Mo \rightarrow {}^{95m+g}_{43}Tc$ (Fig. 482, p. 282)

E [MeV]	ΔE [MeV]	σ [mb]	Δσ [mb]	Ref.
5.00		19.000		Izu 91*
5.50		55.000		Izu 91*
6.50		90.000		Izu 91*
7.00		190.000		Izu 91*
7.50		225.000		Izu 91*
8.00		275.000		Izu 91*
8.50		295.000		Izu 91*
9.00		340.000		Izu 91*
9.50		380.000		Izu 91*
10.50		500.000		Izu 91*
12.00		510.000		Izu 91*
13.00		510.000		Izu 91*
14.00		365.000		Izu 91*
14.50		280.000		Izu 91*
15.00		260.000		Izu 91*
16.00		170.000		Izu 91*
16.50		110.000		Izu 91*
18.00		53.000		Izu 91*
20.50		25.000		Izu 91*
23.00		19.000		Izu 91*
26.00		16.500		Izu 91*
28.00		16.000		Izu 91*

E [MeV]	ΔE [MeV]	σ [mb]	Δσ [mb]	Ref.

892 Reaction $^{96}_{42}Mo \rightarrow {}^{72}_{30}Zn$ (cumulative)

1800		0.600E-2	0.10E-2	Por 64a

893 Reaction $^{96}_{42}Mo \rightarrow {}^{66}_{31}Ga$

1800		6.000	0.20	Por 64a

894 Reaction $^{96}_{42}Mo \rightarrow {}^{67}_{31}Ga$

1800		9.600		Por 64a

895 Reaction $^{96}_{42}Mo \rightarrow {}^{67}_{31}Ga$ (cumulative)

1800		10.500	0.70	Por 64a

896 Reaction $^{96}_{42}Mo \rightarrow {}^{72}_{31}Ga$

1800		0.390E-1		Por 64a
1800		0.220	0.20E-1	Por 64a

897 Reaction $^{96}_{42}Mo \rightarrow {}^{73}_{31}Ga$ (cumulative)

1800		0.400E-1	7.00E-2	Por 64a

898 Reaction $^{96}_{42}Mo \rightarrow {}^{71}_{33}As$

1800		9.400		Por 64a

899 Reaction $^{96}_{42}Mo \rightarrow {}^{71}_{33}As$ (cumulative)

1800		9.700	0.40	Por 64a

900 Reaction $^{96}_{42}Mo \rightarrow {}^{72}_{33}As$

1800		9.50	1.00	Por 64a

901 Reaction $^{96}_{42}Mo \rightarrow {}^{74}_{33}As$

1800		3.30	0.20	Por 64a

902 Reaction $^{96}_{42}Mo \rightarrow {}^{72}_{34}Se$

1800		3.500		Por 64a

903 Reaction $^{96}_{42}Mo \rightarrow {}^{72}_{34}Se$ (cumulative)

1800		3.60	0.50	Por 64a

E [MeV]	ΔE [MeV]	σ [mb]	Δσ [mb]	Ref.

904 Reaction $^{96}_{42}Mo \rightarrow {}^{92m}_{41}Nb$ (Fig. 483, p. 282)

E [MeV]	ΔE [MeV]	σ [mb]	Δσ [mb]	Ref.
18.30		5.000		Lev 91
19.30		9.100		Lev 91
20.30		11.300		Lev 91
21.40		14.600		Lev 91
22.20		17.100		Lev 91
23.10		18.200		Lev 91
24.00		20.000		Lev 91
24.80		21.000		Lev 91
25.70		23.000		Lev 91
26.60		20.000		Lev 91
27.60		20.000		Lev 91
28.50		17.000		Lev 91
29.50		16.200		Lev 91

905 Reaction $^{96}_{42}Mo \rightarrow {}^{95m}_{41}Nb$

E [MeV]	ΔE [MeV]	σ [mb]	Δσ [mb]	Ref.
25.70		0.140		Lev 91
26.60		0.290		Lev 91
27.60		0.470		Lev 91
28.50		0.650		Lev 91
29.50		0.840		Lev 91

906 Reaction $^{96}_{42}Mo \rightarrow {}^{95m+g}_{41}Nb$

E [MeV]	ΔE [MeV]	σ [mb]	Δσ [mb]	Ref.
25.70		0.500		Lev 91
26.60		1.000		Lev 91
27.60		1.500		Lev 91
28.50		2.200		Lev 91
29.50		2.700		Lev 91

907 Reaction $^{96}_{42}Mo \rightarrow {}^{93}_{43}Tc$ (Fig. 484, p. 282)

E [MeV]	ΔE [MeV]	σ [mb]	Δσ [mb]	Ref.
30.00		1.500	3.00	Hog 73
35.00		12.300	0.40	Hog 73
40.00		158.000	1.00	Hog 73
45.00		243.000	5.00	Hog 73
50.00		226.000	3.00	Hog 73
65.00		85.000	1.00	Hog 73

908 Reaction $^{96}_{42}Mo \rightarrow {}^{93m}_{43}Tc$ (Fig. 485, p. 282)

E [MeV]	ΔE [MeV]	σ [mb]	Δσ [mb]	Ref.
30.00		0.900	1.70	Hog 73
35.00		5.100	0.20	Hog 73
40.00		2.400	1.00	Hog 73
45.00		56.000	1.00	Hog 73
50.00		19.900	0.30	Hog 73
65.00		5.900	0.10	Hog 73

E [MeV]	ΔE [MeV]	σ [mb]	Δσ [mb]	Ref.

909 Reaction $^{96}_{42}Mo \rightarrow {}^{93m+g}_{43}Tc$ (Fig. 486, p. 283)

E [MeV]	ΔE [MeV]	σ [mb]	Δσ [mb]	Ref.
30.00		2.300	0.34	Hog 72
35.00		17.400	0.50	Hog 72
40.00		182.000	2.00	Hog 72
45.00		299.000	5.00	Hog 72
50.00		246.000	4.00	Hog 72
65.00		91.000	1.00	Hog 72
80.00		69.000	3.00	Hog 72

910 Reaction $^{96}_{42}Mo \rightarrow {}^{94}_{43}Tc$ (Fig. 487, p. 283)

E [MeV]	ΔE [MeV]	σ [mb]	Δσ [mb]	Ref.
20.00		8.600	1.20	Hog 73
24.00		80.000		Lev 91
24.80		132.000		Lev 91
25.00		42.900	0.20	Hog 73
25.70		202.000		Lev 91
26.60		292.000		Lev 91
27.60		342.000		Lev 91
28.50		367.000		Lev 91
29.50		395.000		Lev 91
30.00		330.300	3.00	Hog 73
35.00		340.000	2.00	Hog 73
40.00		231.000	3.00	Hog 73
45.00		118.000	2.00	Hog 73
50.00		72.000	0.30	Hog 73
65.00		40.000	0.80	Hog 73

911 Reaction $^{96}_{42}Mo \rightarrow {}^{94m}_{43}Tc$ (Fig. 488, p. 283)

E [MeV]	ΔE [MeV]	σ [mb]	Δσ [mb]	Ref.
20.00		5.500	0.30	Hog 73
24.00		57.000		Lev 91
24.80		92.000		Lev 91
25.00		30.300	0.40	Hog 73
25.70		132.000		Lev 91
26.60		220.000		Lev 91
27.60		250.000		Lev 91
28.50		255.000		Lev 91
29.50		277.000		Lev 91
30.00		220.000	3.00	Hog 73
35.00		178.000	2.00	Hog 73
40.00		112.000	3.00	Hog 73
45.00		60.000	10.00	Hog 73
50.00		49.000	1.00	Hog 73
65.00		33.200	0.80	Hog 73

E [MeV]	ΔE [MeV]	σ [mb]	Δσ [mb]	Ref.

912 Reaction $^{96}{}_{42}Mo \rightarrow {}^{94m+g}{}_{43}Tc$ (Fig. 489, p. 283)

E [MeV]	ΔE [MeV]	σ [mb]	Δσ [mb]	Ref.
20.00		14.000	1.20	Hog 72
25.00		72.800	0.50	Hog 72
30.00		551.000	3.00	Hog 72
35.00		518.000	3.00	Hog 72
40.00		343.000	4.00	Hog 72
45.00		178.000	9.00	Hog 72
50.00		122.000	2.00	Hog 72
65.00		73.000	1.00	Hog 72
80.00		57.000	3.00	Hog 72

913 Reaction $^{96}{}_{42}Mo \rightarrow {}^{95}{}_{43}Tc$ (Fig. 490, p. 283)

E [MeV]	ΔE [MeV]	σ [mb]	Δσ [mb]	Ref.
12.10		70.000		Lev 91
12.80		204.000		Lev 91
13.80		465.000		Lev 91
14.80		565.000		Lev 91
15.00		187.000	2.00	Hog 73
15.80		654.000		Lev 91
16.70		685.000		Lev 91
17.50		770.000		Lev 91
18.30		767.000		Lev 91
19.30		822.000		Lev 91
20.00		1111.000	33.00	Hog 73
20.30		844.000		Lev 91
21.40		848.000		Lev 91
22.20		859.000		Lev 91
23.10		820.000		Lev 91
24.00		794.000		Lev 91
24.80		624.000		Lev 91
25.00		928.000	15.00	Hog 73
25.70		568.000		Lev 91
26.60		431.000		Lev 91
27.60		320.000		Lev 91
28.50		239.000		Lev 91
29.50		212.000		Lev 91
30.00		463.000	3.00	Hog 73
35.00		109.000	1.00	Hog 73
40.00		73.800	0.30	Hog 73
45.00		61.900	0.80	Hog 73
50.00		56.000	1.00	Hog 73
65.00		39.400	0.40	Hog 73

914 Reaction $^{96}{}_{42}Mo \rightarrow {}^{95m}{}_{43}Tc$ (Fig. 491, p. 284)

E [MeV]	ΔE [MeV]	σ [mb]	Δσ [mb]	Ref.
12.10		95.000		Lev 91
12.80		165.000		Lev 91

E [MeV]	ΔE [MeV]	σ [mb]	Δσ [mb]	Ref.

914 Reaction $^{96}_{42}Mo \rightarrow {}^{95m}_{43}Tc$ (continued)

E [MeV]	ΔE [MeV]	σ [mb]	Δσ [mb]	Ref.
13.80		262.000		Lev 91
14.80		305.000		Lev 91
15.00		154.000	6.00	Hog 73
15.80		340.000		Lev 91
16.70		350.000		Lev 91
17.50		345.000		Lev 91
18.30		320.000		Lev 91
19.30		317.000		Lev 91
20.00		396.000	33.00	Hog 73
20.30		307.000		Lev 91
21.40		286.000		Lev 91
22.20		287.000		Lev 91
23.10		242.000		Lev 91
24.00		225.000		Lev 91
24.80		189.000		Lev 91
25.00		276.000	7.00	Hog 73
25.70		161.000		Lev 91
26.60		123.000		Lev 91
27.60		91.000		Lev 91
28.50		70.000		Lev 91
29.50		60.000		Lev 91
30.00		105.000	7.00	Hog 73
35.00		25.500	0.30	Hog 73
40.00		29.000	6.00	Hog 73
45.00		21.300	0.50	Hog 73
50.00		21.400	1.30	Hog 73
65.00		17.700	2.10	Hog 73

915 Reaction $^{96}_{42}Mo \rightarrow {}^{95m+g}_{43}Tc$ (Fig. 492, p. 284)

E [MeV]	ΔE [MeV]	σ [mb]	Δσ [mb]	Ref.
15.00		341.000	6.00	Hog 72
20.00		1507.000	48.00	Hog 72
25.00		1205.000	17.00	Hog 72
30.00		568.000	7.00	Hog 72
35.00		144.000	2.00	Hog 72
40.00		103.000	6.00	Hog 72
45.00		84.000	2.00	Hog 72
50.00		78.000	2.00	Hog 72
65.00		57.000	2.00	Hog 72
80.00		51.000	2.00	Hog 72

916 Reaction $^{96}_{42}Mo \rightarrow {}^{96}_{43}Tc$ (Fig. 493, p. 284)

E [MeV]	ΔE [MeV]	σ [mb]	Δσ [mb]	Ref.
4.50		2.310	0.24	Ska 87
5.00		21.600	1.80	Ska 87
5.50		38.000	4.00	Ska 87

E [MeV]	ΔE [MeV]	σ [mb]	Δσ [mb]	Ref.

916 Reaction $^{96}_{42}Mo \rightarrow ^{96}_{43}Tc$ (continued)

E [MeV]	ΔE [MeV]	σ [mb]	Δσ [mb]	Ref.
6.00		110.000	9.00	Ska 87
6.50		189.000	15.00	Ska 87
7.00		277.000	22.00	Ska 87
7.50		380.000	30.00	Ska 87
7.70		500.000		Lev 91
8.00		438.000	35.00	Ska 87
8.50		496.000	40.00	Ska 87
8.60		661.000		Lev 91
9.00		545.000	46.00	Ska 87
9.50		776.000		Lev 91
10.00		498.000	6.00	Hog 72
10.40		912.000		Lev 91
11.30		970.000		Lev 91
12.10		955.000		Lev 91
12.80		851.000		Lev 91
13.80		490.000		Lev 91
14.80		302.000		Lev 91
15.00		522.000	6.00	Hog 72
15.80		174.000		Lev 91
16.70		117.000		Lev 91
17.50		92.000		Lev 91
18.30		65.000		Lev 91
20.00		235.000	1.20	Hog 72
25.00		78.000	8.00	Hog 72
30.00		50.000	6.00	Hog 72
35.00		28.000	9.00	Hog 72
40.00		21.100	1.50	Hog 72
45.00		19.200	0.90	Hog 72
50.00		18.000	0.40	Hog 72
65.00		15.500	0.20	Hog 72
80.00		14.600	0.40	Hog 72

917 Reaction $^{97}_{42}Mo \rightarrow ^{92m}_{41}Nb$

E [MeV]	ΔE [MeV]	σ [mb]	Δσ [mb]	Ref.
24.00		1.500		Lev 91
24.80		2.500		Lev 91
25.70		3.700		Lev 91
26.60		5.300		Lev 91
27.60		9.500		Lev 91
28.50		12.400		Lev 91
29.50		15.000		Lev 91

918 Reaction $^{97}_{42}Mo \rightarrow ^{96}_{41}Nb$

E [MeV]	ΔE [MeV]	σ [mb]	Δσ [mb]	Ref.
24.00		0.330		Lev 91
24.80		0.440		Lev 91

E [MeV]	ΔE [MeV]	σ [mb]	Δσ [mb]	Ref.

918 Reaction $^{97}_{42}Mo \rightarrow ^{96}_{41}Nb$ (continued)

E [MeV]	ΔE [MeV]	σ [mb]	Δσ [mb]	Ref.
25.70		0.570		Lev 91
26.60		0.710		Lev 91
27.60		1.000		Lev 91
28.50		1.300		Lev 91
29.50		1.600		Lev 91

919 Reaction $^{97}_{42}Mo \rightarrow ^{95}_{43}Tc$

E [MeV]	ΔE [MeV]	σ [mb]	Δσ [mb]	Ref.
20.30		61.000		Lev 91
21.40		178.000		Lev 91
22.20		322.000		Lev 91
23.10		374.000		Lev 91
24.00		454.000		Lev 91
24.80		570.000		Lev 91
25.70		652.000		Lev 91
26.60		657.000		Lev 91
27.60		668.000		Lev 91
28.50		731.000		Lev 91
29.50		709.000		Lev 91

920 Reaction $^{97}_{42}Mo \rightarrow ^{95m}_{43}Tc$

E [MeV]	ΔE [MeV]	σ [mb]	Δσ [mb]	Ref.
21.40		42.000		Lev 91
22.20		77.000		Lev 91
23.10		92.000		Lev 91
24.00		117.000		Lev 91
24.80		147.000		Lev 91
25.70		165.000		Lev 91
26.60		170.000		Lev 91
27.60		160.000		Lev 91
28.50		175.000		Lev 91
29.50		165.000		Lev 91

921 Reaction $^{97}_{42}Mo \rightarrow ^{96}_{43}Tc$ **(Fig. 494, p. 284)**

E [MeV]	ΔE [MeV]	σ [mb]	Δσ [mb]	Ref.
12.10		371.000		Lev 91
12.80		758.000		Lev 91
13.80		977.000		Lev 91
14.80		1070.000		Lev 91
15.80		1140.000		Lev 91
16.70		1150.000		Lev 91
17.50		1200.000		Lev 91
18.30		1200.000		Lev 91
19.30		1230.000		Lev 91
20.30		1122.000		Lev 91
21.40		976.000		Lev 91

921 Reaction $^{97}_{42}Mo \rightarrow {}^{96}_{43}Tc$ (continued)

E [MeV]	ΔE [MeV]	σ [mb]	Δσ [mb]	Ref.
22.20		871.000		Lev 91
23.10		692.000		Lev 91
24.00		549.000		Lev 91
24.80		457.000		Lev 91
25.70		380.000		Lev 91
26.60		316.000		Lev 91
27.60		257.000		Lev 91
28.50		186.000		Lev 91
29.50		126.000		Lev 91

922 Reaction $^{98}_{42}Mo \rightarrow {}^{95}_{41}Nb$ (Fig. 495, p. 284)

E [MeV]	ΔE [MeV]	σ [mb]	Δσ [mb]	Ref.
15.90		12.000		Lev 84
17.00		13.000		Lev 84
17.90		13.000		Lev 84
18.60		11.000		Lev 84
19.40		11.500		Lev 84
20.20		10.000		Lev 84
20.80		9.700		Lev 84
21.50		9.000		Lev 84
22.20		8.000		Lev 84
23.20		7.000		Lev 84
24.40		6.500		Lev 84
25.20		6.000		Lev 84
26.10		5.900		Lev 84
27.20		5.500		Lev 84
28.20		5.000		Lev 84

923 Reaction $^{98}_{42}Mo \rightarrow {}^{95m}_{41}Nb$ (Fig. 496, p. 285)

E [MeV]	ΔE [MeV]	σ [mb]	Δσ [mb]	Ref.
8.60		0.950		Lev 91
9.50		1.900		Lev 91
10.40		2.000		Lev 91
11.30		2.800		Lev 91
12.10		2.800		Lev 91
12.80		2.900		Lev 91
13.80		3.400		Lev 91
14.80		3.600		Lev 91
15.80		3.300		Lev 91
16.70		3.700		Lev 91
17.50		3.200		Lev 91
18.30		3.000		Lev 91
19.30		2.900		Lev 91
20.30		2.600		Lev 91

E [MeV]	ΔE [MeV]	σ [mb]	Δσ [mb]	Ref.

924 Reaction $^{98}_{42}Mo \rightarrow {}^{95m+g}_{41}Nb$ (Fig. 497, p. 285)

E [MeV]	ΔE [MeV]	σ [mb]	Δσ [mb]	Ref.
7.70		1.500		Lev 91
8.60		1.600		Lev 91
9.50		3.400		Lev 91
10.40		3.600		Lev 91
11.30		5.600		Lev 91
12.10		6.200		Lev 91
12.80		6.600		Lev 91
13.80		9.100		Lev 91
14.80		10.800		Lev 91
15.80		10.700		Lev 91
16.70		11.500		Lev 91
17.50		10.400		Lev 91
18.30		10.200		Lev 91
19.30		9.800		Lev 91
20.30		9.300		Lev 91
21.40		9.300		Lev 91
22.20		8.100		Lev 91
23.10		7.000		Lev 91
24.00		6.400		Lev 91
24.80		6.300		Lev 91
25.70		5.800		Lev 91
26.60		5.700		Lev 91
27.60		5.600		Lev 91
28.50		5.300		Lev 91
29.50		5.200		Lev 91

925 Reaction $^{98}_{42}Mo \rightarrow {}^{96}_{41}Nb$

E [MeV]	ΔE [MeV]	σ [mb]	Δσ [mb]	Ref.
25.70		0.340		Lev 91
26.60		0.440		Lev 91
27.60		0.610		Lev 91
28.50		0.690		Lev 91
29.50		0.810		Lev 91

926 Reaction $^{98}_{42}Mo \rightarrow {}^{97}_{41}Nb$

E [MeV]	ΔE [MeV]	σ [mb]	Δσ [mb]	Ref.
24.80		0.190		Lev 91
25.70		0.280		Lev 91
26.60		0.390		Lev 91
27.60		0.650		Lev 91
28.50		0.840		Lev 91
29.50		0.970		Lev 91

E [MeV]	ΔE [MeV]	σ [mb]	Δσ [mb]	Ref.

927 Reaction $^{98}_{42}Mo \rightarrow ^{96}_{43}Tc$ (Fig. 498, p. 285)

E [MeV]	ΔE [MeV]	σ [mb]	Δσ [mb]	Ref.
20.30		43.000		Lev 91
21.40		132.000		Lev 91
22.20		245.000		Lev 91
23.10		389.000		Lev 91
24.00		589.000		Lev 91
24.80		631.000		Lev 91
25.70		732.000		Lev 91
26.60		724.000		Lev 91
27.60		813.000		Lev 91
28.50		891.000		Lev 91
29.50		955.000		Lev 91

928 Reaction $^{98}_{42}Mo \rightarrow ^{98}_{43}Tc$ (Fig. 499, p. 285)

E [MeV]	ΔE [MeV]	σ [mb]	Δσ [mb]	Ref.
3.00		1.000		Fly 79*
3.50		10.000		Fly 79*
4.00		20.000		Fly 79*
4.50		35.000		Fly 79*
5.00		65.000		Fly 79*
5.50		115.000		Fly 79*
6.00		200.000		Fly 79*
6.15		260.000		Fly 79*
6.25		205.000		Fly 79*
6.30		275.000		Fly 79*
6.40		220.000		Fly 79*
6.50		200.000		Fly 79*
6.65		300.000		Fly 79*
6.70		270.000		Fly 79*
6.75		310.000		Fly 79*

929 Reaction $^{100}_{42}Mo \rightarrow ^{95}_{41}Nb$

E [MeV]	ΔE [MeV]	σ [mb]	Δσ [mb]	Ref.
25.00		2.000		Lev 84
26.20		7.000		Lev 84
27.35		10.000		Lev 84
28.20		13.000		Lev 84
29.20		17.000		Lev 84

930 Reaction $^{100}_{42}Mo \rightarrow ^{95m}_{41}Nb$

E [MeV]	ΔE [MeV]	σ [mb]	Δσ [mb]	Ref.
25.70		1.400		Lev 91
26.60		1.800		Lev 91
27.60		2.200		Lev 91
28.50		2.600		Lev 91
29.50		3.500		Lev 91

E [MeV]	ΔE [MeV]	σ [mb]	Δσ [mb]	Ref.

931 Reaction $^{100}{}_{42}Mo \rightarrow {}^{95m+g}{}_{41}Nb$

E [MeV]	ΔE [MeV]	σ [mb]	Δσ [mb]	Ref.
23.10		2.700		Lev 91
24.00		4.000		Lev 91
24.80		6.800		Lev 91
25.70		10.000		Lev 91
26.60		13.000		Lev 91
27.60		16.000		Lev 91
28.50		20.000		Lev 91
29.50		26.000		Lev 91

932 Reaction $^{100}{}_{42}Mo \rightarrow {}^{96}{}_{41}Nb$ (Fig. 500, p. 285)

E [MeV]	ΔE [MeV]	σ [mb]	Δσ [mb]	Ref.
15.80		1.700		Lev 91
15.90		1.000		Lev 84*
16.70		3.000		Lev 91
17.00		2.200		Lev 84*
17.50		3.800		Lev 91
17.90		3.800		Lev 84*
18.30		5.600		Lev 91
18.60		4.500		Lev 84*
19.30		7.800		Lev 91
19.40		5.000		Lev 84*
20.10		6.800		Lev 84*
20.30		8.500		Lev 91
20.80		8.000		Lev 84*
21.40		11.200		Lev 91
22.10		10.000		Lev 84*
22.20		11.700		Lev 91
23.10		13.700		Lev 91
23.40		11.500		Lev 84*
24.00		13.800		Lev 91
24.40		13.500		Lev 84*
24.80		12.600		Lev 91
25.40		14.300		Lev 84*
25.70		11.500		Lev 91
26.20		12.800		Lev 84*
26.60		10.200		Lev 91
27.40		12.800		Lev 84*
27.60		10.500		Lev 91
28.50		10.200		Lev 91
29.20		11.000		Lev 84*
29.50		9.800		Lev 91

933 Reaction $^{100}{}_{42}Mo \rightarrow {}^{97}{}_{41}Nb$ (Fig. 501, p. 286)

E [MeV]	ΔE [MeV]	σ [mb]	Δσ [mb]	Ref.
8.60		0.900		Lev 91
9.50		1.500		Lev 91

E [MeV]	ΔE [MeV]	σ [mb]	Δσ [mb]	Ref.

933 Reaction $^{100}_{42}Mo \rightarrow ^{97}_{41}Nb$ (continued)

E [MeV]	ΔE [MeV]	σ [mb]	Δσ [mb]	Ref.
10.40		2.000		Lev 91
11.30		2.700		Lev 91
12.10		3.500		Lev 91
12.80		3.900		Lev 91
13.80		5.200		Lev 91
14.80		5.600		Lev 91
15.75		6.100		Lev 84*
15.80		5.700		Lev 91
16.70		5.500		Lev 91
16.90		6.800		Lev 84*
17.50		5.400		Lev 91
17.80		7.000		Lev 84*
18.30		5.100		Lev 91
18.60		6.800		Lev 84*
19.30		4.900		Lev 91
19.40		6.100		Lev 84*
20.10		5.500		Lev 84*
20.30		4.200		Lev 91
20.80		5.000		Lev 84*
21.40		4.000		Lev 91
21.50		4.200		Lev 84*
22.20		3.800		Lev 91
22.30		4.000		Lev 84*
23.10		3.500		Lev 91
23.30		3.800		Lev 84*
24.00		3.300		Lev 91
24.30		3.500		Lev 84*
24.80		3.200		Lev 91
25.10		3.200		Lev 84*
25.70		2.700		Lev 91
26.00		3.000		Lev 84*
26.60		2.300		Lev 91
27.10		2.500		Lev 84*
27.60		2.200		Lev 91
28.10		2.200		Lev 84*
28.50		2.400		Lev 91
29.50		2.400		Lev 91

934 Reaction $^{100}_{42}Mo \rightarrow ^{99}_{42}Mo$ (Fig. 502, p. 286)

E [MeV]	ΔE [MeV]	σ [mb]	Δσ [mb]	Ref.
12.80		17.000		Lev 91
13.80		35.000		Lev 91
14.80		48.000		Lev 91
15.80		60.000		Lev 91
16.70		74.000		Lev 91
17.50		93.000		Lev 91

E [MeV]	ΔE [MeV]	σ [mb]	Δσ [mb]	Ref.

934 Reaction $^{100}_{42}Mo \rightarrow ^{99}_{42}Mo$ (continued)

E [MeV]	ΔE [MeV]	σ [mb]	Δσ [mb]	Ref.
18.30		110.000		Lev 91
19.30		132.000		Lev 91
20.30		141.000		Lev 91
21.40		158.000		Lev 91
22.20		168.000		Lev 91
23.10		176.000		Lev 91
24.00		180.000		Lev 91
24.80		178.000		Lev 91
25.70		170.000		Lev 91
26.60		174.000		Lev 91
27.60		170.000		Lev 91
28.50		182.000		Lev 91
29.50		178.000		Lev 91
280.0		79.000		Mar 58
380.0		65.000		Mar 58
400.0		76.000		Mar 58
600.0		81.000		Mar 58
800.0		61.000		Mar 58
1000		67.000		Mar 58
1300		76.000		Mar 58
1600		73.000		Mar 58
1900		71.000		Mar 58
2200		73.000		Mar 58
2600		76.000		Mar 58
2900		73.000		Mar 58
3000		82.000		Mar 58
3000		63.000		Mar 58

935 Reaction $^{100}_{42}Mo \rightarrow ^{99m}_{43}Tc$ (Fig. 503, p. 286)

E [MeV]	ΔE [MeV]	σ [mb]	Δσ [mb]	Ref.
8.60		102.000		Lev 91
9.50		160.000		Lev 91
10.40		222.000		Lev 91
11.30		237.000		Lev 91
12.10		275.000		Lev 91
12.80		285.000		Lev 91
13.80		304.000		Lev 91
14.80		305.000		Lev 91
15.80		290.000		Lev 91
16.70		282.000		Lev 91
17.50		267.000		Lev 91
18.30		257.000		Lev 91
19.30		262.000		Lev 91
20.30		182.000		Lev 91
21.40		141.000		Lev 91
22.20		117.000		Lev 91

935 Reaction $^{100}_{42}Mo \rightarrow ^{99m}_{43}Tc$ (continued)

E [MeV]	ΔE [MeV]	σ [mb]	Δσ [mb]	Ref.
23.10		94.000		Lev 91
24.00		79.000		Lev 91
24.80		66.000		Lev 91
25.70		55.000		Lev 91
26.60		42.000		Lev 91
27.60		38.000		Lev 91
28.50		41.000		Lev 91
29.50		36.000		Lev 91

936 Reaction $^{100}_{42}Mo \rightarrow ^{100}_{43}Tc$ (Fig. 504, p. 287)

E [MeV]	ΔE [MeV]	σ [mb]	Δσ [mb]	Ref.
2.00		0.100E-1		Kne 75*
3.00		1.000		Kne 75*
4.00		5.800	1.70	Ska 87
4.00		15.000		Kne 75*
4.50		22.500	4.00	Ska 87
5.00		40.000	7.00	Ska 87
5.00		100.000		Kne 75*
5.50		104.000	17.00	Ska 87
6.00		130.000	2.00	Ska 87
6.00		600.000		Kne 75*
6.50		256.000	40.00	Ska 87
7.00		351.000	55.00	Ska 87
7.00		700.000		Kne 75*
7.50		403.000	63.00	Ska 87
8.00		562.000	63.00	Ska 87
8.00		900.000		Kne 75*
8.50		620.000	100.00	Ska 87
9.00		500.000	80.00	Ska 87
9.00		800.000		Kne 75*
9.50		900.000		Kne 75*
10.00		1000.000		Kne 75*
11.00		1000.000		Kne 75*

937 Reaction $^{96}_{44}Ru \rightarrow ^{72}_{30}Zn$ (cumulative)

E [MeV]	ΔE [MeV]	σ [mb]	Δσ [mb]	Ref.
1800		0.160E-2	0.40E-3	Por 64a

938 Reaction $^{96}_{44}Ru \rightarrow ^{66}_{31}Ga$

E [MeV]	ΔE [MeV]	σ [mb]	Δσ [mb]	Ref.
1800		7.800	0.30	Por 64a

939 Reaction $^{96}_{44}Ru \rightarrow ^{67}_{31}Ga$ (cumulative)

E [MeV]	ΔE [MeV]	σ [mb]	Δσ [mb]	Ref.
1800		11.300	0.70	Por 64a

E [MeV]	ΔE [MeV]	σ [mb]	Δσ [mb]	Ref.
940 Reaction $^{96}_{44}Ru \rightarrow ^{72}_{31}Ga$				
1800		0.800E-1	0.10E-1	Por 64a
941 Reaction $^{96}_{44}Ru \rightarrow ^{73}_{31}Ga$ (cumulative)				
1800		0.800E-2	0.10E-2	Por 64a
942 Reaction $^{96}_{44}Ru \rightarrow ^{71}_{33}As$ (cumulative)				
1800		11.200	1.00	Por 64a
943 Reaction $^{96}_{44}Ru \rightarrow ^{72}_{33}As$				
1800		7.000	1.00	Por 64a
944 Reaction $^{96}_{44}Ru \rightarrow ^{74}_{33}As$				
1800		1.100	0.20	Por 64a
945 Reaction $^{96}_{44}Ru \rightarrow ^{72}_{34}Se$ (cumulative)				
1800		6.500	1.00	Por 64a
946 Reaction $^{100}_{44}Ru \rightarrow ^{100}_{45}Rh$				
4.30		0.30		Bla 51*
4.60		5.50		Bla 51*
4.90		12.00		Bla 51*
5.30		31.00		Bla 51*
5.70		49.00		Bla 51*
5.90		68.00		Bla 51*
6.30		100.00		Bla 51*
947 Reaction $^{101}_{44}Ru \rightarrow ^{101}_{45}Rh$ (Fig. 505, p. 287)				
2.90		0.430		Bla 51*
3.30		3.000		Bla 51*
3.80		6.500		Bla 51*
4.20		14.000		Bla 51*
4.60		28.000		Bla 51*
4.90		39.000		Bla 51*
5.30		50.000		Bla 51*
5.70		75.000		Bla 51*
5.90		105.000		Bla 51*
6.20		130.000		Bla 51*

E [MeV]	ΔE [MeV]	σ [mb]	Δσ [mb]	Ref.
948 Reaction $_{45}Rh \rightarrow {}^{22}{}_{11}Na$				
2600		0.820	0.12	Dit 90
949 Reaction $_{45}Rh \rightarrow {}^{69}{}_{32}Ge$				
600.0		9.460	1.02	Mic 89
950 Reaction $_{45}Rh \rightarrow {}^{74}{}_{33}As$				
600.0		0.940	0.70E-1	Mic 89
951 Reaction $_{45}Rh \rightarrow {}^{75}{}_{34}Se$				
600.0		6.690	0.25	Mic 89
952 Reaction $_{45}Rh \rightarrow {}^{83}{}_{37}Rb$				
600.0		23.400	1.20	Mic 89
953 Reaction $_{45}Rh \rightarrow {}^{84}{}_{37}Rb$				
600.0		2.100	0.10	Mic 89
954 Reaction $_{45}Rh \rightarrow {}^{85}{}_{38}Sr$				
600.0		51.500	3.20	Mic 89
955 Reaction $_{45}Rh \rightarrow {}^{87}{}_{39}Y$				
600.0		41.300	1.60	Mic 89
956 Reaction $_{45}Rh \rightarrow {}^{91}{}_{39}Y$				
600.0		20.800	1.70	Mic 89
957 Reaction $_{45}Rh \rightarrow {}^{88}{}_{40}Zr$				
600.0		38.200	1.50	Mic 89
958 Reaction $_{45}Rh \rightarrow {}^{89}{}_{40}Zr$				
600.0		41.800	1.60	Mic 89
959 Reaction $_{45}Rh \rightarrow {}^{95}{}_{40}Zr$				
600.0		0.200	0.20E-1	Mic 89

E [MeV]	ΔE [MeV]	σ [mb]	Δσ [mb]	Ref.

960 Reaction $_{45}Rh \rightarrow {}^{90}{}_{41}Nb$

E [MeV]	ΔE [MeV]	σ [mb]	Δσ [mb]	Ref.
600.0		38.000	3.50	Mic 89

961 Reaction $_{45}Rh \rightarrow {}^{92m}{}_{41}Nb$

E [MeV]	ΔE [MeV]	σ [mb]	Δσ [mb]	Ref.
600.0		2.190	0.90E-1	Mic 89

962 Reaction $_{45}Rh \rightarrow {}^{95}{}_{41}Nb$

E [MeV]	ΔE [MeV]	σ [mb]	Δσ [mb]	Ref.
600.0		1.180	0.60E-1	Mic 89

963 Reaction $_{45}Rh \rightarrow {}^{94}{}_{43}Tc$ (Fig. 506, p. 287)

E [MeV]	ΔE [MeV]	σ [mb]	Δσ [mb]	Ref.
71.80	1.20	1.100	0.30	Sem 91
79.90	3.30	2.300	0.40	Sem 91
84.90	3.10	7.700	1.30	Sem 91
88.70	2.90	9.400	1.60	Sem 91
94.30	2.70	15.000	2.60	Sem 91
99.00	2.40	19.000	3.20	Sem 91
103.0	1.70	15.600	3.30	Sem 91

964 Reaction $_{45}Rh \rightarrow {}^{95}{}_{43}Tc$ (Fig. 507, p. 287)

E [MeV]	ΔE [MeV]	σ [mb]	Δσ [mb]	Ref.
56.60	2.40	0.500	0.20	Sem 91
61.40	2.20	2.200	0.60	Sem 91
66.40	2.00	7.500	1.30	Sem 91
71.80	1.20	24.000	4.20	Sem 91
79.90	3.30	35.000	6.00	Sem 91
84.90	3.10	40.000	6.00	Sem 91
88.70	2.90	35.000	5.00	Sem 91
94.30	2.70	30.000	5.50	Sem 91
99.00	2.40	27.000	4.50	Sem 91
103.0	1.70	30.000	5.00	Sem 91

965 Reaction $_{45}Rh \rightarrow {}^{95m}{}_{43}Tc$ (Fig. 508, p. 288)

E [MeV]	ΔE [MeV]	σ [mb]	Δσ [mb]	Ref.
66.40	2.00	1.100	0.30	Sem 91
71.80	1.20	2.000	0.60	Sem 91
79.90	3.30	2.600	0.50	Sem 91
84.90	3.10	2.400	0.40	Sem 91
88.70	2.90	2.000	0.30	Sem 91
94.30	2.70	2.300	0.40	Sem 91
99.00	2.40	2.700	0.40	Sem 91
103.0	1.70	2.600	0.60	Sem 91
600.0		3.750	0.15	Mic 89

E [MeV]	ΔE [MeV]	σ [mb]	Δσ [mb]	Ref.

966 Reaction $_{45}$Rh → $^{96}_{43}$Tc (Fig. 509, p. 288)

E [MeV]	ΔE [MeV]	σ [mb]	Δσ [mb]	Ref.
56.60	2.40	4.600	0.80	Sem 91
61.40	2.20	20.000	3.40	Sem 91
66.40	2.00	23.000	4.00	Sem 91
71.80	1.20	20.000	3.50	Sem 91
79.90	3.30	20.000	3.40	Sem 91
84.90	3.10	17.000	2.80	Sem 91
88.70	2.90	19.000	3.00	Sem 91
94.30	2.70	16.000	2.60	Sem 91
99.00	2.40	15.000	2.40	Sem 91
103.0	1.70	17.000	3.00	Sem 91
600.0		23.200	1.20	Mic 89

967 Reaction $_{45}$Rh → $^{97}_{44}$Ru (cumulative) (Fig. 510, p. 288)

E [MeV]	ΔE [MeV]	σ [mb]	Δσ [mb]	Ref.
35.00	0.70	0.100	0.50E-1	Sem 91
37.80	2.00	1.400	0.20	Lag 83
39.20	2.00	5.100	0.70	Lag 83
40.50	2.00	9.400	1.20	Lag 83
41.90	2.10	16.000	2.10	Lag 83
42.00	2.80	12.000	2.00	Sem 91
43.10	2.10	24.000	3.10	Lag 83
44.30	2.10	30.000	3.90	Lag 83
45.50	2.10	39.000	3.10	Lag 83
46.20	2.70	40.000	6.00	Sem 91
46.70	2.10	45.000	5.90	Lag 83
47.90	2.10	51.000	6.60	Lag 83
49.00	2.10	52.000	6.80	Lag 83
50.10	2.10	53.000	6.90	Lag 83
51.80	2.50	60.000	10.00	Sem 91
51.20	2.10	52.000	6.80	Lag 83
52.30	2.10	56.000	7.30	Lag 83
53.40	2.10	51.000	6.60	Lag 83
54.40	2.00	45.000	5.90	Lag 83
55.50	2.00	43.000	5.60	Lag 83
56.50	1.90	38.000	4.90	Lag 83
56.60	2.40	40.000	7.00	Sem 91
57.90	1.90	36.000	4.70	Lag 83
58.50	1.80	29.000	3.80	Lag 83
59.50	1.70	28.000	3.60	Lag 83
60.50	1.70	25.000	3.30	Lag 83
61.40	2.20	25.000	4.50	Sem 91
61.50	1.60	24.000	3.10	Lag 83
62.40	1.50	23.000	3.00	Lag 83
63.40	1.30	22.000	2.90	Lag 83
64.30	1.20	20.000	2.60	Lag 83
65.30	1.00	17.000	2.20	Lag 83

E [MeV]	ΔE [MeV]	σ [mb]	Δσ [mb]	Ref.

967 Reaction $_{45}$Rh → $^{97}_{44}$Ru (continued)

E [MeV]	ΔE [MeV]	σ [mb]	Δσ [mb]	Ref.
66.20	0.80	21.000	2.70	Lag 83
66.40	2.00	22.000	4.00	Sem 91
67.10	0.50	18.000	2.30	Lag 83
71.80	1.20	24.000	4.00	Sem 91
79.90	3.30	40.000	7.00	Sem 91
84.90	3.10	70.000	12.00	Sem 91
88.70	2.90	80.000	13.00	Sem 91
94.30	2.70	95.000	16.00	Sem 91
99.00	2.40	100.000	15.00	Sem 91
103.0	1.70	95.000	16.00	Sem 91
600.0		36.200	1.40	Mic 89

968 Reaction $_{45}$Rh → $^{99}_{45}$Rh (Fig. 511, p. 288)

E [MeV]	ΔE [MeV]	σ [mb]	Δσ [mb]	Ref.
46.20	2.70	3.500	0.70	Sem 91
51.80	2.50	45.000	7.00	Sem 91
56.60	2.40	90.000	14.00	Sem 91
61.40	2.20	65.000	11.00	Sem 91
66.40	2.00	45.000	6.00	Sem 91
71.80	1.20	50.000	7.00	Sem 91
79.90	3.30	50.000	8.00	Sem 91
84.90	3.10	45.000	7.00	Sem 91
88.70	2.90	40.000	6.00	Sem 91
94.30	2.70	40.000	6.00	Sem 91
99.00	2.40	30.000	5.00	Sem 91
103.0	1.70	30.000	6.00	Sem 91
600.0		7.860	0.31	Mic 89

969 Reaction $_{45}$Rh → $^{99m}_{45}$Rh (Fig. 512, p. 289)

E [MeV]	ΔE [MeV]	σ [mb]	Δσ [mb]	Ref.
42.00	2.80	1.100	0.30	Sem 91
46.20	2.70	10.000	1.70	Sem 91
51.80	2.50	90.000	16.00	Sem 91
56.60	2.40	145.000	23.00	Sem 91
61.40	2.20	160.000	25.50	Sem 91
66.40	2.00	100.000	15.00	Sem 91
71.80	1.20	100.000	15.00	Sem 91
79.90	3.30	70.000	10.00	Sem 91
84.90	3.10	70.000	9.00	Sem 91
88.70	2.90	60.000	8.00	Sem 91
94.30	2.70	60.000	9.00	Sem 91
99.00	2.40	70.000	10.00	Sem 91
103.0	1.70	60.000	10.00	Sem 91

970 Reaction $_{45}$Rh → $^{100}{}_{45}$Rh (Fig. 513, p. 289)

E [MeV]	ΔE [MeV]	σ [mb]	Δσ [mb]	Ref.
29.00		2.000		Lag 84*
30.00	1.20	3.000	0.80	Sem 91
30.50		2.000		Lag 84*
32.00		4.000		Lag 84*
33.00		5.500		Lag 84*
35.00		4.000		Lag 84*
35.00	0.70	16.000	2.70	Sem 91
36.00		12.000		Lag 84*
37.50		32.000		Lag 84*
39.00		50.000		Lag 84*
40.50		80.000		Lag 84*
42.00		120.000		Lag 84*
42.00	2.80	100.000	14.00	Sem 91
43.00		160.000		Lag 84*
44.00		190.000		Lag 84*
45.00		200.000		Lag 84*
45.50		220.000		Lag 84*
46.00		230.000		Lag 84*
46.20	2.70	190.000	30.00	Sem 91
47.50		240.000		Lag 84*
48.50		250.000		Lag 84*
49.50		250.000		Lag 84*
51.00		230.000		Lag 84*
51.80	2.50	240.000	42.00	Sem 91
52.00		225.000		Lag 84*
53.00		215.000		Lag 84*
54.00		210.000		Lag 84*
55.50		215.000		Lag 84*
56.00		200.000		Lag 84*
56.60	2.40	170.000	28.00	Sem 91
57.50		190.000		Lag 84*
58.50		180.000		Lag 84*
59.50		160.000		Lag 84*
60.00		140.000		Lag 84*
61.00		125.000		Lag 84*
61.40	2.20	130.000	25.00	Sem 91
62.00		130.000		Lag 84*
64.00		130.000		Lag 84*
65.00		120.000		Lag 84*
66.00		125.000		Lag 84*
66.40	2.00	110.000	18.00	Sem 91
67.00		120.000		Lag 84*
71.80	1.20	130.000	21.00	Sem 91
79.90	3.30	105.000	18.00	Sem 91
84.90	3.10	90.000	13.00	Sem 91
88.70	2.90	80.000	12.00	Sem 91

E [MeV]	ΔE [MeV]	σ [mb]	Δσ [mb]	Ref.

970 Reaction $_{45}Rh \rightarrow {}^{100}{}_{45}Rh$ (continued)

E [MeV]	ΔE [MeV]	σ [mb]	Δσ [mb]	Ref.
94.30	2.70	80.000	13.00	Sem 91
99.00	2.40	90.000	13.00	Sem 91
103.0	1.70	85.000	14.00	Sem 91

971 Reaction $_{45}Rh \rightarrow {}^{101}{}_{45}Rh$ (Fig. 514, p. 289)

E [MeV]	ΔE [MeV]	σ [mb]	Δσ [mb]	Ref.
30.00		120.000		Lag 84*
32.00		150.000		Lag 84*
40.50		170.000		Lag 84*
41.50		160.000		Lag 84*
42.50		160.000		Lag 84*
44.00		150.000		Lag 84*
45.00		130.000		Lag 84*
46.50		130.000		Lag 84*
48.50		170.000		Lag 84*
49.50		115.000		Lag 84*
51.00		130.000		Lag 84*
52.00		150.000		Lag 84*
53.00		140.000		Lag 84*
54.00		110.000		Lag 84*
55.50		120.000		Lag 84*
56.50		110.000		Lag 84*
57.50		100.000		Lag 84*
58.50		110.000		Lag 84*
60.00		100.000		Lag 84*
61.00		95.000		Lag 84*
62.00		75.000		Lag 84*
64.00		80.000		Lag 84*
65.00		95.000		Lag 84*
600.0		11.200	0.40	Mic 89

972 Reaction $_{45}Rh \rightarrow {}^{101m}{}_{45}Rh$ (Fig. 515, p. 290)

E [MeV]	ΔE [MeV]	σ [mb]	Δσ [mb]	Ref.
22.00		1.000		Lag 84*
24.00		1.900		Lag 84*
26.00		22.000		Lag 84*
27.50		110.000		Lag 84*
29.50		290.000		Lag 84*
31.50		400.000		Lag 84*
32.50		440.000		Lag 84*
34.00		540.000		Lag 84*
35.50		330.000		Lag 84*
36.50		400.000		Lag 84*
38.00		350.000		Lag 84*
40.00		330.000		Lag 84*
41.00		300.000		Lag 84*

E [MeV]	ΔE [MeV]	σ [mb]	Δσ [mb]	Ref.

972 Reaction $_{45}Rh \rightarrow {}^{101m}{}_{45}Rh$ (continued)

E [MeV]	ΔE [MeV]	σ [mb]	Δσ [mb]	Ref.
42.00		270.000		Lag 84*
43.50		210.000		Lag 84*
44.00		180.000		Lag 84*
45.00		170.000		Lag 84*
46.00		170.000		Lag 84*
47.50		150.000		Lag 84*
48.50		130.000		Lag 84*
49.50		120.000		Lag 84*
51.00		120.000		Lag 84*
52.00		110.000		Lag 84*
53.00		115.000		Lag 84*
54.00		100.000		Lag 84*
55.50		100.000		Lag 84*
56.50		95.000		Lag 84*
57.50		100.000		Lag 84*
59.00		95.000		Lag 84*
60.50		95.000		Lag 84*
61.50		90.000		Lag 84*
62.50		85.000		Lag 84*
63.50		85.000		Lag 84*
64.50		80.000		Lag 84*
65.00		75.000		Lag 84*
66.00		80.000		Lag 84*
67.00		85.000		Lag 84*
68.00		75.000		Lag 84*
600.0		38.600	2.12	Mic 89

973 Reaction $_{45}Rh \rightarrow {}^{102}{}_{45}Rh$

E [MeV]	ΔE [MeV]	σ [mb]	Δσ [mb]	Ref.
600.0		39.500	2.10	Mic 89

974 Reaction $_{45}Rh \rightarrow {}^{102m}{}_{45}Rh$

E [MeV]	ΔE [MeV]	σ [mb]	Δσ [mb]	Ref.
600.0		18.200	0.70	Mic 89

975 Reaction $_{45}Rh \rightarrow {}^{100}{}_{46}Pd$ (Fig. 516, p. 291)

E [MeV]	ΔE [MeV]	σ [mb]	Δσ [mb]	Ref.
30.50		55.000		Lag 84*
32.00		50.000		Lag 84*
33.50		90.000		Lag 84*
35.00		160.000		Lag 84*
36.50		150.000		Lag 84*
41.00		210.000		Lag 84*
42.00		215.000		Lag 84*
43.00		220.000		Lag 84*
44.50		260.000		Lag 84*
46.00		230.000		Lag 84*

E [MeV]	ΔE [MeV]	σ [mb]	Δσ [mb]	Ref.

975 Reaction $_{45}$Rh → $^{100}{}_{46}$Pd (continued)

E [MeV]	ΔE [MeV]	σ [mb]	Δσ [mb]	Ref.
47.00		205.000		Lag 84*
49.00		190.000		Lag 84*
50.50		170.000		Lag 84*
51.50		150.000		Lag 84*
52.50		120.000		Lag 84*
53.50		130.000		Lag 84*
55.00		110.000		Lag 84*
56.00		105.000		Lag 84*
57.00		100.000		Lag 84*
57.50		100.000		Lag 84*
59.00		90.000		Lag 84*
60.00		95.000		Lag 84*
61.50		85.000		Lag 84*
62.00		90.000		Lag 84*
63.00		75.000		Lag 84*
64.50		70.000		Lag 84*
66.00		65.000		Lag 84*
67.00		65.000		Lag 84*

976 Reaction $_{45}$Rh → $^{101}{}_{46}$Pd (Fig. 517, p. 290)

E [MeV]	ΔE [MeV]	σ [mb]	Δσ [mb]	Ref.
17.40	1.60	0.100	0.40E-1	Sem 91
22.50	1.50	9.000	1.50	Sem 91
23.50		1.800		Lag 84*
25.00		25.000		Lag 84*
26.50	1.40	40.000	5.00	Sem 91
27.00		120.000		Lag 84*
29.00		270.000		Lag 84*
30.50		330.000		Lag 84*
30.80	1.20	400.000	50.00	Sem 91
32.00		380.000		Lag 84*
33.00		410.000		Lag 84*
35.00	0.70	350.000	55.00	Sem 91
35.00		400.000		Lag 84*
36.50		390.000		Lag 84*
38.00		330.000		Lag 84*
39.00		310.000		Lag 84*
41.00		250.000		Lag 84*
42.00		200.000		Lag 84*
42.00	2.80	170.000	25.00	Sem 91
43.00		140.000		Lag 84*
44.50		125.000		Lag 84*
46.00		120.000		Lag 84*
46.20	2.70	80.000	13.00	Sem 91
47.00		110.000		Lag 84*
48.00		90.000		Lag 84*

976 Reaction $_{45}$Rh → $^{101}{}_{46}$Pd (continued)

E [MeV]	ΔE [MeV]	σ [mb]	Δσ [mb]	Ref.
49.00		80.000		Lag 84*
50.50		75.000		Lag 84*
51.50		70.000		Lag 84*
51.80	2.50	60.000	12.00	Sem 91
52.50		70.000		Lag 84*
53.50		65.000		Lag 84*
55.00		60.000		Lag 84*
56.00		60.000		Lag 84*
56.60	2.40	56.000	11.00	Sem 91
57.00		60.000		Lag 84*
57.50		65.000		Lag 84*
59.00		50.000		Lag 84*
60.00		50.000		Lag 84*
61.00		50.000		Lag 84*
61.40	2.20	65.000	9.00	Sem 91
62.00		50.000		Lag 84*
63.00		45.000		Lag 84*
63.50		45.000		Lag 84*
64.50		40.000		Lag 84*
65.50		40.000		Lag 84*
66.40	2.00	45.000	7.00	Sem 91
66.50		40.000		Lag 84*
67.00		35.000		Lag 84*
71.80	1.20	35.000	6.00	Sem 91
79.90	3.30	30.000	5.00	Sem 91
84.90	3.10	35.000	5.00	Sem 91
88.70	2.90	40.000	7.00	Sem 91
94.30	2.70	35.000	6.00	Sem 91
99.00	2.40	40.000	7.00	Sem 91
103.0	1.70	35.000	5.00	Sem 91

977 Reaction $_{45}$Rh → $^{103}{}_{46}$Pd (Fig. 518, p. 291)

E [MeV]	ΔE [MeV]	σ [mb]	Δσ [mb]	Ref.
3.40		2.800		Bla 51*
3.80		7.500		Bla 51*
4.50		20.000		Bla 51*
4.90		39.000		Bla 51*
5.30		60.000		Bla 51*
5.50		80.000		Bla 51*
5.80		105.000		Bla 51*
6.30		150.000		Bla 51*
6.60		200.000		Bla 51*
6.80		230.000		Bla 51*

E [MeV]	ΔE [MeV]	σ [mb]	Δσ [mb]	Ref.

978 Reaction $_{46}Pd \rightarrow {}^{81}_{36}Kr$

E [MeV]	ΔE [MeV]	σ [mb]	Δσ [mb]	Ref.
150.0		0.320E-2	0.12E-2	Lav 84
1050		26.800	2.70	Lav 84
24000		13.800	1.40	Lav 84

979 Reaction $_{46}Pd \rightarrow {}^{85}_{36}Kr$

E [MeV]	ΔE [MeV]	σ [mb]	Δσ [mb]	Ref.
1050		0.570E-1	0.60E-2	Lav 84
24000		0.540E-1	0.11E-1	Lav 84

980 Reaction $_{46}Pd \rightarrow {}^{109}_{46}Pd$ (Fig. 519, p. 291)

E [MeV]	ΔE [MeV]	σ [mb]	Δσ [mb]	Ref.
11.00		1.100		Coh 54*
12.00		3.000		Coh 54*
13.00		5.000		Coh 54*
14.00		10.000		Coh 54*
15.00		20.000		Coh 54*
16.00		35.000		Coh 54*
17.00		50.000		Coh 54*
18.00		75.000		Coh 54*
19.00		95.000		Coh 54*
20.00		120.000		Coh 54*
21.00		140.000		Coh 54*
22.00		160.000		Coh 54*
23.00		190.000		Coh 54*

981 Reaction $^{104}_{46}Pd \rightarrow {}^{104}_{47}Ar$

E [MeV]	ΔE [MeV]	σ [mb]	Δσ [mb]	Ref.
6.00		100.000		Bat 88*
7.00		200.000		Bat 88*
8.00		350.000		Bat 88*
9.00		500.000		Bat 88*

982 Reaction $^{105}_{46}Pd \rightarrow {}^{105}_{47}Ar$

E [MeV]	ΔE [MeV]	σ [mb]	Δσ [mb]	Ref.
5.00		15.000		Bat 88*
6.00		70.000		Bat 88*
7.00		300.000		Bat 88*
8.00		400.000		Bat 88*
9.00		500.000		Bat 88*

983 Reaction $^{108}_{46}Pd \rightarrow {}^{108}_{47}Ar$

E [MeV]	ΔE [MeV]	σ [mb]	Δσ [mb]	Ref.
5.00		25.000		Bat 88*
6.00		100.000		Bat 88*
7.00		250.000		Bat 88*
8.00		350.000		Bat 88*
9.00		450.000		Bat 88*

E [MeV]	ΔE [MeV]	σ [mb]	Δσ [mb]	Ref.

984 Reaction $^{110}_{46}Pd \rightarrow ^{110}_{47}Ar$

E [MeV]	ΔE [MeV]	σ [mb]	Δσ [mb]	Ref.
6.00		120.000		Bat 88*
7.00		250.000		Bat 88*
8.00		400.000		Bat 88*
9.00		500.000		Bat 88*

985 Reaction $_{47}Ag \rightarrow ^{3}_{1}H$

E [MeV]	ΔE [MeV]	σ [mb]	Δσ [mb]	Ref.
2050		160.000		Cur 59
12000		336.000	48.00	Nog 91

986 Reaction $_{47}Ag \rightarrow ^{7}_{4}Be$ (Fig. 520, p. 291)

E [MeV]	ΔE [MeV]	σ [mb]	Δσ [mb]	Ref.
65.00		0.240E-1		Laf 66
85.00		0.280E-1	4.00E-2	Laf 66
335.0		0.100		Mar 51*
340.0		0.100		Bak 58
940.0		2.500	0.60	Dos 61
940.0		2.500		Fow 62
1000		2.550	0.35	Bak 58
1840		11.300	2.80	Dos 61
2200		11.300		Bak 58
3000		12.100		Bak 58
5500		17.400	1.60	Pos 71
300000		18.700	1.30	Por 79
800000		15.800	1.00	Wan 86

987 Reaction $_{47}Ag \rightarrow ^{7}_{4}Be$ (cumulative)

E [MeV]	ΔE [MeV]	σ [mb]	Δσ [mb]	Ref.
11500		14.20	1.00	Eng 74
300000		13.55	0.30E-1	Eng 74a

988 Reaction $_{47}Ag \rightarrow ^{18}_{9}F$ (Fig. 521, p. 292)

E [MeV]	ΔE [MeV]	σ [mb]	Δσ [mb]	Ref.
420.0		0.160E-1	0.80E-2	Mar 52
1000		0.200		Car 58
2000		0.550		Car 58
3000		1.700		Car 58
4500		1.900		Car 58
5900		1.500		Car 58

989 Reaction $_{47}Ag \rightarrow ^{24}_{10}Ne$ (cumulative)

E [MeV]	ΔE [MeV]	σ [mb]	Δσ [mb]	Ref.
1000		0.200E-1		Hud 68
2000		0.900E-1		Hud 68
3000		0.210		Hud 68

E [MeV]	ΔE [MeV]	σ [mb]	Δσ [mb]	Ref.

990 Reaction $_{47}Ag \rightarrow {}^{22}_{11}Na$ (cumulative)

E [MeV]	ΔE [MeV]	σ [mb]	Δσ [mb]	Ref.
3000		1.140		Kat 68
11500		1.600	0.10	Eng 74
29000		2.350		Kat 68
300000		2.370	0.10	Por 79
300000		1.900	0.10	Eng 74a
800000		2.340	0.15	Wan 86

991 Reaction $_{47}Ag \rightarrow {}^{24}_{11}Na$ (Fig. 522, p. 292)

E [MeV]	ΔE [MeV]	σ [mb]	Δσ [mb]	Ref.
400.0		0.165E-1	0.80E-3	Kor 67
700.0		0.100		Cre 63
1000		0.300		Car 58
1000		0.270		Hud 68
2000		1.400		Car 58
2000		1.140		Hud 68
3000		2.700		Car 58
3000		2.060		Hud 68
4500		4.100		Car 58
5900		3.300		Car 58

992 Reaction $_{47}Ag \rightarrow {}^{24}_{11}Na$ (cumulative)

E [MeV]	ΔE [MeV]	σ [mb]	Δσ [mb]	Ref.
3000		2.240		Kat 68
3650		3.800	0.40	Koz 90
11500		3.900	0.40	Eng 74
29000		4.100		Kat 68
300000		5.220	0.14	Por 79
300000		4.700	0.10	Eng 74a
400000		5.150	0.23	Col 81

993 Reaction $_{47}Ag \rightarrow {}^{27}_{12}Mg$ (cumulative)

E [MeV]	ΔE [MeV]	σ [mb]	Δσ [mb]	Ref.
300000		1.540	0.90E-1	Por 79

994 Reaction $_{47}Ag \rightarrow {}^{28}_{12}Mg$

E [MeV]	ΔE [MeV]	σ [mb]	Δσ [mb]	Ref.
3000		0.327		Kat 68
29000		0.594		Kat 68

995 Reaction $_{47}Ag \rightarrow {}^{28}_{12}Mg$ (cumulative)

E [MeV]	ΔE [MeV]	σ [mb]	Δσ [mb]	Ref.
3650		0.500	0.70E-1	Koz 90
11500		0.600	0.10E-1	Eng 74
300000		0.700	0.40E-1	Por 79
300000		0.640	0.50E-1	Eng 74a
400000		0.700	0.20E-1	Col 81

E [MeV]	ΔE [MeV]	σ [mb]	Δσ [mb]	Ref.
996 Reaction $_{47}Ag \rightarrow {}^{29}_{13}Al$ (cumulative)				
300000		2.280	0.13	Por 79
997 Reaction $_{47}Ag \rightarrow {}^{32}_{15}P$				
3000		2.510		Kat 68
998 Reaction $_{47}Ag \rightarrow {}^{38}_{17}Cl$				
11500		1.200	0.10	Eng 74
300000		1.200	0.10	Eng 74a
999 Reaction $_{47}Ag \rightarrow {}^{38}_{17}Cl$ (cumulative)				
300000		1.180	0.40E-1	Por 79
1000 Reaction $_{47}Ag \rightarrow {}^{39}_{17}Cl$ (cumulative)				
300000		0.600	0.80E-1	Por 79
1001 Reaction $_{47}Ag \rightarrow {}^{41}_{18}Ar$ (cumulative)				
11500		0.650	0.30E-1	Eng 74
300000		0.660	0.20E-1	Por 79
300000		0.640	0.30E-1	Eng 74a
1002 Reaction $_{47}Ag \rightarrow {}^{42}_{19}K$				
3000		1.670		Kat 68
29000		0.987		Kat 68
300000		2.240	0.15	Por 79
400000		2.360	0.10	Col 81
1003 Reaction $_{47}Ag \rightarrow {}^{43}_{19}K$				
3000		0.697		Kat 68
29000		0.462		Kat 68
1004 Reaction $_{47}Ag \rightarrow {}^{43}_{19}K$ (cumulative)				
3650		1.100	0.80E-1	Koz 90
300000		1.310	0.50E-1	Por 79
1005 Reaction $_{47}Ag \rightarrow {}^{45}_{20}Ca$				
3000		0.727		Kat 68
29000		1.120		Kat 68

E [MeV]	ΔE [MeV]	σ [mb]	Δσ [mb]	Ref.

1006 Reaction $_{47}Ag \rightarrow {}^{47}_{20}Ca$

E [MeV]	ΔE [MeV]	σ [mb]	Δσ [mb]	Ref.
3000		0.520E-1		Kat 68
29000		0.900E-1		Kat 68

1007 Reaction $_{47}Ag \rightarrow {}^{47}_{20}Ca$ (cumulative)

E [MeV]	ΔE [MeV]	σ [mb]	Δσ [mb]	Ref.
300000		0.900E-1	0.10E-1	Por 79

1008 Reaction $_{47}Ag \rightarrow {}^{43}_{21}Sc$ (cumulative)

E [MeV]	ΔE [MeV]	σ [mb]	Δσ [mb]	Ref.
3000		0.880		Kat 68
29000		1.180		Kat 68
300000		1.870	0.20	Por 79

1009 Reaction $_{47}Ag \rightarrow {}^{44}_{21}Sc$

E [MeV]	ΔE [MeV]	σ [mb]	Δσ [mb]	Ref.
3000		0.860		Kat 68
11500		1.440	0.40E-1	Eng 74
29000		1.120		Kat 68
300000		1.710	0.80E-1	Por 79
300000		1.400	0.10	Eng 74a

1010 Reaction $_{47}Ag \rightarrow {}^{44m}_{21}Sc$

E [MeV]	ΔE [MeV]	σ [mb]	Δσ [mb]	Ref.
3000		1.480		Kat 68
3650		1.900	0.10	Koz 90
11500		1.900	0.10	Eng 74
29000		1.970		Kat 68
300000		2.280	0.70E-1	Por 79
300000		2.400	0.30	Eng 74a
400000		2.220	0.13	Col 81

1011 Reaction $_{47}Ag \rightarrow {}^{46}_{21}Sc$

E [MeV]	ΔE [MeV]	σ [mb]	Δσ [mb]	Ref.
3000		3.120		Kat 68
11500		3.000	0.40	Eng 74
29000		3.440		Kat 68
300000		3.180	0.60E-1	Por 79
300000		2.900	0.30	Eng 74a
800000		3.150	0.22	Wan 86

1012 Reaction $_{47}Ag \rightarrow {}^{47}_{21}Sc$

E [MeV]	ΔE [MeV]	σ [mb]	Δσ [mb]	Ref.
3000		1.670		Kat 68
11500		2.080	0.10E-1	Eng 74
29000		1.700		Kat 68
300000		1.740	0.18	Por 79
300000		1.900	0.20	Eng 74a

E [MeV]	ΔE [MeV]	σ [mb]	Δσ [mb]	Ref.
1013 Reaction $_{47}Ag \rightarrow {}^{48}_{21}Sc$				
3000		0.330		Kat 68
11500		0.550	0.80E-1	Eng 74
29000		0.470		Kat 68
300000		0.550	0.40E-1	Por 79
300000		0.490	0.90E-1	Eng 74a
1014 Reaction $_{47}Ag \rightarrow {}^{49}_{21}Sc$ (cumulative)				
3000		0.110		Kat 68
29000		0.140		Kat 68
1015 Reaction $_{47}Ag \rightarrow {}^{45}_{22}Ti$ (cumulative)				
3000		0.660		Kat 68
29000		0.990		Kat 68
1016 Reaction $_{47}Ag \rightarrow {}^{48}_{23}V$ (cumulative)				
3000		2.030		Kat 68
11500		2.500	0.20	Eng 74
29000		2.780		Kat 68
300000		3.060	0.60E-1	Por 79
300000		2.600	0.20	Eng 74a
400000		3.310	0.16	Col 81
1017 Reaction $_{47}Ag \rightarrow {}^{49}_{23}V$ (cumulative)				
3000		2.640		Kat 68
29000		6.300		Kat 68
1018 Reaction $_{47}Ag \rightarrow {}^{48}_{24}Cr$				
3000		0.450E-1		Kat 68
29000		0.610E-1		Kat 68
1019 Reaction $_{47}Ag \rightarrow {}^{49}_{24}Cr$				
3000		0.510		Kat 68
1020 Reaction $_{47}Ag \rightarrow {}^{49}_{24}Cr$ (cumulative)				
300000		0.630	0.40E-1	Por 79
1021 Reaction $_{47}Ag \rightarrow {}^{51}_{24}Cr$ (cumulative)				
3000		6.490		Kat 68
29000		7.150		Kat 68

1022 Reaction $_{47}Ag \rightarrow {}^{52}_{25}Mn$

E [MeV]	ΔE [MeV]	σ [mb]	Δσ [mb]	Ref.
11500		1.700	0.10	Eng 74
300000		2.190	0.10	Por 79
300000		1.700	0.10	Eng 74a
400000		2.340	0.10	Col 81

1023 Reaction $_{47}Ag \rightarrow {}^{52}_{25}Mn$ ([illegible]nulative)

E [MeV]	ΔE [MeV]	σ [mb]	Δσ [mb]	Ref.
3000		2.470		Kat 68

1024 Reaction $_{47}Ag \rightarrow {}^{52m}_{25}Mn$

E [MeV]	ΔE [MeV]	σ [mb]	Δσ [mb]	Ref.
11500		0.390	0.50E-1	Eng 74
300000		0.400	0.20E-1	Eng 74a

1025 Reaction $_{47}Ag \rightarrow {}^{52m}_{25}Mn$ (cumulative)

E [MeV]	ΔE [MeV]	σ [mb]	Δσ [mb]	Ref.
300000		0.440	0.40E-1	Por 79

1026 Reaction $_{47}Ag \rightarrow {}^{54}_{25}Mn$

E [MeV]	ΔE [MeV]	σ [mb]	Δσ [mb]	Ref.
3000		4.900		Kat 68
3650		4.600	0.40	Koz 90
11500		4.800	0.20	Eng 74
29000		5.550		Kat 68
300000		4.700	0.30	Por 79
300000		4.700	0.20	Eng 74a
800000		6.080	0.47	Wan 86

1027 Reaction $_{47}Ag \rightarrow {}^{56}_{25}Mn$

E [MeV]	ΔE [MeV]	σ [mb]	Δσ [mb]	Ref.
300000		1.070	0.30E-1	Por 79

1028 Reaction $_{47}Ag \rightarrow {}^{55}_{26}Fe$ (cumulative)

E [MeV]	ΔE [MeV]	σ [mb]	Δσ [mb]	Ref.
3000		4.470		Kat 68
29000		5.160		Kat 68

1029 Reaction $_{47}Ag \rightarrow {}^{59}_{26}Fe$ (cumulative)

E [MeV]	ΔE [MeV]	σ [mb]	Δσ [mb]	Ref.
3000		0.520		Kat 68
3650		0.500	0.40E-1	Koz 90
11500		0.600	0.40E-1	Eng 74
29000		0.626		Kat 68
300000		0.590	0.30E-1	Por 79
300000		0.520	0.30E-1	Eng 74a
800000		0.567	0.40E-1	Wan 86

E [MeV]	ΔE [MeV]	σ [mb]	Δσ [mb]	Ref.

1030 Reaction $_{47}Ag \rightarrow {}^{55}{}_{27}Co$

E [MeV]	ΔE [MeV]	σ [mb]	Δσ [mb]	Ref.
3000		0.237		Kat 68
29000		0.207		Kat 68

1031 Reaction $_{47}Ag \rightarrow {}^{55}{}_{27}Co$ (cumulative)

E [MeV]	ΔE [MeV]	σ [mb]	Δσ [mb]	Ref.
3650		0.400	0.70E-1	Koz 90
300000		0.330	0.40E-1	Por 79

1032 Reaction $_{47}Ag \rightarrow {}^{56}{}_{27}Co$

E [MeV]	ΔE [MeV]	σ [mb]	Δσ [mb]	Ref.
3000		2.300		Kat 68
29000		1.800		Kat 68
800000		1.380	0.11	Wan 86

1033 Reaction $_{47}Ag \rightarrow {}^{56}{}_{27}Co$ (cumulative)

E [MeV]	ΔE [MeV]	σ [mb]	Δσ [mb]	Ref.
11500		1.500	0.20	Eng 74
300000		1.490	0.70E-1	Por 79
300000		1.400	0.80E-1	Eng 74a

1034 Reaction $_{47}Ag \rightarrow {}^{57}{}_{27}Co$ (cumulative)

E [MeV]	ΔE [MeV]	σ [mb]	Δσ [mb]	Ref.
3000		4.350		Kat 68
3650		4.200	0.80	Koz 90
11500		4.400	0.70	Eng 74
29000		4.310		Kat 68
300000		4.410	0.18	Por 79
300000		3.900	0.50	Eng 74a

1035 Reaction $_{47}Ag \rightarrow {}^{58}{}_{27}Co$

E [MeV]	ΔE [MeV]	σ [mb]	Δσ [mb]	Ref.
3000		5.800		Kat 68
11500		5.800	0.60	Eng 74
29000		5.510		Kat 68
300000		5.800	0.20	Por 79
300000		6.000	0.70	Eng 74a
800000		5.700	0.52	Wan 86

1036 Reaction $_{47}Ag \rightarrow {}^{60}{}_{27}Co$

E [MeV]	ΔE [MeV]	σ [mb]	Δσ [mb]	Ref.
3000		1.890		Kat 68
11500		1.900	0.40E-1	Eng 74
11500		1.600	0.10	Eng 74
29000		2.050		Kat 68
300000		1.900	0.80E-1	Por 79
300000		1.900	0.10	Eng 74a
800000		1.970	0.15	Wan 86

E [MeV]	ΔE [MeV]	σ [mb]	Δσ [mb]	Ref.
1037 Reaction $_{47}Ag \rightarrow {}^{61}_{27}Co$ (cumulative)				
3000		0.940		Kat 68
29000		0.880		Kat 68
1038 Reaction $_{47}Ag \rightarrow {}^{60}_{29}Cu$ (cumulative)				
11500		0.700	0.30E-1	Eng 74
300000		0.610	0.80E-1	Por 79
300000		0.620	0.20E-1	Eng 74a
1039 Reaction $_{47}Ag \rightarrow {}^{61}_{29}Cu$ (cumulative)				
3000		3.240		Kat 68
29000		2.720		Kat 68
1040 Reaction $_{47}Ag \rightarrow {}^{64}_{29}Cu$				
3000		4.090		Kat 68
29000		3.570		Kat 68
1041 Reaction $_{47}Ag \rightarrow {}^{67}_{29}Cu$ (cumulative)				
3000		0.223		Kat 68
29000		0.191		Kat 68
1042 Reaction $_{47}Ag \rightarrow {}^{65}_{30}Zn$ (cumulative)				
3000		10.350		Kat 68
3650		8.000	0.60	Koz 90
11500		8.800	0.30	Eng 74
29000		8.710		Kat 68
300000		7.390	0.48	Por 79
300000		8.600	0.20	Eng 74a
800000		7.190	0.92	Wan 86
1043 Reaction $_{47}Ag \rightarrow {}^{69m}_{30}Zn$				
300000		0.61	0.60E-1	Por 79
1044 Reaction $_{47}Ag \rightarrow {}^{66}_{31}Ga$ (cumulative)				
3650		3.900	0.30	Koz 90
11500		4.300	0.10	Eng 74
300000		4.120	0.20	Por 79
300000		4.500	0.10	Eng 74a

E [MeV]	ΔE [MeV]	σ [mb]	Δσ [mb]	Ref.

1045 Reaction $_{47}Ag \rightarrow {}^{67}_{31}Ga$ (cumulative)

E [MeV]	ΔE [MeV]	σ [mb]	Δσ [mb]	Ref.
3000		8.080		Kat 68
11500		7.600	0.40	Eng 74
11500		10.900	0.10	Eng 74
29000		6.460		Kat 68
300000		8.700	0.60	Eng 74a

1046 Reaction $_{47}Ag \rightarrow {}^{72}_{31}Ga$

E [MeV]	ΔE [MeV]	σ [mb]	Δσ [mb]	Ref.
3000		0.282		Kat 68
29000		0.261		Kat 68

1047 Reaction $_{47}Ag \rightarrow {}^{67}_{32}Ge$ (cumulative)

E [MeV]	ΔE [MeV]	σ [mb]	Δσ [mb]	Ref.
300000		1.120	0.40E-1	Por 79

1048 Reaction $_{47}Ag \rightarrow {}^{69}_{32}Ge$ (cumulative)

E [MeV]	ΔE [MeV]	σ [mb]	Δσ [mb]	Ref.
300000		0.610	0.60E-1	Por 79

1049 Reaction $_{47}Ag \rightarrow {}^{70}_{33}As$ (cumulative)

E [MeV]	ΔE [MeV]	σ [mb]	Δσ [mb]	Ref.
300000		2.540	0.33	Por 79

1050 Reaction $_{47}Ag \rightarrow {}^{71}_{33}As$ (cumulative)

E [MeV]	ΔE [MeV]	σ [mb]	Δσ [mb]	Ref.
3000		8.120		Kat 68
3650		6.700	0.90	Koz 90
11500		7.000	0.20	Eng 74
29000		6.320		Kat 68
300000		6.320	0.90	Por 79
300000		7.300	0.50	Eng 74a

1051 Reaction $_{47}Ag \rightarrow {}^{72}_{33}As$

E [MeV]	ΔE [MeV]	σ [mb]	Δσ [mb]	Ref.
3000		8.280		Kat 68
3650		5.600	0.50	Koz 90
11500		5.900	0.30	Eng 74
29000		6.660		Kat 68
300000		6.570	0.98	Por 79
300000		6.100	0.50	Eng 74a

1052 Reaction $_{47}Ag \rightarrow {}^{73}_{33}As$ (cumulative)

E [MeV]	ΔE [MeV]	σ [mb]	Δσ [mb]	Ref.
3000		16.200		Kat 68
29000		12.900		Kat 68

E [MeV]	ΔE [MeV]	σ [mb]	Δσ [mb]	Ref.
1053 Reaction $_{47}Ag \rightarrow {}^{74}_{33}As$				
3000		3.010		Kat 68
11500		1.800	0.20	Eng 74
11500		2.700	0.30	Eng 74
29000		2.430		Kat 68
300000		2.420	0.50E-1	Por 79
300000		1.900	0.10	Eng 74a
300000		2.170	0.20E-1	Eng 74a
1054 Reaction $_{47}Ag \rightarrow {}^{72}_{34}Se$				
3000		3.430		Kat 68
29000		2.470		Kat 68
1055 Reaction $_{47}Ag \rightarrow {}^{72}_{34}Se$ (cumulative)				
11500		2.200	0.10	Eng 74
300000		1.800	0.20	Eng 74a
1056 Reaction $_{47}Ag \rightarrow {}^{73}_{34}Se$ (cumulative)				
3650		4.100	0.40	Koz 90
11500		3.800	0.30	Eng 74
300000		5.870	0.10	Por 79
300000		4.300	0.60	Eng 74a
1057 Reaction $_{47}Ag \rightarrow {}^{75}_{34}Se$ (cumulative)				
3000		19.000		Kat 68
3650		10.700	0.90	Koz 90
11500		9.800	0.50	Eng 74
29000		13.500		Kat 68
300000		11.600	0.20	Por 79
300000		9.400	0.40	Eng 74a
800000		11.100	0.80	Wan 86
1058 Reaction $_{47}Ag \rightarrow {}^{74m}_{35}Br$				
300000		1.450	0.11	Por 79
1059 Reaction $_{47}Ag \rightarrow {}^{75}_{35}Br$ (cumulative)				
300000		5.660	0.10	Por 79

E [MeV]	ΔE [MeV]	σ [mb]	Δσ [mb]	Ref.

1060 Reaction $_{47}Ag \rightarrow {}^{76}_{35}Br$

E [MeV]	ΔE [MeV]	σ [mb]	Δσ [mb]	Ref.
3650		7.700	0.70	Koz 90
11500		8.100	0.20	Eng 74
300000		6.790	1.05	Por 79
300000		6.800	0.90	Eng 74a

1061 Reaction $_{47}Ag \rightarrow {}^{77}_{35}Br$ (cumulative)

E [MeV]	ΔE [MeV]	σ [mb]	Δσ [mb]	Ref.
11500		8.900	0.60	Eng 74
300000		9.110	0.23	Por 79
300000		8.100	0.90	Eng 74a

1062 Reaction $_{47}Ag \rightarrow {}^{76}_{36}Kr$ (cumulative)

E [MeV]	ΔE [MeV]	σ [mb]	Δσ [mb]	Ref.
3650		2.100	0.60	Koz 90
300000		1.760	0.19	Por 79

1063 Reaction $_{47}Ag \rightarrow {}^{77}_{36}Kr$ (cumulative)

E [MeV]	ΔE [MeV]	σ [mb]	Δσ [mb]	Ref.
300000		5.520	0.31	Por 79

1064 Reaction $_{47}Ag \rightarrow {}^{79}_{36}Kr$ (cumulative)

E [MeV]	ΔE [MeV]	σ [mb]	Δσ [mb]	Ref.
300000		10.400	0.50	Por 79

1065 Reaction $_{47}Ag \rightarrow {}^{81}_{36}Kr$

E [MeV]	ΔE [MeV]	σ [mb]	Δσ [mb]	Ref.
150.0		0.320E-2	0.12E-2	Lav 84
600.0		10.500	1.10	Lav 84
1050		22.200	2.20	Lav 84
3000		17.000	1.50	Hud 70
24000		12.700	3.60	Lav 84
29000		13.600	1.00	Hud 70

1066 Reaction $_{47}Ag \rightarrow {}^{85}_{36}Kr$

E [MeV]	ΔE [MeV]	σ [mb]	Δσ [mb]	Ref.
600.0		0.950E-2	0.10E-2	Lav 84
1050		0.210E-1	0.20E-2	Lav 84
3000		0.300E-1	0.10E-1	Hud 70
24000		0.180E-1	0.50E-2	Lav 84
29000		0.230E-1	0.40E-2	Hud 70

1067 Reaction $_{47}Ag \rightarrow {}^{79}_{37}Rb$ (cumulative)

E [MeV]	ΔE [MeV]	σ [mb]	Δσ [mb]	Ref.
11500		2.000	0.30	Eng 74
300000		2.500	0.10	Eng 74a

E [MeV]	ΔE [MeV]	σ [mb]	Δσ [mb]	Ref.

1068 Reaction $_{47}Ag \rightarrow {}^{81}_{37}Rb$ (cumulative)

E [MeV]	ΔE [MeV]	σ [mb]	Δσ [mb]	Ref.
3650		13.900	1.10	Koz 90
11500		16.000	1.30	Eng 74
300000		14.000	0.70	Por 79
300000		14.500	2.10	Eng 74a

1069 Reaction $_{47}Ag \rightarrow {}^{82m}_{37}Rb$

E [MeV]	ΔE [MeV]	σ [mb]	Δσ [mb]	Ref.
3650		4.600	0.70	Koz 90
11500		4.800	0.10	Eng 74
300000		5.600	0.28	Por 79
300000		3.800	0.70	Eng 74a

1070 Reaction $_{47}Ag \rightarrow {}^{83}_{37}Rb$

E [MeV]	ΔE [MeV]	σ [mb]	Δσ [mb]	Ref.
3000		4.940		Kat 68
29000		3.600		Kat 68

1071 Reaction $_{47}Ag \rightarrow {}^{83}_{37}Rb$ (cumulative)

E [MeV]	ΔE [MeV]	σ [mb]	Δσ [mb]	Ref.
3000		22.000		Kat 68
3650		14.100	0.50	Koz 90
29000		15.100		Kat 68
11500		12.300	0.40	Eng 74
300000		14.700	0.40	Por 79
300000		11.200	0.40	Eng 74a
800000		14.700	1.10	Wan 86

1072 Reaction $_{47}Ag \rightarrow {}^{84}_{37}Rb$

E [MeV]	ΔE [MeV]	σ [mb]	Δσ [mb]	Ref.
3000		2.260		Kat 68
11500		1.310	0.20	Eng 74
29000		1.520		Kat 68
300000		1.280	0.30E-1	Eng 74a

1073 Reaction $_{47}Ag \rightarrow {}^{84m}_{37}Rb$

E [MeV]	ΔE [MeV]	σ [mb]	Δσ [mb]	Ref.
11500		0.710	0.20E-1	Eng 74
300000		1.180	0.60E-1	Por 79
300000		1.030	0.30E-1	Eng 74a

1074 Reaction $_{47}Ag \rightarrow {}^{84m+g}_{37}Rb$

E [MeV]	ΔE [MeV]	σ [mb]	Δσ [mb]	Ref.
300000		1.500	0.30E-1	Por 79

E [MeV]	ΔE [MeV]	σ [mb]	Δσ [mb]	Ref.

1075 Reaction $_{47}Ag \rightarrow {}^{82}_{38}Sr$ (cumulative)

E [MeV]	ΔE [MeV]	σ [mb]	Δσ [mb]	Ref.
3000		8.500		Kat 68
11500		7.100	0.50	Eng 74
29000		6.360		Kat 68
300000		7.380	0.34	Por 79
300000		7.200	0.80	Eng 74a
800000		6.300	0.44	Wan 86

1076 Reaction $_{47}Ag \rightarrow {}^{83}_{38}Sr$ (cumulative)

E [MeV]	ΔE [MeV]	σ [mb]	Δσ [mb]	Ref.
3000		17.200		Kat 68
29000		11.380		Kat 68
300000		10.900	0.30E-1	Por 79

1077 Reaction $_{47}Ag \rightarrow {}^{85}_{38}Sr$

E [MeV]	ΔE [MeV]	σ [mb]	Δσ [mb]	Ref.
3000		19.500		Kat 68
29000		15.800		Kat 68

1078 Reaction $_{47}Ag \rightarrow {}^{84}_{39}Y$

E [MeV]	ΔE [MeV]	σ [mb]	Δσ [mb]	Ref.
11500		4.500	0.70	Eng 74
300000		5.100	0.30	Eng 74a

1079 Reaction $_{47}Ag \rightarrow {}^{84}_{39}Y$ (cumulative)

E [MeV]	ΔE [MeV]	σ [mb]	Δσ [mb]	Ref.
300000		4.770	0.14	Por 79

1080 Reaction $_{47}Ag \rightarrow {}^{85}_{39}Y$ (cumulative)

E [MeV]	ΔE [MeV]	σ [mb]	Δσ [mb]	Ref.
300000		7.760	0.80E-1	Por 79

1081 Reaction $_{47}Ag \rightarrow {}^{86m}_{39}Y$

E [MeV]	ΔE [MeV]	σ [mb]	Δσ [mb]	Ref.
3650		6.300	0.40	Koz 90
11500		5.800	0.20	Eng 74
300000		7.050	0.14	Por 79
300000		6.900	0.20	Eng 74a

1082 Reaction $_{47}Ag \rightarrow {}^{86m+g}_{39}Y$

E [MeV]	ΔE [MeV]	σ [mb]	Δσ [mb]	Ref.
300000		9.570	0.30	Por 79

1083 Reaction $_{47}Ag \rightarrow {}^{87}_{39}Y$

E [MeV]	ΔE [MeV]	σ [mb]	Δσ [mb]	Ref.
300000		0.570	0.32	Por 79

E [MeV]	ΔE [MeV]	σ [mb]	Δσ [mb]	Ref.

1084 Reaction $_{47}Ag \rightarrow {}^{87}_{39}Y$ (cumulative)

E [MeV]	ΔE [MeV]	σ [mb]	Δσ [mb]	Ref.
3650		15.400	1.10	Koz 90
11500		17.600	0.70	Eng 74
300000		15.400	1.40	Eng 74a

1085 Reaction $_{47}Ag \rightarrow {}^{87m}_{39}Y$ (cumulative)

E [MeV]	ΔE [MeV]	σ [mb]	Δσ [mb]	Ref.
3650		15.300	1.20	Koz 90
11500		15.000	1.30	Eng 74
300000		15.900	0.80	Por 79
300000		15.000	0.50	Eng 74a

1086 Reaction $_{47}Ag \rightarrow {}^{88}_{39}Y$

E [MeV]	ΔE [MeV]	σ [mb]	Δσ [mb]	Ref.
11500		3.900	0.20	Eng 74
300000		3.110	0.20	Por 79
300000		3.000	0.70	Eng 74a

1087 Reaction $_{47}Ag \rightarrow {}^{90m}_{39}Y$

E [MeV]	ΔE [MeV]	σ [mb]	Δσ [mb]	Ref.
3650		0.500	0.10	Koz 90
300000		0.370	0.30E-1	Por 79

1088 Reaction $_{47}Ag \rightarrow {}^{85}_{40}Zr$ (cumulative)

E [MeV]	ΔE [MeV]	σ [mb]	Δσ [mb]	Ref.
300000		1.060	0.10	Por 79

1089 Reaction $_{47}Ag \rightarrow {}^{86}_{40}Zr$ (cumulative)

E [MeV]	ΔE [MeV]	σ [mb]	Δσ [mb]	Ref.
11500		5.000	0.30	Eng 74
300000		5.930	0.36	Por 79
300000		4.500	0.10	Eng 74a

1090 Reaction $_{47}Ag \rightarrow {}^{88}_{40}Zr$ (cumulative)

E [MeV]	ΔE [MeV]	σ [mb]	Δσ [mb]	Ref.
3000		20.100		Kat 68
3650		15.200	0.90	Koz 90
11500		14.600	1.20	Eng 74
29000		15.200		Kat 68
300000		14.700	0.20	Por 79
300000		13.400	0.50	Eng 74a
800000		14.600	1.00	Wan 86

1091 Reaction $_{47}Ag \rightarrow {}^{89}_{40}Zr$ (cumulative)

E [MeV]	ΔE [MeV]	σ [mb]	Δσ [mb]	Ref.
3000		22.600		Kat 68
3650		14.400	1.10	Koz 90

E [MeV]	ΔE [MeV]	σ [mb]	Δσ [mb]	Ref.
1091 Reaction $_{47}Ag \rightarrow {}^{89}_{40}Zr$ (cumulative) (continued)				
11500		15.700	0.90	Eng 74
29000		17.500		Kat 68
300000		15.800	0.40	Por 79
300000		14.900	0.60	Eng 74a
1092 Reaction $_{47}Ag \rightarrow {}^{88}_{41}Nb$ (cumulative)				
300000		2.650	0.18	Por 79
1093 Reaction $_{47}Ag \rightarrow {}^{89}_{41}Nb$				
3650		1.400	0.20	Koz 90
300000		0.830	0.20E-1	Por 79
1094 Reaction $_{47}Ag \rightarrow {}^{89}_{41}Nb$ (cumulative)				
11500		1.600	0.20	Eng 74
300000		1.400	0.40	Eng 74a
1095 Reaction $_{47}Ag \rightarrow {}^{89m}_{41}Nb$				
300000		9.500	1.00	Por 79
1096 Reaction $_{47}Ag \rightarrow {}^{90}_{41}Nb$				
300000		13.200	0.60	Por 79
1097 Reaction $_{47}Ag \rightarrow {}^{90}_{41}Nb$ (cumulative)				
3650		14.800	1.20	Koz 90
11500		16.000	1.20	Eng 74
300000		14.100	0.70	Eng 74a
1098 Reaction $_{47}Ag \rightarrow {}^{92m}_{41}Nb$				
11500		0.640	0.50E-1	Eng 74
300000		0.570	0.60E-1	Eng 74a
1099 Reaction $_{47}Ag \rightarrow {}^{95}_{41}Nb$				
11500		0.430	0.40E-1	Eng 74
300000		0.530	0.10	Eng 74a
1100 Reaction $_{47}Ag \rightarrow {}^{96}_{41}Nb$				
300000		0.570	0.14	Por 79

E [MeV]	ΔE [MeV]	σ [mb]	Δσ [mb]	Ref.

1101 Reaction $_{47}Ag \rightarrow {}^{90}_{42}Mo$ (cumulative)

E [MeV]	ΔE [MeV]	σ [mb]	Δσ [mb]	Ref.
11500		3.300	0.60	Eng 74
300000		3.320	0.16	Por 79
300000		2.700	0.40	Eng 74a

1102 Reaction $_{47}Ag \rightarrow {}^{93m}_{42}Mo$

E [MeV]	ΔE [MeV]	σ [mb]	Δσ [mb]	Ref.
3650		2.700	0.60	Koz 90
11500		3.900	0.80	Eng 74
300000		2.990	0.11	Por 79
300000		3.600	0.60	Eng 74a

1103 Reaction $_{47}Ag \rightarrow {}^{92}_{43}Tc$ (cumulative)

E [MeV]	ΔE [MeV]	σ [mb]	Δσ [mb]	Ref.
300000		2.010	0.13	Por 79

1104 Reaction $_{47}Ag \rightarrow {}^{93}_{43}Tc$ (cumulative)

E [MeV]	ΔE [MeV]	σ [mb]	Δσ [mb]	Ref.
300000		6.560	0.17	Por 79

1105 Reaction $_{47}Ag \rightarrow {}^{94}_{43}Tc$

E [MeV]	ΔE [MeV]	σ [mb]	Δσ [mb]	Ref.
3650		7.000	0.50	Koz 90
300000		7.650	0.39	Por 79

1106 Reaction $_{47}Ag \rightarrow {}^{94}_{43}Tc$ (cumulative)

E [MeV]	ΔE [MeV]	σ [mb]	Δσ [mb]	Ref.
11500		6.620	0.10E-1	Eng 74
11500		7.800	0.20	Eng 74
300000		6.500	0.30	Eng 74a

1107 Reaction $_{47}Ag \rightarrow {}^{94m}_{43}Tc$ (cumulative)

E [MeV]	ΔE [MeV]	σ [mb]	Δσ [mb]	Ref.
3650		2.900	0.40	Koz 90
300000		2.130	0.21	Por 79

1108 Reaction $_{47}Ag \rightarrow {}^{95}_{43}Tc$ (cumulative)

E [MeV]	ΔE [MeV]	σ [mb]	Δσ [mb]	Ref.
11500		15.700	1.30	Eng 74
300000		14.700	0.40	Eng 74a

1109 Reaction $_{47}Ag \rightarrow {}^{95m}_{43}Tc$

E [MeV]	ΔE [MeV]	σ [mb]	Δσ [mb]	Ref.
11500		0.750	0.10	Eng 74
300000		0.740	0.80E-1	Eng 74a

E [MeV]	ΔE [MeV]	σ [mb]	Δσ [mb]	Ref.

1110 Reaction $_{47}Ag \rightarrow {}^{96}_{43}Tc$

E [MeV]	ΔE [MeV]	σ [mb]	Δσ [mb]	Ref.
11500		7.100	0.50	Eng 74
300000		7.380	0.19	Por 79
300000		7.000	0.50	Eng 74a

1111 Reaction $_{47}Ag \rightarrow {}^{95}_{44}Ru$ (cumulative)

E [MeV]	ΔE [MeV]	σ [mb]	Δσ [mb]	Ref.
11500		5.700	0.70	Eng 74
300000		5.400	0.70	Por 79
300000		4.800	0.40	Eng 74a

1112 Reaction $_{47}Ag \rightarrow {}^{97}_{44}Ru$ (cumulative)

E [MeV]	ΔE [MeV]	σ [mb]	Δσ [mb]	Ref.
3000		19.500		Kat 68
3650		19.100	1.60	Koz 90
11500		14.200	1.00	Eng 74
29000		19.100		Kat 68
300000		15.700	0.60	Por 79
300000		14.400	1.60	Eng 74a

1113 Reaction $_{47}Ag \rightarrow {}^{103}_{44}Ru$ (cumulative)

E [MeV]	ΔE [MeV]	σ [mb]	Δσ [mb]	Ref.
3000		0.810		Kat 68
11500		1.260	0.10E-1	Eng 74
29000		0.920		Kat 68
300000		0.750	0.30E-1	Por 79
300000		1.290	0.50E-1	Eng 74a
800000		0.810	0.55E-1	Wan 86

1114 Reaction $_{47}Ag \rightarrow {}^{96}_{45}Rh$ (cumulative)

E [MeV]	ΔE [MeV]	σ [mb]	Δσ [mb]	Ref.
300000		0.750	0.40E-1	Por 79

1115 Reaction $_{47}Ag \rightarrow {}^{97m}_{45}Rh$

E [MeV]	ΔE [MeV]	σ [mb]	Δσ [mb]	Ref.
300000		3.940	0.15	Por 79

1116 Reaction $_{47}Ag \rightarrow {}^{99}_{45}Rh$

E [MeV]	ΔE [MeV]	σ [mb]	Δσ [mb]	Ref.
300000		2.910	0.11	Por 79

1117 Reaction $_{47}Ag \rightarrow {}^{99m}_{45}Rh$

E [MeV]	ΔE [MeV]	σ [mb]	Δσ [mb]	Ref.
300000		8.420	0.46	Por 79

E [MeV]	ΔE [MeV]	σ [mb]	Δσ [mb]	Ref.
1118 Reaction $_{47}Ag \rightarrow {}^{100}_{45}Rh$				
3650		9.900	1.00	Koz 90
11500		11.100	0.70	Eng 74
300000		12.400	0.30	Por 79
300000		9.900	0.80	Eng 74a
1119 Reaction $_{47}Ag \rightarrow {}^{101}_{45}Rh$				
3000		3.240		Kat 68
11500		3.800	0.10	Eng 74
29000		2.890		Kat 68
300000		4.400	0.10	Eng 74a
1120 Reaction $_{47}Ag \rightarrow {}^{101m}_{45}Rh$ (cumulative)				
11500		18.600	1.30	Eng 74
300000		21.200	1.00	Eng 74a
1121 Reaction $_{47}Ag \rightarrow {}^{102}_{45}Rh$				
11500		1.200	0.30	Eng 74
300000		1.200	0.10	Por 79
300000		2.880	0.46	Eng 74a
1122 Reaction $_{47}Ag \rightarrow {}^{102m}_{45}Rh$				
11500		4.500	0.30	Eng 74
300000		5.600	1.00	Por 79
300000		5.600	0.10	Eng 74a
1123 Reaction $_{47}Ag \rightarrow {}^{106m}_{45}Rh$ (cumulative)				
3650		1.500	0.20	Koz 90
1124 Reaction $_{47}Ag \rightarrow {}^{99}_{46}Pd$ (cumulative)				
11500		1.640	0.10E-1	Eng 74
300000		2.940	0.11	Por 79
300000		1.500	0.10	Eng 74a
1125 Reaction $_{47}Ag \rightarrow {}^{100}_{46}Pd$ (cumulative)				
3000		6.670		Kat 68
11500		4.300	0.30	Eng 74
11500		5.200	0.10	Eng 74
29000		5.630		Kat 68
300000		5.780	0.25	Por 79
300000		5.100	0.40	Eng 74a

E [MeV]	ΔE [MeV]	σ [mb]	Δσ [mb]	Ref.

1126 Reaction $_{47}Ag \rightarrow {}^{101}_{46}Pd$ (cumulative)

E [MeV]	ΔE [MeV]	σ [mb]	Δσ [mb]	Ref.
3000		16.200		Kat 68
3650		11.300	0.90	Koz 90
29000		13.400		Kat 68
300000		12.900	0.40	Por 79

1127 Reaction $_{47}Ag \rightarrow {}^{103}_{46}Pd$

E [MeV]	ΔE [MeV]	σ [mb]	Δσ [mb]	Ref.
3000		18.400		Kat 68
29000		17.800		Kat 68

1128 Reaction $_{47}Ag \rightarrow {}^{103}_{46}Pd$ (cumulative)

E [MeV]	ΔE [MeV]	σ [mb]	Δσ [mb]	Ref.
3000		27.300		Kat 68
29000		25.900		Kat 68

1129 Reaction $_{47}Ag \rightarrow {}^{102}_{47}Ag$ (cumulative)

E [MeV]	ΔE [MeV]	σ [mb]	Δσ [mb]	Ref.
300000		2.710	0.12	Por 79

1130 Reaction $_{47}Ag \rightarrow {}^{103}_{47}Ag$ (cumulative)

E [MeV]	ΔE [MeV]	σ [mb]	Δσ [mb]	Ref.
3000		9.600		Kat 68
11500		7.700	0.50	Eng 74
11500		12.400	0.40	Eng 74
29000		8.080		Kat 68
300000		13.000	0.80	Por 79
300000		13.100	0.10	Eng 74a

1131 Reaction $_{47}Ag \rightarrow {}^{104}_{47}Ag$

E [MeV]	ΔE [MeV]	σ [mb]	Δσ [mb]	Ref.
3650		9.900	1.20	Koz 90
11500		9.400	1.00	Eng 74
300000		11.700	1.60	Eng 74a

1132 Reaction $_{47}Ag \rightarrow {}^{104m}_{47}Ag$ (cumulative)

E [MeV]	ΔE [MeV]	σ [mb]	Δσ [mb]	Ref.
11500		5.000	6.00	Eng 74
300000		1.940	0.42	Por 79
300000		4.700	0.30	Eng 74a

1133 Reaction $_{47}Ag \rightarrow {}^{105}_{47}Ag$ (cumulative)

E [MeV]	ΔE [MeV]	σ [mb]	Δσ [mb]	Ref.
3000		30.100		Kat 68
3650		28.700	2.30	Koz 90
11500		25.000	1.70	Eng 74
29000		28.400		Kat 68

1133 Reaction $_{47}Ag \rightarrow {}^{105}{}_{47}Ag$ (cumulative) (continued)

E [MeV]	ΔE [MeV]	σ [mb]	Δσ [mb]	Ref.
300000		25.000	0.90	Por 79
300000		24.300	2.10	Eng 74a
800000		23.500	1.70	Wan 86

1134 Reaction $_{47}Ag \rightarrow {}^{106}{}_{47}Ag$ (Fig. 523, p. 292)

E [MeV]	ΔE [MeV]	σ [mb]	Δσ [mb]	Ref.
9.00		1.000		Lin 59*
10.40		1.900		Lin 59*
11.90		3.900		Lin 59*
13.40		0.600		Coh 54*
14.00		6.000		Lin 59*
14.40		2.800		Coh 54*
14.50		4.300		Coh 54*
15.00		10.000		Coh 54*
15.90		13.000		Lin 59*
16.00		20.000		Coh 54*
16.70		20.000		Lin 59*
17.00		40.000		Coh 54*
17.50		32.000		Lin 59*
18.00		50.000		Coh 54*
18.50		41.000		Lin 59*
19.00		75.000		Coh 54*
19.10		48.000		Lin 59*
19.80		61.000		Lin 59*
20.00		100.000		Coh 54*
20.10		61.000		Lin 59*
21.00		100.000		Coh 54*
22.00		110.000		Coh 54*
23.00		115.000		Coh 54*
3000		12.400		Kat 68

1135 Reaction $_{47}Ag \rightarrow {}^{106m}{}_{47}Ag$ (Fig. 524, p. 292)

E [MeV]	ΔE [MeV]	σ [mb]	Δσ [mb]	Ref.
15.70		1.000		Lin 59*
16.00		5.500		Lin 59*
16.80		10.000		Lin 59*
17.50		18.000		Lin 59*
18.00		32.000		Lin 59*
18.50		28.000		Lin 59*
19.00		41.000		Lin 59*
19.40		43.000		Lin 59*
20.00		60.000		Lin 59*
3650		10.200	0.80	Koz 90
11500		9.400	0.70	Eng 74
300000		11.400	0.60	Por 79
300000		9.700	0.60	Eng 74a

1136 Reaction $_{47}Ag \rightarrow {}^{107}_{48}Cd$ (Fig. 525, p. 293)

E [MeV]	ΔE [MeV]	σ [mb]	Δσ [mb]	Ref.
1.90		0.600E-1		Win 62
2.51		0.190E-1	0.20E-2	Col 74
3.12		0.520	0.40E-1	Col 74
3.20		0.500		Win 62
3.65		1.820	0.13	Col 74
3.84		2.520	0.17	Col 74
4.16		5.160	0.35	Col 74
4.90		19.000		Win 62
4.94		24.400	1.60	Col 74
5.00		24.000		Win 62
5.10		38.000		Win 62
6.00		98.000		Win 62
6.06		101.000	7.00	Col 74
6.10		115.000		Win 62
6.40		133.000		Win 62
6.90		179.000		Win 62
7.10		209.000		Win 62
7.20		216.000	15.00	Col 74
8.00		296.000		Win 62
8.10		324.000		Win 62
8.17		331.000	20.00	Col 74
8.70		375.000		Win 62
8.90		418.000		Win 62
9.00		430.500	9.50	Win 62
9.18		521.000	36.00	Col 74
9.35		495.000	47.00	Col 74
9.45		489.000		Win 62
9.85		521.000		Win 62
10.16		521.000		Win 62
10.30		606.000	41.00	Col 74
10.31		644.000	55.00	Col 74
10.33		562.000		Win 62
10.35		576.000		Win 62
10.43		590.000		Win 62
10.45		589.000		Win 62
10.50		581.000		Win 62
11.33		664.000	46.00	Col 74
11.34		680.000	47.00	Col 74
12.29		548.000	37.00	Col 74
12.43		496.000	33.00	Col 74
12.58		491.000	34.00	Col 74
12.62		518.000	35.00	Col 74
12.95		460.000	31.00	Col 74
12.99		449.000	31.00	Col 74
13.27		394.000	27.00	Col 74
13.44		371.000	30.50	Col 74

E [MeV]	ΔE [MeV]	σ [mb]	Δσ [mb]	Ref.

1136 Reaction $_{47}Ag \rightarrow {}^{107}_{48}Cd$ (continued)

E [MeV]	ΔE [MeV]	σ [mb]	Δσ [mb]	Ref.
13.46		390.000	26.00	Col 74
13.58		370.000	25.00	Col 74
13.99		314.000	21.00	Col 74
14.39		267.000	18.00	Col 74
15.59		154.000	11.00	Col 74
15.95		144.000	10.00	Col 74
16.19		132.000	9.00	Col 74
16.45		110.000	7.00	Col 74
16.69		104.000	7.00	Col 74
16.92		96.400	6.50	Col 74
18.71		54.800	3.80	Col 74
19.04		51.100	3.70	Col 74
19.27		48.500	3.30	Col 74
19.50		46.000	3.20	Col 74
19.73		47.300	3.20	Col 74
19.96		50.900	3.50	Col 74
22.43		26.300	1.80	Col 74
22.73		293.000	20.00	Col 74
22.92		346.000	25.00	Col 74
23.12		357.000	24.00	Col 74
23.40		392.000	27.00	Col 74

1137 Reaction $^{107}_{47}Ag \rightarrow {}^{107}_{48}Cd$ (Fig. 526, p. 293)

E [MeV]	ΔE [MeV]	σ [mb]	Δσ [mb]	Ref.
3.40		0.700		Bla 51*
3.80		3.000		Bla 51*
4.10		5.700		Bla 51*
4.30		8.000		Bla 51*
4.40		10.000		Bla 51*
4.50		12.000		Bla 51*
4.60		14.000		Bla 51*
4.70		18.000		Bla 51*
4.80		20.000		Bla 51*
5.00		30.000		Bla 51*
5.30		35.000		Bla 51*
5.40		40.000		Bla 51*
5.60		50.000		Bla 51*
5.70		60.000		Bla 51*
5.90		80.000		Bla 51*
6.10		85.000		Bla 51*
6.30		95.000		Bla 51*
6.40		110.000		Bla 51*
6.70		140.000		Bla 51b*

E [MeV]	ΔE [MeV]	σ [mb]	Δσ [mb]	Ref.

1138 Reaction $^{109}_{47}Ag \rightarrow ^{109}_{48}Cd$ (Fig. 527, p. 293)

E [MeV]	ΔE [MeV]	σ [mb]	Δσ [mb]	Ref.
4.20		6.000		Bla 51*
4.40		8.800		Bla 51*
4.60		12.000		Bla 51*
4.80		18.000		Bla 51*
5.00		22.000		Bla 51*
5.00		24.000		Win 62
5.00		35.000		Win 62
5.30		30.000		Bla 51*
5.40		39.000		Bla 51*
5.60		42.000		Bla 51*
5.70		55.000		Bla 51*
5.90		63.000		Bla 51*
6.00		95.000		Win 62
6.10		78.000		Bla 51*
6.30		83.000		Bla 51*
6.40		100.000		Bla 51*
6.50		130.000		Win 62
6.70		120.000		Bla 51b*
6.90		146.000		Win 62
8.00		280.000		Win 62
8.70		332.000		Win 62
8.90		365.000		Win 62
9.10		362.000		Win 62
9.40		353.000		Win 62
9.85		342.000		Win 62
10.20		320.000		Win 62
10.40		305.000		Win 62
10.50		305.000		Win 62
140.0		3.220	0.15	Rea 65
210.0		3.000	0.15	Rea 65
290.0		1.700	0.15	Rea 65
380.0		1.450	0.15	Rea 65

1139 Reaction $_{48}Cd \rightarrow ^{22}_{11}Na$

E [MeV]	ΔE [MeV]	σ [mb]	Δσ [mb]	Ref.
400.0		0.465E-2	0.40E-3	Kor 67

1140 Reaction $_{48}Cd \rightarrow ^{24}_{11}Na$

E [MeV]	ΔE [MeV]	σ [mb]	Δσ [mb]	Ref.
400.0		0.126E-1	0.40E-3	Kor 67

1141 Reaction $_{48}Cd \rightarrow ^{24}_{11}Na$ (cumulative)

E [MeV]	ΔE [MeV]	σ [mb]	Δσ [mb]	Ref.
400000		7.190	0.45	Col 81

E [MeV]	ΔE [MeV]	σ [mb]	Δσ [mb]	Ref.
1142 Reaction $_{48}Cd \rightarrow {}^{28}_{12}Mg$ (cumulative)				
400000		1.610	0.10	Col 81
1143 Reaction $_{48}Cd \rightarrow {}^{32}_{15}P$				
550.0		0.264E-1	0.14E-2	Kor 70
1144 Reaction $_{48}Cd \rightarrow {}^{33}_{15}P$				
550.0		0.251E-1	0.31E-2	Kor 70
1145 Reaction $_{48}Cd \rightarrow {}^{42}_{19}K$				
400000		2.070	0.56	Col81
1146 Reaction $_{48}Cd \rightarrow {}^{44m}_{21}Sc$				
400000		1.390	0.50E-1	Col 81
1147 Reaction $_{48}Cd \rightarrow {}^{48}_{21}Sc$				
400000		0.870	0.30E-1	Col 81
1148 Reaction $_{48}Cd \rightarrow {}^{48}_{23}V$				
400000		1.610	0.13	Col 81
1149 Reaction $_{48}Cd \rightarrow {}^{52}_{25}Mn$				
400000		0.880	0.40E-1	Col 81
1150 Reaction $_{48}Cd \rightarrow {}^{109(m1+m2+g)}_{49}In$ (Fig. 528, p. 293)				
13.00		1.800	0.30	Nor 90
15.00		25.900	2.10	Nor 90
16.90		69.300	5.60	Nor 90
18.60		93.800	7.50	Nor 90
21.40		109.900	8.80	Nor 90
23.90		133.600	10.70	Nor 90
26.20		153.000	12.30	Nor 90
28.40		140.700	11.30	Nor 90
30.50		127.500	10.20	Nor 90
32.40		116.800	9.40	Nor 90
34.30		117.700	9.40	Nor 90
36.10		122.900	9.90	Nor 90
37.90		132.000	10.60	Nor 90
39.50		137.000	11.00	Nor 90

E [MeV]	ΔE [MeV]	σ [mb]	Δσ [mb]	Ref.

1150 Reaction $_{48}Cd \rightarrow {}^{109(m1+m2+g)}{}_{49}In$ (continued)

E [MeV]	ΔE [MeV]	σ [mb]	Δσ [mb]	Ref.
42.30		146.300	11.80	Nor 90
44.90		150.200	12.10	Nor 90
47.40		143.500	11.50	Nor 90
50.80		126.200	10.10	Nor 90
54.00		109.800	8.80	Nor 90
57.10		100.200	8.00	Nor 90
60.10		96.600	7.70	Nor 90
62.90		95.900	7.70	Nor 90
65.70		98.600	7.90	Nor 90
69.10		82.600	6.70	Nor 90
72.40		73.400	5.90	Nor 90
75.60		66.100	5.30	Nor 90
78.70		57.900	4.70	Nor 90
81.80		52.900	4.30	Nor 90
84.70		49.000	3.90	Nor 90
87.60		45.300	3.60	Nor 90
90.40		43.600	3.50	Nor 90
93.10		39.900	3.20	Nor 90
95.80		38.300	3.10	Nor 90
98.50		34.900	2.80	Nor 90

1151 Reaction $_{48}Cd \rightarrow {}^{110m}{}_{49}In$ (Fig. 529, p. 294)

E [MeV]	ΔE [MeV]	σ [mb]	Δσ [mb]	Ref.
8.00		1.000	0.10	Nor 90
10.70		6.000	0.50	Nor 90
13.00		16.000	1.30	Nor 90
15.00		29.500	2.40	Nor 90
16.90		39.700	3.20	Nor 90
18.60		44.200	3.60	Nor 90
21.40		50.400	4.10	Nor 90
23.90		57.000	4.60	Nor 90
26.20		81.300	6.50	Nor 90
28.40		101.100	8.10	Nor 90
30.50		111.900	9.00	Nor 90
32.40		120.700	9.70	Nor 90
34.30		125.400	10.10	Nor 90
36.10		124.500	10.00	Nor 90
37.90		116.800	9.40	Nor 90
39.50		102.600	8.20	Nor 90
42.30		83.800	6.80	Nor 90
44.90		77.800	6.30	Nor 90
47.40		80.900	6.50	Nor 90
50.80		86.700	7.00	Nor 90
54.00		88.300	7.10	Nor 90
57.10		81.000	6.50	Nor 90
60.10		68.300	5.50	Nor 90

E [MeV]	ΔE [MeV]	σ [mb]	Δσ [mb]	Ref.
1151 Reaction $_{48}Cd \rightarrow {}^{110m}{}_{49}In$ (continued)				
62.90		57.200	4.60	Nor 90
65.70		48.700	3.90	Nor 90
69.10		40.900	3.30	Nor 90
72.40		36.500	2.90	Nor 90
75.60		33.500	2.70	Nor 90
78.70		31.000	2.50	Nor 90
81.80		28.000	2.30	Nor 90
84.70		26.500	2.10	Nor 90
87.60		24.100	1.90	Nor 90
90.40		22.900	1.80	Nor 90
93.10		20.600	1.70	Nor 90
95.80		19.500	1.60	Nor 90
98.50		17.300	1.40	Nor 90
1152 Reaction $_{48}Cd \rightarrow {}^{111}{}_{49}In$				
400.0		6.370	0.13	Nie 69
1153 Reaction $_{48}Cd \rightarrow {}^{111m+g}{}_{49}In$ (Fig. 530, p. 294)				
4.10		1.200	0.10	Nor 90
8.00		24.100	1.90	Nor 90
10.70		72.300	5.80	Nor 90
13.00		133.700	10.70	Nor 90
15.00		203.400	16.30	Nor 90
16.90		238.400	19.10	Nor 90
18.60		256.700	20.60	Nor 90
21.40		294.700	23.60	Nor 90
23.90		315.800	25.30	Nor 90
26.20		261.600	20.90	Nor 90
28.40		204.500	16.40	Nor 90
30.50		186.400	14.90	Nor 90
32.40		189.800	15.20	Nor 90
34.30		204.800	16.40	Nor 90
36.10		221.600	17.70	Nor 90
37.90		232.800	18.60	Nor 90
39.50		240.500	19.20	Nor 90
42.30		237.800	19.10	Nor 90
44.90		207.100	16.60	Nor 90
47.40		168.700	13.50	Nor 90
50.80		126.400	10.10	Nor 90
54.00		101.600	8.10	Nor 90
57.10		88.200	7.10	Nor 90
60.10		81.600	6.50	Nor 90
62.90		76.500	6.10	Nor 90
65.70		72.100	5.80	Nor 90

E [MeV]	ΔE [MeV]	σ [mb]	Δσ [mb]	Ref.

1153 Reaction $_{48}Cd \rightarrow {}^{111m+g}{}_{49}In$ (continued)

E [MeV]	ΔE [MeV]	σ [mb]	Δσ [mb]	Ref.
69.10		64.500	5.20	Nor 90
72.40		57.400	4.60	Nor 90
75.60		52.900	4.30	Nor 90
78.70		47.400	3.80	Nor 90
81.80		44.700	3.60	Nor 90
84.70		42.100	3.40	Nor 90
87.60		39.500	3.20	Nor 90
90.40		36.900	3.00	Nor 90
93.10		35.200	2.80	Nor 90
95.80		33.600	2.70	Nor 90
98.50		31.200	2.50	Nor 90

1154 Reaction $_{48}Cd \rightarrow {}^{114m}{}_{49}In$ (Fig. 531, p. 294)

E [MeV]	ΔE [MeV]	σ [mb]	Δσ [mb]	Ref.
4.10		0.700	0.20	Nor 90
8.00		21.300	1.70	Nor 90
10.70		70.100	5.60	Nor 90
13.00		60.200	4.80	Nor 90
15.00		29.700	2.40	Nor 90
16.90		16.100	1.30	Nor 90
18.60		11.800	1.00	Nor 90
21.40		26.800	2.20	Nor 90
23.90		50.900	4.10	Nor 90
26.20		62.400	5.00	Nor 90
28.40		64.400	5.20	Nor 90
30.50		57.800	4.60	Nor 90
32.40		47.000	3.80	Nor 90
34.30		34.800	2.80	Nor 90
36.10		26.100	2.10	Nor 90
37.90		20.300	1.70	Nor 90
39.50		17.000	1.40	Nor 90
42.30		13.100	1.10	Nor 90
44.90		10.300	0.80	Nor 90
47.40		8.900	0.70	Nor 90
50.80		7.900	0.70	Nor 90
54.00		7.100	0.60	Nor 90
57.10		6.700	0.60	Nor 90
60.10		6.400	0.50	Nor 90
62.90		5.500	0.50	Nor 90
65.70		5.200	0.40	Nor 90
69.10		5.100	0.40	Nor 90
72.40		4.800	0.40	Nor 90
75.60		4.400	0.40	Nor 90
78.70		3.900	0.30	Nor 90
81.80		4.300	0.40	Nor 90
84.70		4.000	0.30	Nor 90

E [MeV]	ΔE [MeV]	σ [mb]	Δσ [mb]	Ref.

1154 Reaction $_{48}Cd \rightarrow {}^{114m}{}_{49}In$ (continued)

E [MeV]	ΔE [MeV]	σ [mb]	Δσ [mb]	Ref.
87.60		3.700	0.30	Nor 90
90.40		3.500	0.30	Nor 90
93.10		3.200	0.30	Nor 90
95.80		3.200	0.30	Nor 90
98.50		3.000	0.30	Nor 90

1155 Reaction $^{106}{}_{48}Cd \rightarrow {}^{105}{}_{47}Ag$

E [MeV]	ΔE [MeV]	σ [mb]	Δσ [mb]	Ref.
400.0		21.000	2.00	Nie 70

1156 Reaction $^{106}{}_{48}Cd \rightarrow {}^{105}{}_{48}Cd$

E [MeV]	ΔE [MeV]	σ [mb]	Δσ [mb]	Ref.
400.0		30.400	0.90	Nie 70

1157 Reaction $^{108}{}_{48}Cd \rightarrow {}^{107}{}_{48}Cd$

E [MeV]	ΔE [MeV]	σ [mb]	Δσ [mb]	Ref.
400.0		101.700	12.80	Nie 70

1158 Reaction $^{108}{}_{48}Cd \rightarrow {}^{107}{}_{49}In$

E [MeV]	ΔE [MeV]	σ [mb]	Δσ [mb]	Ref.
400.0		6.250	0.84	Nie 70

1159 Reaction $^{110}{}_{48}Cd \rightarrow {}^{109}{}_{48}Cd$

E [MeV]	ΔE [MeV]	σ [mb]	Δσ [mb]	Ref.
12.60		1.000	2.00	Oto 66
13.50		5.000	5.00	Oto 66
14.70		3.000	2.00	Oto 66
400.0		44.800	4.70	Nie 70

1160 Reaction $^{110}{}_{48}Cd \rightarrow {}^{109}{}_{49}In$ **(Fig. 532, p. 294)**

E [MeV]	ΔE [MeV]	σ [mb]	Δσ [mb]	Ref.
12.60		6.000	1.00	Oto 66
13.50		77.000	6.00	Oto 66
13.70		105.000		Ska 75
14.70		320.000	30.00	Oto 66
15.20		371.000		Ska 75
17.00		511.000		Ska 75
17.70		558.000		Ska 75
18.30		604.000		Ska 75
19.00		638.000		Ska 75
19.70		668.000		Ska 75
20.30		660.000		Ska 75
20.90		672.000		Ska 75
400.0		3.580	0.54	Nie 70

1161 Reaction $^{110}_{48}Cd \rightarrow {}^{110}_{49}In$ (Fig. 533, p. 295)

E [MeV]	ΔE [MeV]	σ [mb]	Δσ [mb]	Ref.
4.50		0.400		Bla 51*
4.60		2.000		Bla 51*
4.90		10.000		Bla 51*
5.00		17.000		Bla 51*
5.25		25.000		Bla 51*
5.50		41.000		Bla 51*
5.70		55.000		Bla 51*
5.90		65.000		Bla 51*
6.30		100.000		Bla 51*
6.40		105.000		Bla 51*
6.50		108.000		Ska 75
6.70		160.000		Bla 51*
7.30		168.000		Ska 75
7.30		300.000	10.00	Oto 66
8.00		250.000		Ska 75
8.50		312.000		Ska 75
9.00		340.000		Ska 75
9.10		500.000		Ska 75
11.20		650.000	30.00	Oto 66
11.80		597.000		Ska 75
13.10		760.000	50.00	Oto 66
13.70		544.000		Ska 75
14.70		420.000	30.00	Oto 66
15.20		262.000		Ska 75
17.00		146.000		Ska 75
18.30		102.000		Ska 75
19.70		73.000		Ska 75
20.90		64.000		Ska 75

1162 Reaction $^{110}_{48}Cd \rightarrow {}^{110m}_{49}In$

E [MeV]	ΔE [MeV]	σ [mb]	Δσ [mb]	Ref.
7.30		17.000	2.00	Oto 66
9.10		41.000	5.00	Oto 66
11.20		120.000	10.00	Oto 66
13.10		130.000	10.00	Oto 66
14.70		170.000	20.00	Oto 66

1163 Reaction $^{111}_{48}Cd \rightarrow {}^{108}_{49}In$

E [MeV]	ΔE [MeV]	σ [mb]	Δσ [mb]	Ref.
35.50		90.000		Mar 85*
38.50		240.000		Mar 85*
41.50		330.000		Mar 85*
44.20		360.000		Mar 85*

E [MeV]	ΔE [MeV]	σ [mb]	Δσ [mb]	Ref.

1164 Reaction $^{111}_{48}Cd \rightarrow ^{109m+g}_{49}In$ (Fig. 534, p. 295)

E [MeV]	ΔE [MeV]	σ [mb]	Δσ [mb]	Ref.
22.00		75.000		Mar 85*
24.00		300.000		Mar 85*
25.50		530.000		Mar 85*
27.80		800.000		Mar 85*
29.50		830.000		Mar 85*
32.30		850.000		Mar 85*
35.50		650.000		Mar 85*
38.50		390.000		Mar 85*
41.50		280.000		Mar 85*
44.20		200.000		Mar 85*

1165 Reaction $^{111}_{48}Cd \rightarrow ^{110}_{49}In$ (Fig. 535, p. 295)

E [MeV]	ΔE [MeV]	σ [mb]	Δσ [mb]	Ref.
11.80		17.000		Ska 75
13.10		270.000		Oto 66
13.50		300.000		Mar 85*
13.70		332.000		Ska 75
14.70		590.000		Oto 66
15.20		628.000		Ska 75
15.50		650.000		Mar 85*
17.00		711.000		Ska 75
17.20		850.000		Mar 85*
18.30		790.000		Ska 75
19.20		900.000		Mar 85*
19.70		917.000		Ska 75
20.90		718.000		Ska 75
22.00		950.000		Mar 85*
24.00		700.000		Mar 85*
25.50		550.000		Mar 85*
27.80		280.000		Mar 85*
29.50		210.000		Mar 85*
32.30		140.000		Mar 85*
35.50		110.000		Mar 85*
38.50		90.000		Mar 85*
41.50		85.000		Mar 85*
44.20		75.000		Mar 85*

1166 Reaction $^{111}_{48}Cd \rightarrow ^{111}_{49}In$ (Fig. 536, p. 295)

E [MeV]	ΔE [MeV]	σ [mb]	Δσ [mb]	Ref.
2.10		0.400E-1		Win 62
3.00		0.380		Bla 51*
3.30		0.900		Win 62
3.35		1.000		Bla 51*
3.55		2.300		Bla 51*
3.75		4.500		Bla 51*
4.10		8.000		Bla 51*

1166 Reaction $^{111}_{48}Cd \rightarrow ^{111}_{49}In$ (continued)

E [MeV]	ΔE [MeV]	σ [mb]	Δσ [mb]	Ref.
4.30		4.600		Ska 75
4.30		4.900	3.00	Oto 66
4.35		12.000		Bla 51*
4.40		12.000		Win 62
4.50		20.000		Bla 51*
4.70		30.000		Bla 51*
4.85		39.000		Bla 51*
5.10		47.000		Bla 51*
5.20		275.000		Ska 75
5.30		62.000		Bla 51*
5.40		45.000		Win 62
5.40		75.000		Bla 51*
5.60		90.000		Bla 51*
5.75		100.000		Bla 51*
5.90		66.600		Ska 75
5.90		120.000		Bla 51*
6.10		130.000		Bla 51*
6.30		180.000		Bla 51*
6.40		134.000		Win 62
6.40		200.000		Bla 51*
6.70		134.000		Ska 75
6.70		270.000		Bla 51*
6.70		210.000	10.00	Oto 66
7.30		226.000		Ska 75
7.40		262.000		Win 62
8.00		300.000		Ska 75
8.40		366.000		Win 62
8.50		377.000		Ska 75
9.00		430.000		Ska 75
9.50		508.000		Win 62
9.75		526.000		Win 62
9.80		489.000		Ska 75
10.10		540.000		Win 62
11.80		696.000		Ska 75
12.00		810.000		Blo 55
13.10		530.000	70.00	Oto 66
13.50		550.000		Mar 85*
13.70		480.000		Ska 75
14.70		390.000	90.00	Oto 66
15.20		250.000		Ska 75
15.50		330.000		Mar 85*
17.00		140.000		Ska 75
17.20		125.000		Mar 85*
18.30		82.000		Ska 75
19.20		95.000		Mar 85*
19.70		72.000		Ska 75

E [MeV]	ΔE [MeV]	σ [mb]	Δσ [mb]	Ref.

1166 Reaction $^{111}_{48}Cd \rightarrow ^{111}_{49}In$ (continued)

E [MeV]	ΔE [MeV]	σ [mb]	Δσ [mb]	Ref.
20.90		61.000		Ska 75
22.00		63.000		Mar 85*
24.00		66.000		Mar 85*
25.50		52.000		Mar 85*
27.80		47.000		Mar 85*
29.50		33.000		Mar 85*
32.30		34.000		Mar 85*
35.50		30.000		Mar 85*
38.50		23.000		Mar 85*
41.50		24.000		Mar 85*
44.20		22.000		Mar 85*
70.00		6.550	0.45	Nie 69
100.0		5.960	0.34	Nie 69
150.0		3.110	0.16	Nie 69
200.0		1.960	0.15	Nie 69
300.0		1.270	0.60E-1	Nie 69
400.0		0.920	0.10E-1	Nie 69

1167 Reaction $^{112}_{48}Cd \rightarrow ^{111}_{47}Ag$

E [MeV]	ΔE [MeV]	σ [mb]	Δσ [mb]	Ref.
400.0		16.200	0.40	Nie 70

1168 Reaction $^{112}_{48}Cd \rightarrow ^{110}_{49}In$

E [MeV]	ΔE [MeV]	σ [mb]	Δσ [mb]	Ref.
22.80		29.000	2.00	Oto 66
28.60		390.000	30.00	Oto 66
34.50		320.000	20.00	Oto 66

1169 Reaction $^{112}_{48}Cd \rightarrow ^{110m}_{49}In$

E [MeV]	ΔE [MeV]	σ [mb]	Δσ [mb]	Ref.
22.80		19.000	3.00	Oto 66
28.60		310.000	40.00	Oto 66
34.50		380.000	50.00	Oto 66

1170 Reaction $^{112}_{48}Cd \rightarrow ^{111}_{49}In$ (Fig. 537, p. 296)

E [MeV]	ΔE [MeV]	σ [mb]	Δσ [mb]	Ref.
12.60		5.300	0.80	Oto 66
13.20		200.000		Ska 75
13.50		340.000	20.00	Oto 66
14.50		490.000	20.00	Oto 66
14.70		560.000	30.00	Oto 66
15.20		597.000		Ska 75
17.00		823.000		Ska 75
17.30		840.000	20.00	Oto 66
18.30		913.000		Ska 75
19.70		970.000		Ska 75

1170 Reaction $^{112}_{48}Cd \rightarrow ^{111}_{49}In$ (continued)

E [MeV]	ΔE [MeV]	σ [mb]	Δσ [mb]	Ref.
20.80		870.000	40.00	Oto 66
20.90		1070.000		Ska 75
22.80		1000.000	30.00	Oto 66
25.80		620.000	30.00	Oto 66
28.60		340.000	10.00	Oto 66
31.50		166.000	7.00	Oto 66
34.50		123.000	4.00	Oto 66
37.40		93.000	4.00	Oto 66
50.00		77.800	12.80	Nie 69
70.00		29.700	1.90	Nie 69
100.00		20.400	0.60	Nie 69
150.00		10.800	0.10	Nie 69
200.00		8.830	0.18	Nie 69
300.00		5.860	0.11	Nie 69
400.00		4.050	0.22	Nie 69
400.00		4.060	0.70E-1	Nie 70

1171 Reaction $^{112}_{48}Cd \rightarrow ^{112}_{49}In$ (Figs. 538, 539, p. 296)

E [MeV]	ΔE [MeV]	σ [mb]	Δσ [mb]	Ref.
3.20		0.200E-1		Bla 51*
3.40		0.150		Bla 51*
3.60		0.400		Bla 51*
3.80		0.800		Bla 51*
4.10		1.400		Bla 51*
4.40		2.700		Bla 51*
4.50		4.000		Bla 51*
4.70		6.000		Bla 51*
4.70		17.000		Ska 75
4.90		10.000		Bla 51*
5.10		13.000		Bla 51*
5.20		53.000		Ska 75
5.40		25.000		Bla 51*
5.60		30.000		Bla 51*
5.80		40.000		Bla 51*
5.90		45.000		Bla 51*
5.90		96.000		Ska 75
6.10		60.000		Bla 51*
6.30		70.000		Bla 51*
6.40		80.000		Bla 51*
6.40		163.000		Ska 75
6.70		45.000		Bla51
6.80		220.000		Ska 75
7.30		321.000		Ska 75
7.80		375.000		Ska 75
8.20		483.000		Ska 75
8.70		544.000		Ska 75

E [MeV]	ΔE [MeV]	σ [mb]	Δσ [mb]	Ref.

1171 Reaction $^{112}_{48}Cd \rightarrow ^{112}_{49}In$ (continued)

E [MeV]	ΔE [MeV]	σ [mb]	Δσ [mb]	Ref.
9.00		600.000		Ska 75
12.50		680.000		Ska 75
12.60		400.000	60.00	Oto 66
13.50		300.000	50.00	Oto 66
14.00		426.000		Ska 75
14.70		220.000	40.00	Oto 66
15.60		295.000		Ska 75
17.00		190.000		Ska 75
17.30		60.000	20.00	Oto 66
18.30		135.000		Ska 75
19.70		110.000		Ska 75
20.90		96.000		Ska 75
22.80		40.000	10.00	Oto 66

1172 Reaction $^{112}_{48}Cd \rightarrow ^{112m}_{49}In$ (Fig. 540, p. 296)

E [MeV]	ΔE [MeV]	σ [mb]	Δσ [mb]	Ref.
6.70		67.000		Bla 51
12.60		300.000	40.00	Oto 66
13.50		280.000	50.00	Oto 66
14.70		180.000	40.00	Oto 66
17.30		110.000	10.00	Oto 66
22.80		41.000	6.00	Oto 66

1173 Reaction $^{113}_{48}Cd \rightarrow ^{112}_{47}Ag$

E [MeV]	ΔE [MeV]	σ [mb]	Δσ [mb]	Ref.
400.0		14.200	0.30	Nie 70

1174 Reaction $^{113}_{48}Cd \rightarrow ^{111}_{49}In$ (Fig. 541, p. 297)

E [MeV]	ΔE [MeV]	σ [mb]	Δσ [mb]	Ref.
3.30		4.000		Zai 90
10.60	1.60	3.000		Zai 90
15.50	1.10	14.000		Zai 90
19.60	0.90	37.000		Zai 90
23.00	0.70	263.000		Zai 90
26.10	0.80	518.000		Zai 90
29.00	0.60	629.000		Zai 90
31.70	0.60	682.000		Zai 90
34.20	0.60	606.000		Zai 90
36.60	0.50	461.000		Zai 90
38.80	0.50	326.000		Zai 90
41.10	0.50	245.000		Zai 90
43.30	0.50	186.000		Zai 90
45.40	0.50	159.000		Zai 90
47.30	0.40	138.000		Zai 90
49.30	0.40	122.000		Zai 90
51.20	0.40	111.000		Zai 90

E [MeV]	ΔE [MeV]	σ [mb]	Δσ [mb]	Ref.

1174 Reaction $^{113}_{48}Cd \rightarrow ^{111}_{49}In$ (continued)

E [MeV]	ΔE [MeV]	σ [mb]	Δσ [mb]	Ref.
53.00	0.40	102.000		Zai 90
54.80	0.40	101.000		Zai 90
56.50	0.40	92.000		Zai 90
58.20	0.40	85.000		Zai 90
59.90	0.40	82.000		Zai 90
61.50	0.30	78.000		Zai 90
63.00	0.30	78.000		Zai 90
70.00		39.500	1.70	Nie 69
100.0		26.700	1.20	Nie 69
150.0		16.700	0.30	Nie 69
200.0		12.300	0.10	Nie 69
300.0		8.150	0.10	Nie 69
400.0		5.770	0.21	Nie 69

1175 Reaction $^{113}_{48}Cd \rightarrow ^{112}_{49}In$ (Fig. 542, p. 297)

E [MeV]	ΔE [MeV]	σ [mb]	Δσ [mb]	Ref.
12.50		423.000		Ska 75
14.00		755.000		Ska 75
15.60		910.000		Ska 75
17.00		1105.000		Ska 75
18.30		1056.000		Ska 75
19.70		984.000		Ska 75
20.90		917.000		Ska 75

1176 Reaction $^{113}_{48}Cd \rightarrow ^{113m}_{49}In$ (Fig. 543, p. 297)

E [MeV]	ΔE [MeV]	σ [mb]	Δσ [mb]	Ref.
4.70		4.60		Ska 75
5.00		11.30		Ska 75
5.90		29.00		Ska 75
6.50		59.00		Ska 75
7.30		85.00		Ska 75
7.80		109.00		Ska 75
8.50		127.00		Ska 75
9.00		140.00		Ska 75
10.90		140.00		Ska 75
13.50		56.00		Ska 75
15.20		26.00		Ska 75
16.70		16.00		Ska 75
18.30		14.00		Ska 75
19.50		13.00		Ska 75
20.90		14.00		Ska 75

1177 Reaction $^{114}_{48}Cd \rightarrow ^{113}_{47}Ag$

E [MeV]	ΔE [MeV]	σ [mb]	Δσ [mb]	Ref.
400.0		13.700	0.10	Nie 70

E [MeV]	ΔE [MeV]	σ [mb]	Δσ [mb]	Ref.

1178 Reaction $^{114}{}_{48}Cd \rightarrow {}^{111}{}_{49}In$ (Fig. 544, p. 297)

E [MeV]	ΔE [MeV]	σ [mb]	Δσ [mb]	Ref.
7.50	1.70	1.000		Zai 90
13.20	1.20	3.000		Zai 90
17.60	1.00	5.000		Zai 90
21.40	0.90	6.000		Zai 90
24.60	0.70	8.000		Zai 90
27.60	0.70	10.000		Zai 90
30.40	0.60	22.000		Zai 90
33.00	0.60	98.000		Zai 90
35.40	0.60	243.000		Zai 90
37.70	0.60	385.000		Zai 90
40.00	0.60	462.000		Zai 90
42.20	0.60	500.000		Zai 90
44.30	0.50	486.000		Zai 90
46.30	0.50	451.000		Zai 90
48.30	0.50	367.000		Zai 90
50.20	0.50	309.000		Zai 90
52.10	0.50	257.000		Zai 90
54.00	0.40	207.000		Zai 90
55.60	0.40	178.000		Zai 90
57.30	0.40	160.000		Zai 90
59.00	0.40	142.000		Zai 90
60.70	0.40	129.000		Zai 90
62.30	0.40	119.000		Zai 90
70.00		74.400	0.30	Nie 69
100.0		44.300	3.50	Nie 69
150.0		20.700	0.80	Nie 69
200.0		14.500	0.80	Nie 69
300.0		10.100	0.10	Nie 69
400.0		7.120	0.21	Nie 69

1179 Reaction $^{114}{}_{48}Cd \rightarrow {}^{113m}{}_{49}In$ (Fig. 545, p. 297)

E [MeV]	ΔE [MeV]	σ [mb]	Δσ [mb]	Ref.
9.80		18.000	2.70	Ska 75
10.90		74.000	11.10	Ska 75
11.80		88.000	13.20	Ska 75
13.70		201.000	30.20	Ska 75
14.50		183.000	27.50	Ska 75
16.20		204.000	30.60	Ska 75
16.70		179.000	26.90	Ska 75
18.30		190.000	28.50	Ska 75
19.00		162.000	24.30	Ska 75
20.30		190.000	28.50	Ska 75
20.90		145.000	21.80	Ska 75

E [MeV]	ΔE [MeV]	σ [mb]	Δσ [mb]	Ref.

1180 Reaction $^{114}_{48}Cd \rightarrow ^{114}_{49}In$ (Fig. 546, p. 298)

E [MeV]	ΔE [MeV]	σ [mb]	Δσ [mb]	Ref.
3.60		0.200		Bla 51*
3.80		0.400		Bla 51*
4.10		0.800		Bla 51*
4.30		1.600		Bla 51*
4.50		3.000		Bla 51*
4.70		4.000		Bla 51*
4.70		26.000		Win 62
4.80		6.000		Bla 51*
5.10		9.000		Bla 51*
5.30		12.000		Bla 51*
5.40		16.000		Bla 51*
5.60		20.000		Bla 51*
5.70		25.000		Bla 51*
5.80		91.000		Win 62
5.90		30.000		Bla 51*
5.90		94.000		Win 62
6.10		35.000		Bla 51*
6.30		40.000		Bla 51*
6.40		50.000		Bla 51*
6.70		86.000		Bla 51
6.80		156.000	16.00	Win 62
7.80		221.000	6.50	Win 62
8.90		268.000		Win 62
9.90		275.000		Win 62
10.00		264.000		Win 62

1181 Reaction $^{114}_{48}Cd \rightarrow ^{114m}_{49}In$ (Fig. 547, p. 298)

E [MeV]	ΔE [MeV]	σ [mb]	Δσ [mb]	Ref.
3.80		<0.100		Ska 75
4.40		3.000		Win 62
5.20		14.000		Win 62
5.40		15.000		Win 62
5.90		29.000		Ska 75
6.30		57.000		Win 62
6.40		54.000		Win 62
6.70		56.000		Bla 51
7.40		119.000		Win 62
7.50		135.000		Win 62
7.50		39.000		Zai 90
7.70		150.000		Ska 75
8.40		208.000		Win 62
9.00		270.000		Ska 75
9.50		308.000		Win 62
9.60		328.000		Win 62
9.75		321.000		Win 62
10.10		340.000		Win 62

E [MeV]	ΔE [MeV]	σ [mb]	Δσ [mb]	Ref.

1181 Reaction $^{114}_{48}Cd \rightarrow ^{114m}_{49}In$ (continued)

E [MeV]	ΔE [MeV]	σ [mb]	Δσ [mb]	Ref.
10.30		347.000		Win 62
11.80		226.000		Ska 75
12.00		305.000		Blo 55
13.20	1.20	178.000		Zai 90
17.60	1.00	90.000		Zai 90
19.00		41.000		Ska 75
20.90		41.000		Ska 75
21.40	0.90	38.000		Zai 90
24.60	0.70	29.000		Zai 90
27.60	0.70	26.000		Zai 90
30.40	0.60	22.000		Zai 90
33.00	0.60	18.000		Zai 90
35.40	0.50	15.000		Zai 90
37.70	0.40	16.000		Zai 90
40.00	0.60	14.000		Zai 90
42.20	0.60	15.000		Zai 90
44.30	0.50	12.000		Zai 90
46.30	0.30	11.000		Zai 90
48.30	0.50	11.000		Zai 90
50.20	0.50	10.000		Zai 90
52.10	0.50	11.000		Zai 90
54.00	0.40	12.000		Zai 90
55.60	0.50	9.000		Zai 90
57.30	0.40	9.000		Zai 90
59.00	0.40	9.000		Zai 90
60.70	0.40	8.000		Zai 90
62.30	0.50	8.000		Zai 90
70.00		4.290	9.00E-1	Nie 69
100.0		3.450	9.00E-1	Nie 69
150.0		2.020	8.00E-1	Nie 69
200.0		1.410	8.00E-1	Nie 69
300.0		0.980	3.00E-1	Nie 69
400.0		0.830	5.00E-1	Nie 69

1182 Reaction $^{114}_{48}Cd \rightarrow ^{115m}_{49}In$

E [MeV]	ΔE [MeV]	σ [mb]	Δσ [mb]	Ref.
3.00		0.710E-2		Ska 87a
4.00		0.370E-1		Ska 87a
5.00		0.119		Ska 87a
6.00		0.320		Ska 87a
7.00		0.370		Ska 87a
8.00		0.420		Ska 87a

E [MeV]	ΔE [MeV]	σ [mb]	Δσ [mb]	Ref.

1183 Reaction $^{116}_{48}Cd \rightarrow ^{111}_{49}In$ (Fig. 548, p. 298)

E [MeV]	ΔE [MeV]	σ [mb]	Δσ [mb]	Ref.
70.00		129.000	1.00	Nie 69
100.0		45.200	5.50	Nie 69
150.0		14.900	0.80	Nie 69
200.0		11.600	0.40	Nie 69
300.0		7.400	0.46	Nie 69
400.0		6.010	0.50E-1	Nie 69

1184 Reaction $^{116}_{48}Cd \rightarrow ^{114m}_{49}In$ (Fig. 549, p. 298)

E [MeV]	ΔE [MeV]	σ [mb]	Δσ [mb]	Ref.
70.00		28.600	0.30	Nie 69
100.0		19.300	2.00	Nie 69
150.0		9.150	0.46	Nie 69
200.0		7.590	0.18	Nie 69
300.0		4.890	0.20	Nie 69
400.0		4.260	0.30E-1	Nie 69

1185 Reaction $^{116}_{48}Cd \rightarrow ^{115m}_{49}In$ (Fig. 550, p. 299)

E [MeV]	ΔE [MeV]	σ [mb]	Δσ [mb]	Ref.
11.50		195.000		Ska 75
13.50		289.000		Ska 75
15.20		319.000		Ska 75
16.70		287.000		Ska 75
18.30		279.000		Ska 75
19.50		243.000		Ska 75
20.90		210.000		Ska 75

1186 Reaction $^{116}_{48}Cd \rightarrow ^{116m}_{49}In$ (Fig. 551, p. 299)

E [MeV]	ΔE [MeV]	σ [mb]	Δσ [mb]	Ref.
4.50		1.500		Ska 75
5.20		6.700		Ska 75
5.90		20.000		Ska 75
6.70		37.000		Ska 75
7.30		61.000		Ska 75
8.00		85.000		Ska 75
8.50		123.000		Ska 75
9.00		147.000		Ska 75
11.50		90.000		Ska 75
13.50		37.000		Ska 75
15.20		24.000		Ska 75
16.70		18.000		Ska 75
18.30		16.000		Ska 75
19.50		14.000		Ska 75
20.90		15.000		Ska 75

1187 Reaction $_{49}In \rightarrow ^{3}_{1}H$

E [MeV]	ΔE [MeV]	σ [mb]	Δσ [mb]	Ref.
12000		288.000	40.00	Nog 91

E [MeV]	ΔE [MeV]	σ [mb]	Δσ [mb]	Ref.

1188 Reaction $_{49}In \rightarrow {}^{7}_{4}Be$

E [MeV]	ΔE [MeV]	σ [mb]	Δσ [mb]	Ref.
6200		14.100	0.40	Net 60

1189 Reaction $_{49}In \rightarrow {}^{13}_{7}N$

E [MeV]	ΔE [MeV]	σ [mb]	Δσ [mb]	Ref.
940.0		0.560E-1		Dos 61
1840		0.190		Dos 61
2900		0.270		Dos 61

1190 Reaction $_{49}In \rightarrow {}^{22}_{11}Na$

E [MeV]	ΔE [MeV]	σ [mb]	Δσ [mb]	Ref.
400.0		0.544E-2	0.54E-3	Kor 67

1191 Reaction $_{49}In \rightarrow {}^{24}_{11}Na$

E [MeV]	ΔE [MeV]	σ [mb]	Δσ [mb]	Ref.
400.0		0.131E-1	0.20E-3	Kor 67

1192 Reaction $_{49}In \rightarrow {}^{32}_{15}P$

E [MeV]	ΔE [MeV]	σ [mb]	Δσ [mb]	Ref.
550.0		0.229E-1	0.61E-2	Kor 70c

1193 Reaction $_{49}In \rightarrow {}^{33}_{15}P$

E [MeV]	ΔE [MeV]	σ [mb]	Δσ [mb]	Ref.
550.0		0.165E-1	0.30E-3	Kor 70c

1194 Reaction $_{49}In \rightarrow {}^{43}_{21}Sc$

E [MeV]	ΔE [MeV]	σ [mb]	Δσ [mb]	Ref.
28500		0.570	0.60E-1	Chu 77

1195 Reaction $_{49}In \rightarrow {}^{44}_{21}Sc$

E [MeV]	ΔE [MeV]	σ [mb]	Δσ [mb]	Ref.
28500		0.950	0.90E-1	Chu 77

1196 Reaction $_{49}In \rightarrow {}^{44m}_{21}Sc$

E [MeV]	ΔE [MeV]	σ [mb]	Δσ [mb]	Ref.
28500		1.620	0.16	Chu 77

1197 Reaction $_{49}In \rightarrow {}^{46}_{21}Sc$

E [MeV]	ΔE [MeV]	σ [mb]	Δσ [mb]	Ref.
28500		2.960	0.30	Chu 77

1198 Reaction $_{49}In \rightarrow {}^{47}_{21}Sc$

E [MeV]	ΔE [MeV]	σ [mb]	Δσ [mb]	Ref.
28500		1.590	0.16	Chu 77

1199 Reaction $_{49}In \rightarrow {}^{48}_{21}Sc$

E [MeV]	ΔE [MeV]	σ [mb]	Δσ [mb]	Ref.
28500		0.480	0.50E-1	Chu 77

E [MeV]	ΔE [MeV]	σ [mb]	Δσ [mb]	Ref.
1200 Reaction $_{49}In \rightarrow {}^{65}_{28}Ni$ (cumulative)				
2900		0.180	0.20E-1	Kau 63
1201 Reaction $_{49}In \rightarrow {}^{66}_{28}Ni$ (cumulative)				
2900		0.520E-1	0.10E-1	Kau 63
1202 Reaction $_{49}In \rightarrow {}^{67}_{29}Cu$ (cumulative)				
2900		0.290	0.30E-1	Kau 63
1203 Reaction $_{49}In \rightarrow {}^{72}_{30}Zn$ (cumulative)				
2900		0.870E-2	0.90E-3	Kau 63
1204 Reaction $_{49}In \rightarrow {}^{66}_{31}Ga$				
2900		3.500	0.20	Kau 63
1205 Reaction $_{49}In \rightarrow {}^{67}_{31}Ga$				
2900		4.000	0.50	Kau 63
1206 Reaction $_{49}In \rightarrow {}^{72}_{31}Ga$				
2900		0.230	0.20E-1	Kau 63
1207 Reaction $_{49}In \rightarrow {}^{73}_{31}Ga$ (cumulative)				
2900		0.870E-1	0.90E-1	Kau 63
1208 Reaction $_{49}In \rightarrow {}^{66}_{32}Ge$ (cumulative)				
2900		0.160	0.20E-1	Kau 63
1209 Reaction $_{49}In \rightarrow {}^{67}_{32}Ge$ (cumulative)				
2900		0.680	0.12	Kau 63
1210 Reaction $_{49}In \rightarrow {}^{71}_{33}As$ (cumulative)				
2900		6.200	0.80	Kau 63
1211 Reaction $_{49}In \rightarrow {}^{72}_{33}As$				
2900		7.300	0.80	Kau 63

E [MeV]	ΔE [MeV]	σ [mb]	Δσ [mb]	Ref.
1212 Reaction $_{49}In \rightarrow {}^{74}_{33}As$				
2900		2.700	0.30	Kau 63
1213 Reaction $_{49}In \rightarrow {}^{72}_{34}Se$				
2900		2.800	0.50	Kau 63
1214 Reaction $_{49}In \rightarrow {}^{101}_{47}Ag$				
200.0		2.240	0.10	Pan 66
400.0		5.150	0.18	Pan 66
1215 Reaction $_{49}In \rightarrow {}^{115}_{47}Ag$				
2900		≤ 0.600E-2		Por 62
30000		≤ 0.300E-3		Por 62
1216 Reaction $_{49}In \rightarrow {}^{107}_{48}Cd$				
1000		26.000		Net 60
2000		32.000		Net 60
4100		26.000	2.00	Net 60
6200		29.000	3.00	Net 60
1217 Reaction $_{49}In \rightarrow {}^{109}_{48}Cd$				
1000		35.000		Net 60
2000		51.000	2.00	Net 60
4100		44.000	3.00	Net 60
6200		29.000	3.00	Net 60
1218 Reaction $_{49}In \rightarrow {}^{115}_{48}Cd$				
1000		0.300E-1		Net 60
2000		0.600E-1	0.10E-1	Net 60
2900		0.480E-1		Por 62
4100		0.660E-1	0.10E-2	Net 60
6200		0.710E-1	0.40E-2	Net 60
30000		0.230E-1		Por 62
1219 Reaction $_{49}In \rightarrow {}^{115m}_{48}Cd$				
1000		0.130		Net 60
2000		0.145	0.10E-2	Net 60
2900		0.130		Por 62
4100		0.150	0.20E-1	Net 60
6200		0.710E-1	0.40E-2	Net 60
30000		0.100		Por 62

E [MeV]	ΔE [MeV]	σ [mb]	Δσ [mb]	Ref.

1220 Reaction $_{49}In \rightarrow {}^{109}_{49}In$

E [MeV]	ΔE [MeV]	σ [mb]	Δσ [mb]	Ref.
1000		10.000		Net 60
2000		11.000		Net 60
2900		7.900		Por 62
30000		7.400		Por 62

1221 Reaction $_{49}In \rightarrow {}^{110m}_{49}In$

E [MeV]	ΔE [MeV]	σ [mb]	Δσ [mb]	Ref.
1000		12.000		Net 60
2000		17.000		Net 60
2900		8.800		Por 62
30000		9.500		Por 62

1222 Reaction $_{49}In \rightarrow {}^{111}_{49}In$

E [MeV]	ΔE [MeV]	σ [mb]	Δσ [mb]	Ref.
1000		16.000		Net 60
2000		21.000		Net 60
2900		14.400		Por 62
4100		17.000		Net 60
6200		26.000		Net 60
30000		12.900		Por 62

1223 Reaction $_{49}In \rightarrow {}^{113m}_{49}In$

E [MeV]	ΔE [MeV]	σ [mb]	Δσ [mb]	Ref.
1000		1.900E-1		Net 60
2000		2.500E-1		Net 60
2900		3.800E-1		Por 62
30000		3.300E-1		Por 62

1224 Reaction $_{49}In \rightarrow {}^{114m}_{49}In$

E [MeV]	ΔE [MeV]	σ [mb]	Δσ [mb]	Ref.
1000		42.000		Net 60
2000		49.000	1.00	Net 60
2900		40.000		Por 62
4100		57.000	1.00	Net 60
6200		61.000	9.00	Net 60
30000		40.000		Por 62

1225 Reaction $_{49}In \rightarrow {}^{115m}_{49}In$

E [MeV]	ΔE [MeV]	σ [mb]	Δσ [mb]	Ref.
1000		1.700		Net 60
2000		4.100	0.60	Net 60
2900		1.900		Por 62
4100		5.000		Net 60
6200		4.200	0.60	Net 60
30000		1.700		Por 62

E [MeV]	ΔE [MeV]	σ [mb]	Δσ [mb]	Ref.

1226 Reaction $_{49}In \rightarrow {}^{109}{}_{50}Sn$ (Fig. 552, p. 299)

E [MeV]	ΔE [MeV]	σ [mb]	Δσ [mb]	Ref.
56.60		2.800	0.70	Nor 90
59.60		3.100	0.60	Nor 90
62.40		4.800	0.80	Nor 90
65.20		3.900	0.60	Nor 90
68.80		6.300	1.60	Nor 90
72.10		6.300	1.20	Nor 90
75.30		16.600	3.50	Nor 90
78.40		4.600	0.50	Nor 90
81.40		16.700	3.40	Nor 90
84.40		24.500	2.50	Nor 90
87.20		32.000	4.10	Nor 90
90.00		28.000	3.70	Nor 90
92.80		21.900	2.20	Nor 90
95.50		28.100	3.50	Nor 90
98.10		21.800	2.20	Nor 90

1227 Reaction $_{49}In \rightarrow {}^{110}{}_{50}Sn$ (Fig. 553, p. 299)

E [MeV]	ΔE [MeV]	σ [mb]	Δσ [mb]	Ref.
33.70		0.700	0.10	Nor 90
36.30		3.600	0.30	Nor 90
38.80		8.000	0.60	Nor 90
41.80		12.400	1.00	Nor 90
44.40		15.500	1.30	Nor 90
46.90		15.300	1.20	Nor 90
50.30		12.900	1.00	Nor 90
53.50		11.000	0.90	Nor 90
56.60		16.500	1.30	Nor 90
59.60		33.500	2.70	Nor 90
62.40		58.400	4.70	Nor 90
65.20		79.500	6.40	Nor 90
68.80		94.500	7.70	Nor 90
72.10		93.700	7.60	Nor 90
75.30		84.500	6.80	Nor 90
78.40		71.200	5.70	Nor 90
81.40		61.300	4.90	Nor 90
84.40		53.200	4.30	Nor 90
87.20		45.900	3.70	Nor 90
90.00		40.400	3.30	Nor 90
92.80		37.000	3.00	Nor 90
95.50		33.800	2.70	Nor 90
98.10		31.400	2.50	Nor 90

1228 Reaction $_{49}In \rightarrow {}^{111}{}_{50}Sn$ (Fig. 554, p. 299)

E [MeV]	ΔE [MeV]	σ [mb]	Δσ [mb]	Ref.
24.60		8.700	1.30	Nor 90
27.80		17.700	1.90	Nor 90

E [MeV]	ΔE [MeV]	σ [mb]	Δσ [mb]	Ref.

1228 Reaction $_{49}In \rightarrow {}^{111}{}_{50}Sn$ (continued)

E [MeV]	ΔE [MeV]	σ [mb]	Δσ [mb]	Ref.
30.90		27.500	2.60	Nor 90
33.70		25.800	2.50	Nor 90
36.30		20.100	2.00	Nor 90
38.80		14.600	1.50	Nor 90
41.80		9.200	2.60	Nor 90
44.40		24.000	3.60	Nor 90
46.90		68.000	7.00	Nor 90
50.30		131.900	12.20	Nor 90
53.50		176.000	15.50	Nor 90
56.60		191.500	16.70	Nor 90
59.60		179.400	15.50	Nor 90
62.40		161.500	14.00	Nor 90
65.20		126.400	11.10	Nor 90
68.80		91.700	11.00	Nor 90
72.10		68.400	7.70	Nor 90
75.30		52.300	7.70	Nor 90
78.40		48.800	6.00	Nor 90
81.40		39.200	6.90	Nor 90
84.40		47.400	5.70	Nor 90
87.20		33.500	6.00	Nor 90
90.00		28.100	6.50	Nor 90
92.80		34.300	4.90	Nor 90
95.50		21.700	5.00	Nor 90
98.10		30.500	4.80	Nor 90

1229 Reaction $_{49}In \rightarrow {}^{113(0.91m+g)}{}_{50}Sn$ (Fig. 555, p. 300)

E [MeV]	ΔE [MeV]	σ [mb]	Δσ [mb]	Ref.
5.80		5.800	1.10	Nor 90
11.10		24.700	2.70	Nor 90
14.90		13.200	1.80	Nor 90
18.10		4.700	1.30	Nor 90
20.90		118.500	10.20	Nor 90
24.60		553.000	44.90	Nor 90
27.80		779.500	62.90	Nor 90
30.90		776.900	62.70	Nor 90
33.70		681.100	55.00	Nor 90
36.30		502.000	40.80	Nor 90
38.80		346.200	28.20	Nor 90
41.80		239.500	19.30	Nor 90
44.40		171.300	13.80	Nor 90
46.90		136.000	11.00	Nor 90
50.30		111.500	9.00	Nor 90
53.50		97.400	7.80	Nor 90
56.60		85.200	6.90	Nor 90
59.60		78.600	6.30	Nor 90
62.40		72.000	5.80	Nor 90

E [MeV]	ΔE [MeV]	σ [mb]	Δσ [mb]	Ref.

1229 Reaction $_{49}In \rightarrow {}^{113(0.91m+g)}{}_{50}Sn$ (continued)

E [MeV]	ΔE [MeV]	σ [mb]	Δσ [mb]	Ref.
65.20		66.900	5.40	Nor 90
68.80		60.500	5.00	Nor 90
72.10		58.400	4.80	Nor 90
75.30		55.400	4.60	Nor 90
78.40		50.900	4.20	Nor 90
81.40		48.600	4.00	Nor 90
84.40		46.200	3.80	Nor 90
87.20		42.900	3.50	Nor 90
90.00		41.700	3.40	Nor 90
92.80		41.000	3.30	Nor 90
95.50		37.400	3.10	Nor 90
98.10		36.200	3.00	Nor 90

1230 Reaction $^{115}{}_{49}In \rightarrow {}^{115}{}_{48}Cd$ (Fig. 556, p. 300)

E [MeV]	ΔE [MeV]	σ [mb]	Δσ [mb]	Ref.
130.0		0.310E-1		Ryb 65
200.0		0.290E-1	0.20E-2	Ryb 65
300.0		0.460E-1	0.90E-2	Ryb 65
400.0		0.550E-1	0.90E-2	Ryb 65
500.0		0.650E-1	0.13E-1	Ryb 65
660.0		0.780E-1	0.12E-1	Ryb 65
1000		0.300E-1		Net 60
2000		0.600E-1	0.10E-1	Net 60
2900		0.480E-1		Por 62
4100		0.660E-1	0.10E-2	Net 60
6200		0.710E-1	0.40E-2	Net 60
30000		0.230E-1		Por 62

1231 Reaction $^{115}{}_{49}In \rightarrow {}^{115m}{}_{48}Cd$ (Fig. 557, p. 300)

E [MeV]	ΔE [MeV]	σ [mb]	Δσ [mb]	Ref.
130.0		0.650E-1	0.12E-1	Ryb 65
200.0		0.680E-1	0.40E-2	Ryb 65
300.0		0.113	0.18E-1	Ryb 65
400.0		0.139	0.37E-1	Ryb 65
500.0		0.142	0.27E-1	Ryb 65
660.0		0.161	0.16E-1	Ryb 65
1000		0.130		Net 60
2000		0.145	0.10E-2	Net 60
2900		0.130		Por 62
4100		0.150	0.20E-1	Net 60
6200		0.147	0.60E-2	Net 60
30000		0.100		Por 62

E [MeV]	ΔE [MeV]	σ [mb]	Δσ [mb]	Ref.
1232 Reaction $_{50}Sn \rightarrow {}^{3}_{1}H$				
150		25.000	3.00	Bru 62b
6200		440.000		Cur 59
12000		295.000	42.00	Nog 91
1233 Reaction $_{50}Sn \rightarrow {}^{18}_{9}F$				
660.0		1.700	0.50	Lav 62
1234 Reaction $_{50}Sn \rightarrow {}^{22}_{11}Na$				
400.0		0.510E-2	0.13E-2	Kor 67
1235 Reaction $_{50}Sn \rightarrow {}^{24}_{11}Na$				
400.0		0.113E-1	0.50E-1	Kor 67
660.0		4.500	0.80	Lav 62
1236 Reaction $_{50}Sn \rightarrow {}^{32}_{15}P$				
550.0		0.150E-1		Kor 70c
1237 Reaction $_{50}Sn \rightarrow {}^{33}_{15}P$				
550.0		0.900E-2		Kor 70c
1238 Reaction $_{50}Sn \rightarrow {}^{120}_{49}In$				
350.0		0.500E-2		Mar 51
1239 Reaction $_{50}Sn \rightarrow {}^{121}_{49}In$				
350.0		0.100E-2		Mar 51
1240 Reaction $_{50}Sn \rightarrow {}^{123}_{49}In$				
350.0		0.700E-2		Mar 51
1241 Reaction $_{50}Sn \rightarrow {}^{124}_{49}In$				
350.0		0.400E-2		Mar 51
1242 Reaction $_{50}Sn \rightarrow {}^{126}_{49}In$				
350.0		0.600E-2		Mar 51

E [MeV]	ΔE [MeV]	σ [mb]	Δσ [mb]	Ref.

1243 Reaction $^{112}_{50}Sn \rightarrow {}^{113}_{51}Sb$

E [MeV]	ΔE [MeV]	σ [mb]	Δσ [mb]	Ref.
5.00		2.010		Ska 87a
7.00		26.300		Ska 87a
9.00		31.500		Ska 87a

1244 Reaction $^{116}_{50}Sn \rightarrow {}^{116}_{51}Sb$ (Fig. 558, p. 300)

E [MeV]	ΔE [MeV]	σ [mb]	Δσ [mb]	Ref.
5.80		9.000		Klu 70*
6.00		13.000		Klu 70*
6.50		40.000		Klu 70*
6.75		60.000		Klu 70*
7.00		70.000		Klu 70*
7.20		90.000		Klu 70*
7.50		100.000		Klu 70*
7.60		110.000		Klu 70*
8.00		120.000		Klu 70*
8.15		130.000		Klu 70*

1245 Reaction $^{117}_{50}Sn \rightarrow {}^{117}_{51}Sb$ (Fig. 559, p. 300)

E [MeV]	ΔE [MeV]	σ [mb]	Δσ [mb]	Ref.
2.829		0.140E-1		Joh 77
2.900		0.100		Klu 70*
2.944		0.270E-1		Joh 77
3.062		0.510E-1		Joh 77
3.181		0.890E-1		Joh 77
3.302		0.155		Joh 77
3.425		0.258		Joh 77
3.550		0.412		Joh 77
3.677		0.636		Joh 77
3.700		1.000		Bla 51*
3.800		5.500		Klu 70*
3.806		0.950		Joh 77
3.938		1.388		Joh 77
4.000		2.100		Bla 51*
4.071		2.039		Joh 77
4.206		2.923		Joh 77
4.300		3.900		Bla 51*
4.343		4.144		Joh 77
4.482		7.045		Joh 77
4.500		7.800		Bla 51*
4.600		16.000		Klu 70*
4.623		7.743		Joh 77
4.700		10.500		Bla 51*
4.767		10.290		Joh 77
4.911		13.680		Joh 77
5.000		17.000		Bla 51*
5.059		17.900		Joh 77

E [MeV]	ΔE [MeV]	σ [mb]	Δσ [mb]	Ref.

1245 Reaction $^{117}_{50}Sn \rightarrow ^{117}_{51}Sb$ (continued)

E [MeV]	ΔE [MeV]	σ [mb]	Δσ [mb]	Ref.
5.178		21.840		Joh 77
5.250		22.000		Bla 51*
5.298		26.670		Joh 77
5.400		32.000		Bla 51*
5.418		32.010		Joh 77
5.540		38.250		Joh 77
5.600		40.000		Bla 51*
5.664		46.310		Joh 77
5.750		47.000		Bla 51*
5.789		53.790		Joh 77
5.916		62.310		Joh 77
6.000		53.000		Bla 51*
6.043		72.250		Joh 77
6.100		60.000		Klu 70*
6.172		84.290		Joh 77
6.200		60.000		Bla 51*
6.302		94.000		Joh 77
6.400		68.000		Bla 51*
6.433		106.700		Joh 77
6.565		119.900		Joh 77
6.699		132.700		Joh 77
6.700		90.000		Bla 51
6.700		90.000		Klu 70*
6.901		157.600		Joh 77
7.071		173.600		Joh 77
7.200		130.000		Klu 70*
7.900		180.000		Klu 70*
8.400		200.000		Klu 70*

1246 Reaction $^{118}_{50}Sn \rightarrow ^{118}_{51}Sb$ (Fig. 560, p. 301)

E [MeV]	ΔE [MeV]	σ [mb]	Δσ [mb]	Ref.
4.50		2.000		Klu 70*
4.53		1.380		Joh 77
4.57		2.500		Joh 77
4.64		4.080		Joh 77
4.79		7.590		Joh 77
4.93		11.540		Joh 77
5.08		15.730		Joh 77
5.10		21.000		Klu 70*
5.23		21.000		Joh 77
5.32		24.610		Joh 77
5.44		30.260		Joh 77
5.50		42.000		Klu 70*
5.56		36.610		Joh 77
5.68		44.350		Joh 77
5.70		60.000		Klu 70*

1246 Reaction $^{118}_{50}Sn \rightarrow ^{118}_{51}Sb$ (continued)

E [MeV]	ΔE [MeV]	σ [mb]	Δσ [mb]	Ref.
5.81		52.820		Joh 77
5.90		70.000		Klu 70*
5.94		62.290		Joh 77
6.06		72.050		Joh 77
6.15		80.000		Klu 70*
6.32		94.050		Joh 77
6.45		105.900		Joh 77
6.50		95.000		Klu 70*
6.58		118.200		Joh 77
6.62		122.000		Joh 77
6.70		75.000		Bla 51
6.70		105.000		Klu 70*
6.72		132.100		Joh 77
6.78		139.600		Joh 77
6.92		154.600		Joh 77
7.00		120.000		Klu 70*
7.40		170.000		Klu 70*
7.60		190.000		Klu 70*
8.00		200.000		Klu 70*
8.40		210.000		Klu 70*

1247 Reaction $^{119}_{50}Sn \rightarrow ^{119}_{51}Sb$ (Fig. 561, p. 301)

E [MeV]	ΔE [MeV]	σ [mb]	Δσ [mb]	Ref.
2.62		0.900E-2		Joh 77
2.84		0.300E-1		Joh 77
3.07		0.860E-1		Joh 77
3.19		0.141		Joh 77
3.31		0.226		Joh 77
3.43		0.347		Joh 77
3.56		0.532		Joh 77
3.69		0.803		Joh 77
3.82		1.184		Joh 77
3.95		1.705		Joh 77
4.08		2.449		Joh 77
4.22		3.447		Joh 77
4.35		4.799		Joh 77
4.49		6.660		Joh 77
4.63		11.500		Joh 77
4.77		11.780		Joh 77
4.92		15.350		Joh 77
5.07		19.950		Joh 77
5.19		24.240		Joh 77
5.31		29.240		Joh 77
5.43		35.200		Joh 77
5.50		38.000		Klu 70*
5.55		41.600		Joh 77

E [MeV]	ΔE [MeV]	σ [mb]	Δσ [mb]	Ref.

1247 Reaction $^{119}_{50}Sn \rightarrow {}^{119}_{51}Sb$ (continued)

E [MeV]	ΔE [MeV]	σ [mb]	Δσ [mb]	Ref.
5.67		49.110		Joh 77
5.80		58.490		Joh 77
5.92		66.310		Joh 77
6.05		76.150		Joh 77
6.18		87.130		Joh 77
6.20		63.000		Klu 70*
6.31		98.790		Joh 77
6.44		112.000		Joh 77
6.57		123.500		Joh 77
6.67		134.100		Joh 77
6.70		80.000		Klu 70*
6.81		148.300		Joh 77
6.94		164.400		Joh 77
7.08		186.400		Joh 77
7.40		105.000		Klu 70*
7.90		120.000		Klu 70*
8.50		140.000		Klu 70*

1248 Reaction $^{120}_{50}Sn \rightarrow {}^{117m}_{49}In$ (Fig. 562, p. 301)

E [MeV]	ΔE [MeV]	σ [mb]	Δσ [mb]	Ref.
3.50		0.850E-2		Nee 63*
5.50		0.200E-1		Nee 63*
10.50		0.100		Nee 63*
14.00		0.200		Nee 63*
21.00		0.700		Nee 63*
25.00		0.900		Nee 63*
26.00		1.300		Nee 63*
30.00		1.200		Nee 63*
31.00		1.800		Nee 63*
38.00		1.900		Nee 63*
46.00		2.800		Nee 63*
50.00		2.300		Nee 63*
55.00		2.300		Nee 63*
60.00		2.300		Nee 63*
70.00		2.100		Nee 63*

1249 Reaction $^{120}_{50}Sn \rightarrow {}^{120}_{51}Sb$ (Fig. 563, p. 301)

E [MeV]	ΔE [MeV]	σ [mb]	Δσ [mb]	Ref.
3.50		0.180		Klu 70*
3.60		0.600E-1		Bla 51*
3.68		0.596		Joh 77
3.80		0.600		Bla 51*
3.81		1.043		Joh 77
3.94		1.602		Joh 77
4.00		1.400		Bla 51*
4.00		1.800		Klu 70*

1249 Reaction $^{120}_{50}Sn \rightarrow ^{120}_{51}Sb$ (continued)

E [MeV]	ΔE [MeV]	σ [mb]	Δσ [mb]	Ref.
4.07		2.358		Joh 77
4.21		3.380		Joh 77
4.25		4.300		Klu 70*
4.30		4.700		Bla 51*
4.35		4.761		Joh 77
4.49		6.528		Bla 51*
4.50		8.000		Joh 77
4.55		9.000		Klu 70*
4.63		8.817		Joh 77
4.70		12.000		Bla 51*
4.77		11.920		Joh 77
4.90		21.000		Klu 70*
5.00		20.000		Bla 51*
5.06		20.330		Joh 77
5.20		30.000		Bla 51*
5.20		38.000		Klu 70*
5.21		25.900		Joh 77
5.30		29.730		Joh 77
5.36		32.400		Joh 77
5.40		40.000		Bla 51*
5.42		35.300		Joh 77
5.50		53.000		Klu 70*
5.55		41.720		Joh 77
5.60		50.000		Bla 51*
5.67		49.130		Joh 77
5.79		57.080		Joh 77
5.80		62.000		Bla 51*
5.80		70.000		Klu 70*
5.92		66.160		Joh 77
5.95		68.390		Joh 77
6.00		80.000		Bla 51*
6.05		76.030		Joh 77
6.18		86.950		Joh 77
6.25		95.000		Bla 51*
6.30		105.000		Klu 70*
6.31		98.670		Joh 77
6.35		120.000		Bla 51*
6.44		110.700		Joh 77
6.57		124.000		Joh 77
6.70		170.000		Bla 51*
6.70		137.800		Joh 77
6.80		130.000		Klu 70*
6.84		152.100		Joh 77
6.97		167.700		Joh 77
7.30		180.000		Klu 70*
8.20		220.000		Klu 70*

1250 Reaction $^{122}_{50}Sn \rightarrow ^{122}_{51}Sb$ (Fig. 564, p. 302)

E [MeV]	ΔE [MeV]	σ [mb]	Δσ [mb]	Ref.
2.638		0.130E-1		Joh 77
2.860		0.430E-1		Joh 77
2.974		0.700E-1		Joh 77
3.090		0.120		Joh 77
3.209		0.195		Joh 77
3.329		0.297		Joh 77
3.451		0.459		Joh 77
3.500		1.000		Klu 70*
3.576		0.683		Joh 77
3.700		1.800		Bla 51*
3.703		1.019		Joh 77
3.831		1.449		Joh 77
3.961		2.094		Joh 77
4.000		3.800		Bla 51*
4.094		2.947		Joh 77
4.229		4.154		Joh 77
4.300		7.300		Bla 51*
4.365		5.686		Joh 77
4.500		11.000		Bla 51*
4.500		12.000		Klu 70*
4.645		10.180		Joh 77
4.700		18.000		Bla 51*
4.787		13.330		Joh 77
4.932		17.320		Joh 77
5.000		27.000		Bla 51*
5.078		22.220		Joh 77
5.200		39.000		Bla 51*
5.227		27.950		Joh 77
5.317		31.960		Joh 77
5.350		47.000		Klu 70*
5.400		47.000		Bla 51*
5.440		38.050		Joh 77
5.562		44.780		Joh 77
5.600		60.000		Bla 51*
5.684		52.460		Joh 77
5.750		80.000		Bla 51*
5.809		60.730		Joh 77
5.935		70.360		Joh 77
6.000		100.000		Bla 51*
6.000		105.000		Klu 70*
6.062		79.840		Joh 77
6.191		90.700		Joh 77
6.200		120.000		Bla 51*
6.320		103.100		Joh 77
6.400		140.000		Bla 51*
6.451		114.700		Joh 77

E [MeV]	ΔE [MeV]	σ [mb]	Δσ [mb]	Ref.

1250 Reaction $^{122}_{50}Sn \rightarrow {}^{122}_{51}Sb$ (continued)

E [MeV]	ΔE [MeV]	σ [mb]	Δσ [mb]	Ref.
6.583		127.700		Joh 77
6.700		195.000		Bla 51
6.716		141.900		Joh 77
6.750		200.000		Klu 70*
6.851		156.400		Joh 77
6.986		171.600		Joh 77
7.500		380.000		Klu 70*
8.400		420.000		Klu 70*

1251 Reaction $^{124}_{50}Sn \rightarrow {}^{124}_{51}Sb$ (Fig. 565, p. 302)

E [MeV]	ΔE [MeV]	σ [mb]	Δσ [mb]	Ref.
2.609		0.130E-1		Joh 77
2.948		0.700E-1		Joh 77
3.064		0.115		Joh 77
3.183		0.187		Joh 77
3.304		0.294		Joh 77
3.427		0.451		Joh 77
3.552		0.689		Joh 77
3.679		1.029		Joh 77
3.808		1.499		Joh 77
3.939		2.148		Joh 77
4.072		3.046		Joh 77
4.207		4.272		Joh 77
4.344		5.830		Joh 77
4.483		7.820		Joh 77
4.624		10.450		Joh 77
4.767		13.600		Joh 77
4.912		17.690		Joh 77
5.059		22.660		Joh 77
5.178		27.250		Joh 77
5.298		32.630		Joh 77
5.420		38.600		Joh 77
5.543		45.430		Joh 77
5.666		53.260		Joh 77
5.791		61.320		Joh 77
5.918		70.890		Joh 77
6.045		80.820		Joh 77
6.174		91.650		Joh 77
6.304		102.700		Joh 77
6.435		115.100		Joh 77
6.567		128.300		Joh 77
6.700		141.800		Joh 77
6.835		155.700		Joh 77
6.971		170.800		Joh 77
7.107		187.100		Joh 77

E [MeV]	ΔE [MeV]	σ [mb]	Δσ [mb]	Ref.

1252 Reaction $_{51}Sb \rightarrow {}^{7}_{4}Be$

E [MeV]	ΔE [MeV]	σ [mb]	Δσ [mb]	Ref.
660.0		250.000	140.0	Lav 62

1253 Reaction $_{51}Sb \rightarrow {}^{18}_{9}F$

E [MeV]	ΔE [MeV]	σ [mb]	Δσ [mb]	Ref.
660.0		2.800	1.30	Lav 62

1254 Reaction $_{51}Sb \rightarrow {}^{24}_{11}Na$

E [MeV]	ΔE [MeV]	σ [mb]	Δσ [mb]	Ref.
660.0		14.000	4.00	Lav 62

1255 Reaction $_{51}Sb \rightarrow {}^{28}_{12}Mg$

E [MeV]	ΔE [MeV]	σ [mb]	Δσ [mb]	Ref.
660.0		0.700	0.10	Lav 62

1256 Reaction $_{51}Sb \rightarrow {}^{31}_{14}Si$

E [MeV]	ΔE [MeV]	σ [mb]	Δσ [mb]	Ref.
660.0		0.600	0.40	Lav 62

1257 Reaction $_{51}Sb \rightarrow {}^{32}_{15}P$

E [MeV]	ΔE [MeV]	σ [mb]	Δσ [mb]	Ref.
660.0		0.630	0.38	Lav 62

1258 Reaction $_{51}Sb \rightarrow {}^{117}_{52}Te$ (Fig. 566, p. 302)

E [MeV]	ΔE [MeV]	σ [mb]	Δσ [mb]	Ref.
43.80	0.60	0.200		Lag 90a
44.90	0.50	3.100		Lag 90a
46.00	0.50	6.800		Lag 90a
47.10	0.50	7.700		Lag 90a
48.20	0.50	10.900		Lag 90a
49.20	0.50	14.600		Lag 90a
50.30	0.50	18.400		Lag 90a
51.30	0.50	20.000		Lag 90a
52.30	0.50	23.700		Lag 90a
53.40	0.50	26.200		Lag 90a
54.40	0.50	27.500		Lag 90a
55.40	0.50	31.700		Lag 90a
56.50	0.50	31.300		Lag 90a
57.50	0.50	29.900		Lag 90a
58.50	0.50	28.100		Lag 90a
59.50	0.50	18.900		Lag 90a

1259 Reaction $_{51}Sb \rightarrow {}^{118}_{52}Te$ (Fig. 567, p. 303)

E [MeV]	ΔE [MeV]	σ [mb]	Δσ [mb]	Ref.
31.20	0.60	0.100		Lag 90a
32.60	0.60	0.200		Lag 90a
33.90	0.60	0.400		Lag 90a

E [MeV]	ΔE [MeV]	σ [mb]	Δσ [mb]	Ref.

1259 Reaction $_{51}Sb \rightarrow {}^{118}{}_{52}Te$ (continued)

E [MeV]	ΔE [MeV]	σ [mb]	Δσ [mb]	Ref.
35.20	0.60	1.700		Lag 90a
36.50	0.60	3.300		Lag 90a
37.70	0.50	14.500		Lag 90a
39.00	0.60	37.100		Lag 90a
40.20	0.60	54.900		Lag 90a
41.50	0.60	220.200		Lag 90a
42.60	0.50	372.400		Lag 90a
43.80	0.50	475.600		Lag 90a
44.90	0.50	552.400		Lag 90a
46.00	0.50	632.700		Lag 90a
47.10	0.50	660.000		Lag 90a
48.20	0.50	663.100		Lag 90a
49.20	0.50	609.100		Lag 90a
50.30	0.50	590.900		Lag 90a
51.30	0.50	530.000		Lag 90a
52.30	0.50	462.200		Lag 90a
53.40	0.50	418.100		Lag 90a
54.40	0.50	327.700		Lag 90a
55.40	0.50	299.000		Lag 90a
56.50	0.50	310.200		Lag 90a
57.50	0.50	295.900		Lag 90a
58.50	0.50	294.700		Lag 90a
59.50	0.50	305.600		Lag 90a
60.50	0.40	312.100		Lag 90a
61.40	0.50	359.300		Lag 90a
62.20	0.40	367.400		Lag 90a
63.00	0.40	378.800		Lag 90a
63.90	0.50	376.500		Lag 90a
64.70	0.40	367.600		Lag 90a
65.60	0.40	355.800		Lag 90a
66.40	0.40	333.900		Lag 90a
67.10	0.30	300.000		Lag 90a

1260 Reaction $_{51}Sb \rightarrow {}^{119}{}_{52}Te$ (Fig. 568, p. 303)

E [MeV]	ΔE [MeV]	σ [mb]	Δσ [mb]	Ref.
25.40	0.80	2.400		Lag 90a
26.90	0.70	4.400		Lag 90a
28.40	0.70	16.400		Lag 90a
29.80	0.60	16.300		Lag 90a
31.20	0.60	25.600		Lag 90a
32.60	0.60	31.300		Lag 90a
33.90	0.60	95.100		Lag 90a
35.20	0.60	172.200		Lag 90a
36.50	0.60	325.900		Lag 90a
37.70	0.50	386.800		Lag 90a
39.00	0.60	377.900		Lag 90a

E [MeV]	ΔE [MeV]	σ [mb]	Δσ [mb]	Ref.

1260 Reaction $_{51}Sb \rightarrow {}^{119}_{52}Te$ (continued)

E [MeV]	ΔE [MeV]	σ [mb]	Δσ [mb]	Ref.
40.20	0.60	277.500		Lag 90a
41.50	0.60	279.500		Lag 90a
42.60	0.50	214.900		Lag 90a
43.80	0.40	192.200		Lag 90a
44.90	0.50	184.600		Lag 90a
46.00	0.50	167.000		Lag 90a
47.10	0.50	151.700		Lag 90a
48.20	0.50	117.300		Lag 90a
49.20	0.50	107.900		Lag 90a
50.30	0.50	118.100		Lag 90a
51.30	0.50	130.300		Lag 90a
52.30	0.50	145.000		Lag 90a
53.40	0.50	171.700		Lag 90a
54.40	0.50	183.800		Lag 90a
55.40	0.50	177.100		Lag 90a
56.50	0.50	163.200		Lag 90a
57.50	0.50	147.200		Lag 90a
58.50	0.50	137.200		Lag 90a
59.50	0.50	131.700		Lag 90a
60.50	0.40	101.400		Lag 90a
61.40	0.40	90.600		Lag 90a
62.20	0.40	91.600		Lag 90a
63.00	0.40	65.100		Lag 90a
63.90	0.40	60.800		Lag 90a
64.70	0.40	59.800		Lag 90a
65.60	0.40	42.500		Lag 90a
66.40	0.40	38.100		Lag 90a
67.10	0.30	31.400		Lag 90a

1261 Reaction $_{51}Sb \rightarrow {}^{119m}_{52}Te$ (Fig. 569, p. 303)

E [MeV]	ΔE [MeV]	σ [mb]	Δσ [mb]	Ref.
25.40	0.80	1.700		Lag 90a
26.90	0.70	2.700		Lag 90a
28.40	0.70	5.400		Lag 90a
29.80	0.60	9.200		Lag 90a
31.20	0.60	17.300		Lag 90a
32.60	0.60	30.300		Lag 90a
33.90	0.60	54.400		Lag 90a
35.20	0.60	101.500		Lag 90a
36.50	0.60	156.600		Lag 90a
37.70	0.50	206.800		Lag 90a
39.00	0.60	226.700		Lag 90a
40.20	0.60	214.300		Lag 90a
41.50	0.60	205.000		Lag 90a
42.60	0.50	177.600		Lag 90a
43.80	0.40	129.400		Lag 90a

1261 Reaction $_{51}Sb \rightarrow {}^{119m}{}_{52}Te$ (continued)

E [MeV]	ΔE [MeV]	σ [mb]	Δσ [mb]	Ref.
44.90	0.50	103.400		Lag 90a
46.00	0.50	81.600		Lag 90a
47.10	0.50	74.600		Lag 90a
48.20	0.50	73.900		Lag 90a
49.20	0.50	80.100		Lag 90a
50.30	0.50	91.900		Lag 90a
51.30	0.50	104.300		Lag 90a
52.30	0.50	110.900		Lag 90a
53.40	0.50	121.700		Lag 90a
54.40	0.50	114.200		Lag 90a
55.40	0.50	109.100		Lag 90a
56.50	0.50	118.000		Lag 90a
57.50	0.50	109.100		Lag 90a
58.50	0.50	104.400		Lag 90a
59.50	0.50	99.000		Lag 90a
60.50	0.40	93.600		Lag 90a
61.40	0.40	96.800		Lag 90a
62.20	0.40	95.000		Lag 90a
63.00	0.40	93.500		Lag 90a
63.90	0.40	90.500		Lag 90a
64.70	0.40	71.800		Lag 90a
65.60	0.40	57.100		Lag 90a
66.40	0.40	47.600		Lag 90a
67.10	0.30	29.100		Lag 90a

1262 Reaction $_{51}Sb \rightarrow {}^{121}{}_{52}Te$ (Fig. 570, p. 303)

E [MeV]	ΔE [MeV]	σ [mb]	Δσ [mb]	Ref.
25.40	0.80	16.200		Lag 90a
26.90	0.70	22.900		Lag 90a
28.40	0.70	40.200		Lag 90a
29.80	0.60	63.000		Lag 90a
31.20	0.60	67.200		Lag 90a
32.60	0.60	54.000		Lag 90a
33.90	0.60	67.300		Lag 90a
35.20	0.60	110.000		Lag 90a
36.50	0.60	135.600		Lag 90a
37.70	0.50	146.500		Lag 90a
39.00	0.60	138.000		Lag 90a
40.20	0.60	98.500		Lag 90a
41.50	0.60	87.800		Lag 90a
42.60	0.50	66.600		Lag 90a
43.80	0.40	48.500		Lag 90a
44.90	0.50	38.500		Lag 90a
46.00	0.50	33.600		Lag 90a
47.10	0.50	32.300		Lag 90a
48.20	0.50	24.000		Lag 90a

1262 Reaction $_{51}Sb \rightarrow {}^{121}_{52}Te$ (continued)

E [MeV]	ΔE [MeV]	σ [mb]	Δσ [mb]	Ref.
49.20	0.50	21.400		Lag 90a
50.30	0.50	25.800		Lag 90a
51.30	0.50	23.000		Lag 90a
52.30	0.50	18.500		Lag 90a
53.40	0.50	19.200		Lag 90a
54.40	0.50	15.500		Lag 90a
55.40	0.50	14.000		Lag 90a
56.50	0.50	14.700		Lag 90a
57.50	0.50	13.800		Lag 90a
58.50	0.50	14.800		Lag 90a
59.50	0.50	16.100		Lag 90a

1263 Reaction $_{51}Sb \rightarrow {}^{121m}_{52}Te$ (Fig. 571, p. 303)

E [MeV]	ΔE [MeV]	σ [mb]	Δσ [mb]	Ref.
25.40	0.80	12.300		Lag 90a
26.90	0.70	17.200		Lag 90a
28.40	0.70	31.100		Lag 90a
29.80	0.60	49.700		Lag 90a
31.20	0.60	59.900		Lag 90a
32.60	0.60	56.600		Lag 90a
33.90	0.60	120.600		Lag 90a
35.20	0.60	132.900		Lag 90a
36.50	0.60	155.300		Lag 90a
37.70	0.50	177.500		Lag 90a
39.00	0.60	172.000		Lag 90a
40.20	0.60	132.700		Lag 90a
41.50	0.60	125.400		Lag 90a
42.60	0.50	101.000		Lag 90a
43.80	0.40	71.100		Lag 90a
44.90	0.50	55.200		Lag 90a
46.00	0.50	44.000		Lag 90a
47.10	0.50	44.400		Lag 90a
48.20	0.50	32.700		Lag 90a
49.20	0.50	27.000		Lag 90a
50.30	0.50	25.800		Lag 90a
51.30	0.50	27.600		Lag 90a
52.30	0.50	22.800		Lag 90a
53.40	0.50	22.200		Lag 90a
54.40	0.50	20.200		Lag 90a
55.40	0.50	18.300		Lag 90a
56.50	0.50	19.600		Lag 90a
57.50	0.50	20.600		Lag 90a
58.50	0.50	17.800		Lag 90a
59.50	0.50	18.800		Lag 90a

E [MeV]	ΔE [MeV]	σ [mb]	Δσ [mb]	Ref.

1264 Reaction $^{121}_{51}Sb \rightarrow ^{121}_{52}Te$

E [MeV]	ΔE [MeV]	σ [mb]	Δσ [mb]	Ref.
5.20		23.900	3.70	Bat 84
6.10		51.700	7.90	Bat 84
6.90		80.000	12.00	Bat 84
7.60		120.000	18.00	Bat 84
8.30		183.000	28.00	Bat 84
9.00		207.000	31.00	Bat 84

1265 Reaction $^{121}_{51}Sb \rightarrow ^{121m}_{52}Te$

E [MeV]	ΔE [MeV]	σ [mb]	Δσ [mb]	Ref.
4.20		3.900	0.60	Bat 84
5.20		10.000	1.30	Bat 84
6.10		23.500	3.00	Bat 84
6.90		42.100	5.20	Bat 84
7.60		67.500	8.20	Bat 84
8.30		99.000	12.00	Bat 84
9.00		127.000	15.00	Bat 84

1266 Reaction $^{123}_{51}Sb \rightarrow ^{123}_{52}Te$

E [MeV]	ΔE [MeV]	σ [mb]	Δσ [mb]	Ref.
5.20		16.900	4.50	Bat 84
6.10		35.500	9.60	Bat 84
6.90		61.100	14.80	Bat 84
7.60		78.500	23.30	Bat 84
8.30		139.000	35.00	Bat 84
9.00		162.000	39.00	Bat 84

1267 Reaction $^{123}_{51}Sb \rightarrow ^{123m}_{52}Te$

E [MeV]	ΔE [MeV]	σ [mb]	Δσ [mb]	Ref.
4.20		3.000	0.70	Bat 84
5.20		17.000	2.10	Bat 84
6.10		39.700	4.50	Bat 84
6.90		61.000	7.00	Bat 84
7.60		109.000	12.00	Bat 84
8.30		143.000	16.00	Bat 84
9.00		172.000	19.00	Bat 84

1268 Reaction $_{52}Te \rightarrow ^{121}_{53}I$ (Fig. 572, p. 304)

E [MeV]	ΔE [MeV]	σ [mb]	Δσ [mb]	Ref.
15.20		4.800		Sch 89*
15.80		8.800		Sch 89*
16.20		10.000		Sch 89*
16.50		11.800		Sch 89*
17.00		12.000		Sch 89*
17.50		12.500		Sch 89*
18.00		12.800		Sch 89*
18.50		13.000		Sch 89*

E [MeV]	ΔE [MeV]	σ [mb]	Δσ [mb]	Ref.

1268 Reaction $_{52}Te \rightarrow {}^{121}{}_{53}I$ (continued)

E [MeV]	ΔE [MeV]	σ [mb]	Δσ [mb]	Ref.
19.00		16.000		Sch 89*
19.50		18.000		Sch 89*

1269 Reaction $_{52}Te \rightarrow {}^{123}{}_{53}I$ (Fig. 573, p. 304)

E [MeV]	ΔE [MeV]	σ [mb]	Δσ [mb]	Ref.
9.80		1.400		Sch 89*
10.20		2.000		Sch 89*
10.90		2.800		Sch 89*
11.80		3.200		Sch 89*
12.30		4.000		Sch 89*
12.70		5.200		Sch 89*
13.10		7.000		Sch 89*
13.50		12.000		Sch 89*
14.00		18.000		Sch 89*
14.50		21.000		Sch 89*
15.20		30.000		Sch 89*
15.80		32.000		Sch 89*
16.20		30.000		Sch 89*
16.80		32.000		Sch 89*
17.00		32.000		Sch 89*
17.20		30.000		Sch 89*
17.80		35.000		Sch 89*
18.20		32.000		Sch 89*
18.50		35.000		Sch 89*
18.80		40.000		Sch 89*
19.00		35.000		Sch 89*
19.30		45.000		Sch 89*
19.60		40.000		Sch 89*

1270 Reaction $_{52}Te \rightarrow {}^{124}{}_{53}I$ (Fig. 574, p. 304)

E [MeV]	ΔE [MeV]	σ [mb]	Δσ [mb]	Ref.
10.90		17.000		Sch 89*
11.90		19.000		Sch 89*
12.20		22.000		Sch 89*
12.80		28.000		Sch 89*
13.00		42.000		Sch 89*
13.60		59.000		Sch 89*
14.00		60.000		Sch 89*
14.50		63.000		Sch 89*
15.20		68.000		Sch 89*
15.80		68.000		Sch 89*
16.20		62.000		Sch 89*
16.70		78.000		Sch 89*
17.70		68.000		Sch 89*
17.90		60.000		Sch 89*
18.40		62.000		Sch 89*

E [MeV]	ΔE [MeV]	σ [mb]	Δσ [mb]	Ref.

1270 Reaction $_{52}Te \rightarrow {}^{124}_{53}I$ (continued)

E [MeV]	ΔE [MeV]	σ [mb]	Δσ [mb]	Ref.
18.60		68.000		Sch 89*
19.00		72.000		Sch 89*
19.30		89.000		Sch 89*
19.70		82.000		Sch 89*

1271 Reaction $_{52}Te \rightarrow {}^{125}_{53}I$ (Fig. 575, p. 304)

E [MeV]	ΔE [MeV]	σ [mb]	Δσ [mb]	Ref.
12.20		1.300		Sch 89*
12.70		1.400		Sch 89*
13.00		2.100		Sch 89*
13.50		2.300		Sch 89*
14.00		2.900		Sch 89*
14.50		3.200		Sch 89*
15.20		3.700		Sch 89*
15.80		4.000		Sch 89*
16.20		3.400		Sch 89*
16.70		4.200		Sch 89*
16.90		4.700		Sch 89*
17.20		4.200		Sch 89*
17.50		4.500		Sch 89*
17.80		4.100		Sch 89*
18.50		4.800		Sch 89*
18.70		4.600		Sch 89*
19.00		5.500		Sch 89*
19.20		5.000		Sch 89*
19.60		4.900		Sch 89*

1272 Reaction $_{52}Te \rightarrow {}^{126}_{53}I$ (Fig. 576, p. 304)

E [MeV]	ΔE [MeV]	σ [mb]	Δσ [mb]	Ref.
9.80		45.000		Sch 89*
11.80		80.000		Sch 89*
12.20		72.000		Sch 89*
12.80		50.000		Sch 89*
13.10		45.000		Sch 89*
13.50		48.000		Sch 89*
14.00		22.000		Sch 89*
14.50		17.000		Sch 89*
15.30		14.000		Sch 89*
16.20		9.500		Sch 89*

1273 Reaction $_{52}Te \rightarrow {}^{128}_{53}I$ (Fig. 577, p. 304)

E [MeV]	ΔE [MeV]	σ [mb]	Δσ [mb]	Ref.
8.00		8.500		Sch 89*
9.10		30.000		Sch 89*
9.80		55.000		Sch 89*
10.10		75.000		Sch 89*

E [MeV]	ΔE [MeV]	σ [mb]	Δσ [mb]	Ref.

1273 Reaction $_{52}Te \rightarrow {}^{128}_{53}I$ (continued)

E [MeV]	ΔE [MeV]	σ [mb]	Δσ [mb]	Ref.
10.80		90.000		Sch 89*
11.80		75.000		Sch 89*
12.20		50.000		Sch 89*
12.60		45.000		Sch 89*
13.00		40.000		Sch 89*
13.50		30.000		Sch 89*
15.30		22.000		Sch 89*
16.60		16.000		Sch 89*
17.10		13.000		Sch 89*
17.50		9.000		Sch 89*
18.80		13.000		Sch 89*
19.00		9.500		Sch 89*

1274 Reaction $_{52}Te \rightarrow {}^{130}_{53}I$ **(Fig. 578, p. 305)**

E [MeV]	ΔE [MeV]	σ [mb]	Δσ [mb]	Ref.
8.00		12.000		Sch 89*
9.10		35.000		Sch 89*
9.80		60.000		Sch 89*
10.10		70.000		Sch 89*
10.90		70.000		Sch 89*
11.90		60.000		Sch 89*
12.20		40.000		Sch 89*
12.70		40.000		Sch 89*
13.00		35.000		Sch 89*
13.50		30.000		Sch 89*
14.00		22.000		Sch 89*
14.50		20.000		Sch 89*
15.20		20.000		Sch 89*
15.70		18.000		Sch 89*
16.10		16.000		Sch 89*
16.50		15.000		Sch 89*
16.90		14.000		Sch 89*
17.10		15.000		Sch 89*
17.90		12.000		Sch 89*
19.00		10.000		Sch 89*
19.30		11.000		Sch 89*
19.60		10.000		Sch 89*

1275 Reaction ${}^{122}_{52}Te \rightarrow {}^{121}_{52}Te$

E [MeV]	ΔE [MeV]	σ [mb]	Δσ [mb]	Ref.
300.0		19.700	4.90	Sch 70

1276 Reaction ${}^{122}_{52}Te \rightarrow {}^{121}_{53}I$

E [MeV]	ΔE [MeV]	σ [mb]	Δσ [mb]	Ref.
300.0		3.730	0.75	Sch 70

E [MeV]	ΔE [MeV]	σ [mb]	Δσ [mb]	Ref.

1277 Reaction $^{123}_{52}Te \rightarrow ^{122}_{51}Sb$

E [MeV]	ΔE [MeV]	σ [mb]	Δσ [mb]	Ref.
300.0		11.800	1.90	Sch 70

1278 Reaction $^{123}_{52}Te \rightarrow ^{122}_{53}I$ (Fig. 579, p. 305)

E [MeV]	ΔE [MeV]	σ [mb]	Δσ [mb]	Ref.
12.90		83.000	33.00	Sch 89
13.70		118.000	41.00	Sch 89
14.50		99.000	25.00	Sch 89
15.30		390.000	78.00	Sch 89
15.80		319.000	48.00	Sch 89
16.70		682.000	102.00	Sch 89
17.40		672.000	101.00	Sch 89
18.00		617.000	93.00	Sch 89
19.20		688.000	118.00	Sch 89
19.60		932.000	168.00	Sch 89

1279 Reaction $^{123}_{52}Te \rightarrow ^{123}_{53}I$ (Fig. 580, p. 305)

E [MeV]	ΔE [MeV]	σ [mb]	Δσ [mb]	Ref.
5.90		1.600	0.50	Sch 89
7.50		24.000	6.00	Sch 89
8.10		80.000	16.00	Sch 89
9.30		200.000	30.00	Sch 89
9.90		344.000	38.00	Sch 89
10.20		286.000	31.00	Sch 89
10.40		451.000	50.00	Sch 89
10.70		426.000	47.00	Sch 89
11.70		489.000	54.00	Sch 89
12.20		634.000	70.00	Sch 89
12.60		670.000	74.00	Sch 89
12.90		583.000	64.00	Sch 89
13.70		618.000	68.00	Sch 89
14.50		464.000	51.00	Sch 89
15.30		494.000	54.00	Sch 89
15.80		217.000	24.00	Sch 89
16.70		252.000	28.00	Sch 89
17.40		171.000	21.00	Sch 89
18.00		137.000	16.00	Sch 89
18.70		87.000	10.00	Sch 89
19.20		121.000	15.00	Sch 89
19.60		149.000	19.00	Sch 89

1280 Reaction $^{124}_{52}Te \rightarrow ^{123m}_{52}Te$

E [MeV]	ΔE [MeV]	σ [mb]	Δσ [mb]	Ref.
300.0		15.700	4.40	Sch 70

E [MeV]	ΔE [MeV]	σ [mb]	Δσ [mb]	Ref.

1281 Reaction $^{124}_{52}Te \rightarrow ^{123}_{53}I$ (Fig. 581, p. 305)

E [MeV]	ΔE [MeV]	σ [mb]	Δσ [mb]	Ref.
12.45	0.79	24.500	5.60	Kon 77
14.54	0.72	160.600	6.40	Kon 77
16.35	0.70	427.700	51.10	Kon 77
18.82	0.45	630.900	70.50	Kon 77
20.86	0.41	889.600	69.50	Kon 77
22.65	0.40	1011.400	59.60	Kon 77
24.10	0.37	914.400	45.70	Kon 77
25.36	0.34	1017.000	107.00	Kon 77
26.43	0.34	649.800	68.80	Kon 77
28.19	0.30	436.200	34.90	Kon 77
300.0		5.300	2.10	Sch 70
400.0		3.600	0.20	Reu 69

1282 Reaction $^{124}_{52}Te \rightarrow ^{124}_{53}I$ (Fig. 582, p. 305)

E [MeV]	ΔE [MeV]	σ [mb]	Δσ [mb]	Ref.
9.05	1.10	115.900	18.80	Kon 77
12.45	0.79	308.900	38.80	Kon 77
14.54	0.72	455.500	29.60	Kon 77
16.35	0.70	194.900	16.20	Kon 77
18.82	0.45	84.800	13.90	Kon 77
20.86	0.41	65.600	4.00	Kon 77
22.65	0.40	53.800	5.80	Kon 77
24.10	0.37	30.500	8.20	Kon 77
25.36	0.34	37.300	15.60	Kon 77
26.43	0.34	34.900	6.80	Kon 77
28.19	0.30	43.700	11.40	Kon 77
400.0		1.300	0.10	Reu 69

1283 Reaction $^{125}_{52}Te \rightarrow ^{119}_{51}Sb$

E [MeV]	ΔE [MeV]	σ [mb]	Δσ [mb]	Ref.
120.0		9.100		Zai 62
300.0		6.800		Zai 62
480.0		5.100		Zai 62
660.0		6.800		Zai 62

1284 Reaction $^{125}_{52}Te \rightarrow ^{120}_{51}Sb$

E [MeV]	ΔE [MeV]	σ [mb]	Δσ [mb]	Ref.
120.0		10.600		Zai 62
300.0		7.600		Zai 62
480.0		6.800		Zai 62
660.0		10.200		Zai 62

1285 Reaction $^{125}_{52}Te \rightarrow ^{122}_{51}Sb$

E [MeV]	ΔE [MeV]	σ [mb]	Δσ [mb]	Ref.
120.0		20.000		Zai 62
300.0		17.600		Zai 62

E [MeV]	ΔE [MeV]	σ [mb]	Δσ [mb]	Ref.

1285 Reaction $^{125}_{52}Te \rightarrow ^{122}_{51}Sb$ (continued)

E [MeV]	ΔE [MeV]	σ [mb]	Δσ [mb]	Ref.
480.0		15.400		Zai 62
660.0		22.100		Zai 62

1286 Reaction $^{125}_{52}Te \rightarrow ^{124}_{51}Sb$

E [MeV]	ΔE [MeV]	σ [mb]	Δσ [mb]	Ref.
120.0		9.500		Zai 62
300.0		11.000		Zai 62
300.0		6.390	1.54	Sch 70
480.0		12.600		Zai 62
660.0		20.000		Zai 62

1287 Reaction $^{125}_{52}Te \rightarrow ^{123}_{53}I$

E [MeV]	ΔE [MeV]	σ [mb]	Δσ [mb]	Ref.
120.0		20.000		Zai 62
300.0		2.400		Zai 62
400.0		2.090	0.25	Chu 69
660.0		1.800		Zai 62

1288 Reaction $^{125}_{52}Te \rightarrow ^{124}_{53}I$

E [MeV]	ΔE [MeV]	σ [mb]	Δσ [mb]	Ref.
120.0		13.300		Zai 62
300.0		2.500		Zai 62
300.0		4.410	0.26	Sch 70
400.0		3.120	0.40	Chu 69
480.0		2.200		Zai 62
660.0		2.300		Zai 62

1289 Reaction $^{125}_{52}Te \rightarrow ^{125}_{53}I$

E [MeV]	ΔE [MeV]	σ [mb]	Δσ [mb]	Ref.
120.0		7.200		Zai 62
300.0		1.200		Zai 62
400.0		1.230	0.16	Chu 69

1290 Reaction $^{125}_{52}Te \rightarrow ^{126}_{53}I$

E [MeV]	ΔE [MeV]	σ [mb]	Δσ [mb]	Ref.
120.0		2.200		Zai 62
300.0		0.300		Zai 62
660.0		0.400		Zai 62

1291 Reaction $^{126}_{52}Te \rightarrow ^{119}_{51}Sb$

E [MeV]	ΔE [MeV]	σ [mb]	Δσ [mb]	Ref.
120.0		5.600		Zai 62
200.0		6.400		Zai 62
300.0		6.800		Zai 62
480.0		6.800		Zai 62
660.0		5.700		Zai 62

E [MeV]	ΔE [MeV]	σ [mb]	Δσ [mb]	Ref.

1292 Reaction $^{126}_{52}Te \rightarrow ^{120}_{51}Sb$

E [MeV]	ΔE [MeV]	σ [mb]	Δσ [mb]	Ref.
120.0		9.400		Zai 62
200.0		12.100		Zai 62
300.0		10.200		Zai 62
480.0		9.200		Zai 62
660.0		8.600		Zai 62

1293 Reaction $^{126}_{52}Te \rightarrow ^{122}_{51}Sb$

E [MeV]	ΔE [MeV]	σ [mb]	Δσ [mb]	Ref.
120.0		18.100		Zai 62
200.0		14.000		Zai 62
300.0		21.100		Zai 62
480.0		22.000		Zai 62
660.0		21.600		Zai 62

1294 Reaction $^{126}_{52}Te \rightarrow ^{125}_{51}Sb$

E [MeV]	ΔE [MeV]	σ [mb]	Δσ [mb]	Ref.
300.0		7.800	2.10	Sch 70

1295 Reaction $^{126}_{52}Te \rightarrow ^{125m}_{52}Te$

E [MeV]	ΔE [MeV]	σ [mb]	Δσ [mb]	Ref.
300.0		17.700	1.80	Sch 70

1296 Reaction $^{126}_{52}Te \rightarrow ^{123}_{53}I$

E [MeV]	ΔE [MeV]	σ [mb]	Δσ [mb]	Ref.
120.0		15.600		Zai 62
200.0		5.500		Zai 62
300.0		2.200		Zai 62
400.0		2.500	0.32	Chu 69
480.0		2.000		Zai 62
660.0		1.800		Zai 62

1297 Reaction $^{126}_{52}Te \rightarrow ^{124}_{53}I$

E [MeV]	ΔE [MeV]	σ [mb]	Δσ [mb]	Ref.
120.0		15.400		Zai 62
200.0		5.500		Zai 62
300.0		2.800		Zai 62
400.0		3.520	0.44	Chu 69
480.0		1.900		Zai 62
660.0		2.200		Zai 62

1298 Reaction $^{126}_{52}Te \rightarrow ^{125}_{53}I$

E [MeV]	ΔE [MeV]	σ [mb]	Δσ [mb]	Ref.
120.0		13.000		Zai 62
200.0		4.600		Zai 62
300.0		2.300		Zai 62
300.0		6.300	1.70	Sch 70

E [MeV]	ΔE [MeV]	σ [mb]	Δσ [mb]	Ref.
1298 Reaction $^{126}_{52}Te \rightarrow ^{125}_{53}I$ (continued)				
400.0		3.330	0.41	Chu 69
480.0		1.200		Zai 62
660.0		1.800		Zai 62
1299 Reaction $^{126}_{52}Te \rightarrow ^{126}_{53}I$				
120.0		8.500		Zai 62
200.0		~3.000		Zai 62
300.0		1.100		Zai 62
400.0		0.920	0.12	Chu 69
480.0		0.800		Zai 62
660.0		1.200		Zai 62
1300 Reaction $^{128}_{52}Te \rightarrow ^{127}_{51}Sb$				
300.0		5.430	0.87	Sch 70
1301 Reaction $^{128}_{52}Te \rightarrow ^{127}_{52}Te$				
300.0		27.500	4.20	Sch 70
1302 Reaction $^{128}_{52}Te \rightarrow ^{127m}_{52}Te$				
300.0		19.400	2.00	Sch 70
1303 Reaction $^{128}_{52}Te \rightarrow ^{128}_{53}I$ (Fig. 583, p. 306)				
3.30		0.400E-1		Bla 51*
3.60		0.230		Bla 51*
3.90		0.600		Bla 51*
4.20		1.500		Bla 51*
4.50		2.300		Bla 51*
4.80		4.200		Bla 51*
5.00		9.000		Bla 51*
5.80		27.000		Bla 51*
5.90		34.000		Bla 51*
6.10		42.000		Bla 51*
6.30		50.000		Bla 51*
6.70		90.000		Bla 51
1304 Reaction $^{130}_{52}Te \rightarrow ^{129}_{51}Sb$				
300.0		5.840	0.82	Sch 70

E [MeV]	ΔE [MeV]	σ [mb]	Δσ [mb]	Ref.

1305 Reaction $^{130}_{52}Te \rightarrow ^{129}_{52}Te$

300.0		30.200	1.10	Sch 70

1306 Reaction $^{130}_{52}Te \rightarrow ^{129m}_{52}Te$

300.0		23.200	3.10	Sch 70

1307 Reaction $^{130}_{52}Te \rightarrow ^{130}_{53}I$ (Fig. 584, p. 306)

3.30		0.300E-1		Bla 51*
3.60		0.200		Bla 51*
4.20		1.100		Bla 51*
4.50		2.100		Bla 51*
4.80		4.200		Bla 51*
5.00		8.500		Bla 51*
5.30		12.000		Bla 51*
5.50		19.000		Bla 51*
5.90		38.000		Bla 51*
6.10		47.000		Bla 51*
6.40		70.000		Bla 51*
6.70		95.000		Bla 51
7.50		85.000		Dal 64*
9.00		195.000		Dal 64*
11.50		215.000		Dal 64*
13.00		135.000		Dal 64*
14.00		80.000		Dal 64*
15.50		60.000		Dal 64*
17.00		45.000		Dal 64*
18.50		35.000		Dal 64*
20.00		30.000		Dal 64*
130.0		7.680	0.87	Tre 66
205.0		4.070	0.24	Tre 66
279.0		3.260	0.40E-1	Tre 66

1308 Reaction $^{130}_{52}Te \rightarrow ^{131}_{53}I$ (Fig. 585, p. 306)

6.50		0.500E-1		Dal 64*
7.50		0.200		Dal 64*
9.00		0.500		Dal 64*
11.50		0.800		Dal 64*
12.00		0.850		Dal 64*
14.00		0.700		Dal 64*
15.50		0.500		Dal 64*
17.00		0.450		Dal 64*
18.50		0.350		Dal 64*
20.00		0.300		Dal 64*

3.2 Figures

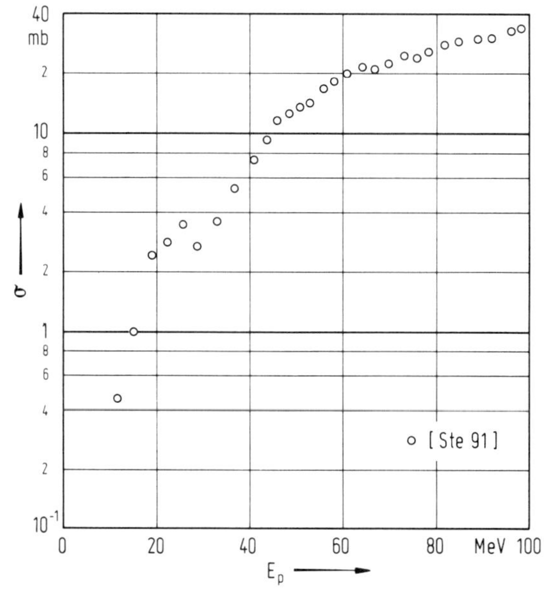

Fig. 329. Reaction $_{36}Kr \rightarrow {}^{77}_{35}Br$ (cumulative)

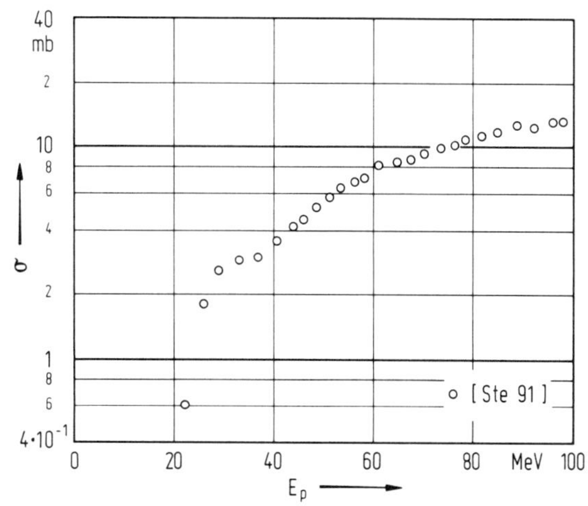

Fig. 330. Reaction $_{36}Kr \rightarrow {}^{82}_{35}Br$

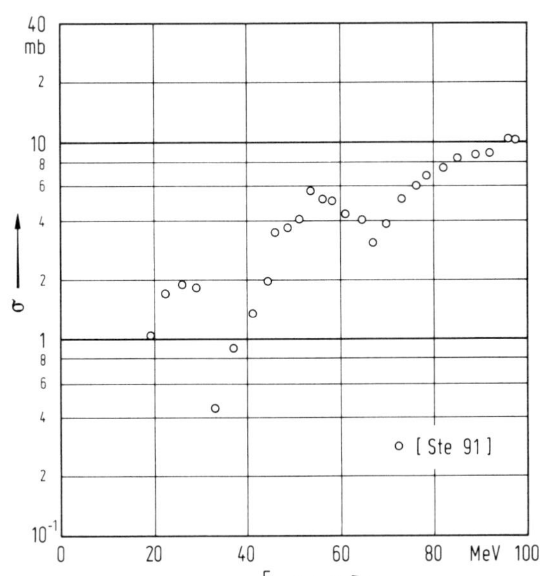

Fig. 331. Reaction $_{36}Kr \rightarrow {}^{77}_{36}Kr$

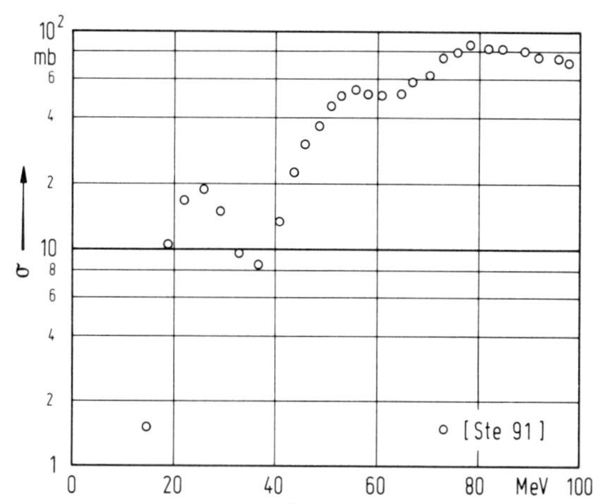

Fig. 332. Reaction $_{36}Kr \rightarrow {}^{79}_{36}Kr$ (cumulative)

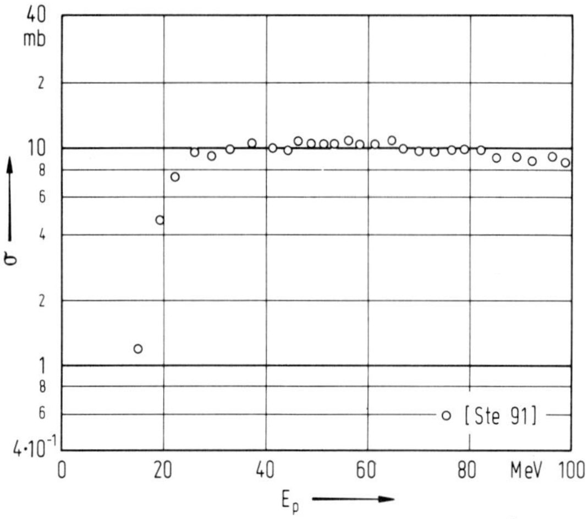

Fig. 333. Reaction $_{36}Kr \rightarrow {}^{85m}{}_{36}Kr$

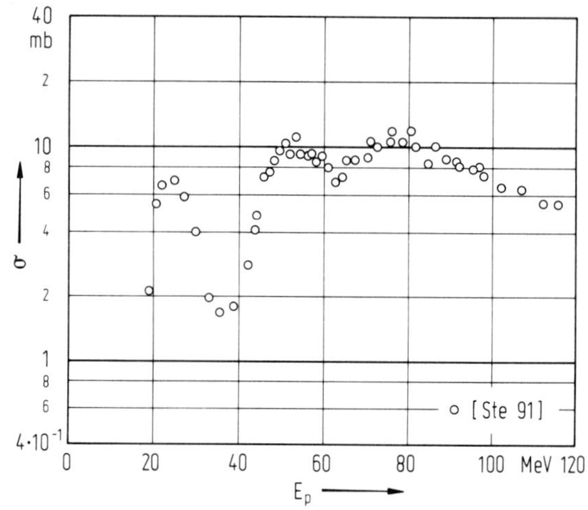

Fig. 334. Reaction $_{36}Kr \rightarrow {}^{79}{}_{37}Rb$

Fig. 336. Reaction $_{36}Kr \rightarrow {}^{81m}{}_{37}Rb$ ▶

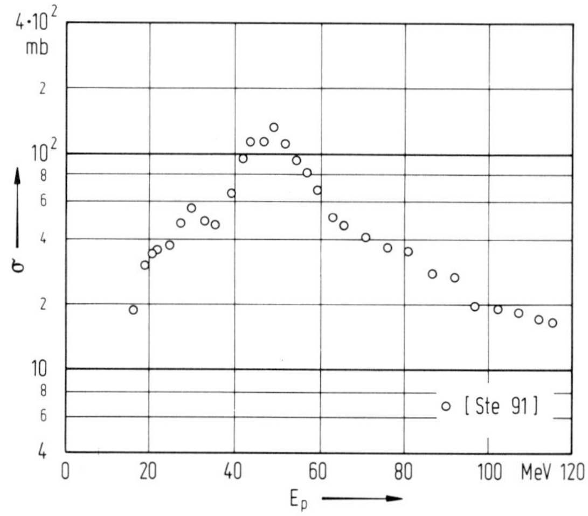

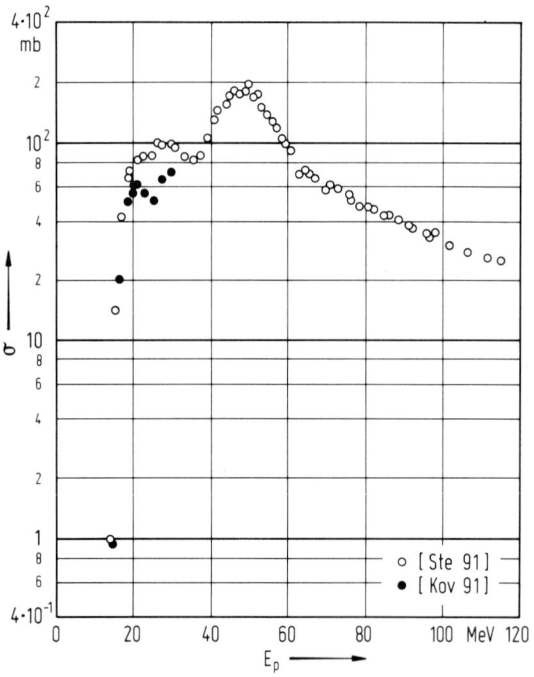

Fig. 335. Reaction $_{36}Kr \rightarrow {}^{81}{}_{37}Rb$ (cumulative)

Fig. 337. Reaction $_{36}Kr \rightarrow {}^{82m}{}_{37}Rb$ ▶

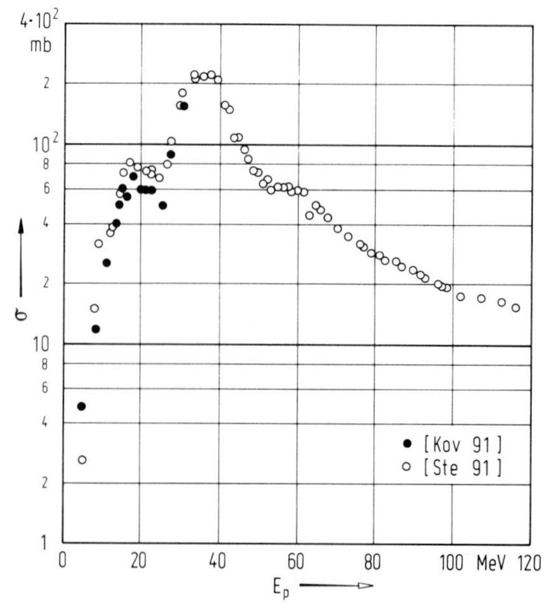

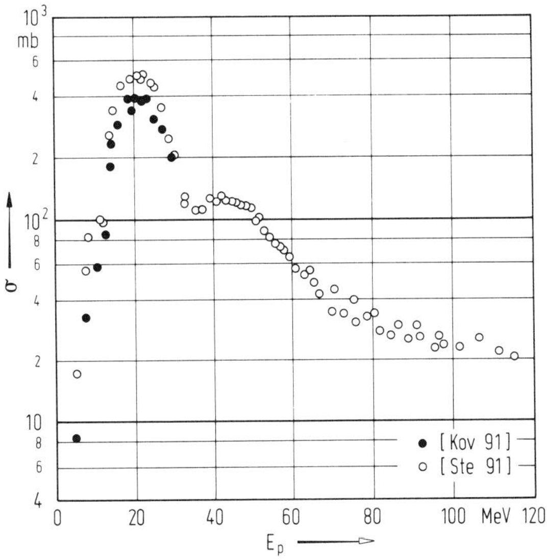

Fig. 338. Reaction $_{36}Kr \rightarrow {}^{83}_{37}Rb$

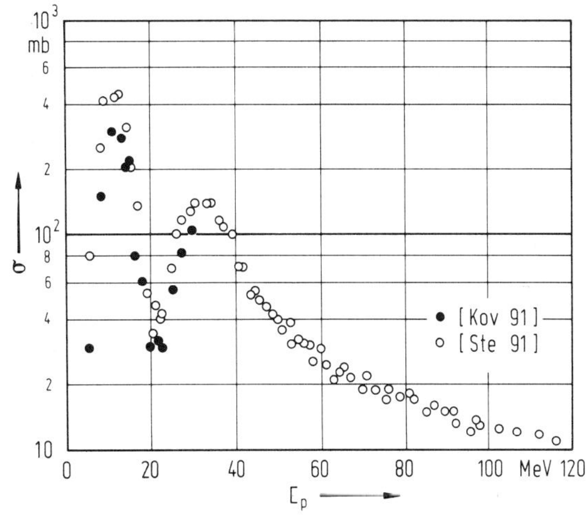

Fig. 339. Reaction $_{36}Kr \rightarrow {}^{84}_{37}Rb$

Fig. 341. Reaction $_{36}Kr \rightarrow {}^{86}_{37}Rb$ ▶

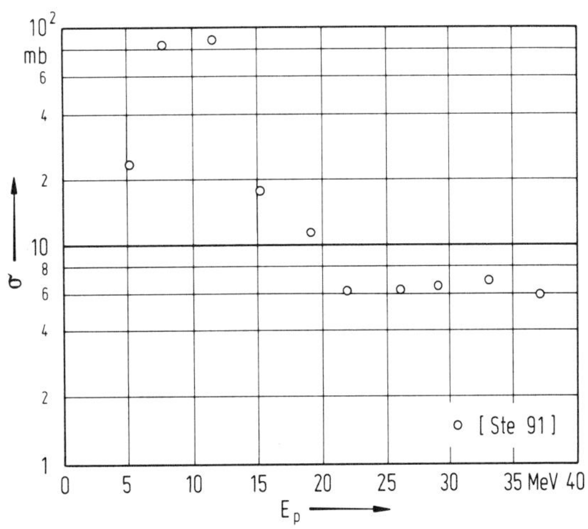

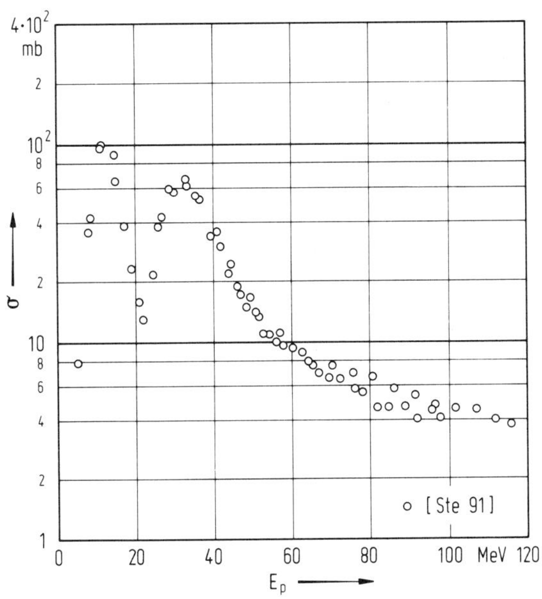

Fig. 340. Reaction $_{36}Kr \rightarrow {}^{84m}_{37}Rb$

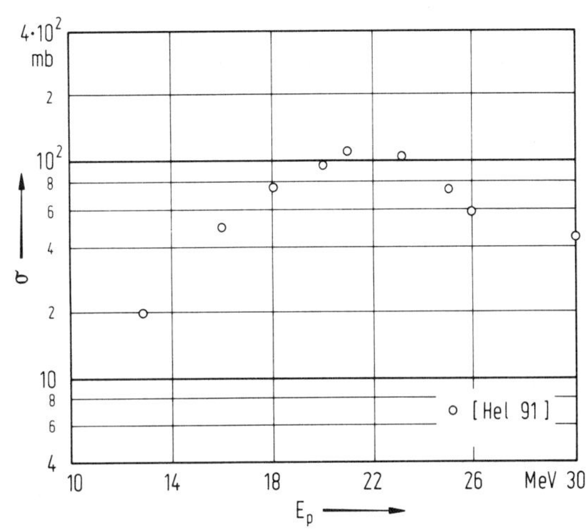

Fig. 342. Reaction $^{78}_{36}Kr \rightarrow {}^{75}_{35}Br$ ▶

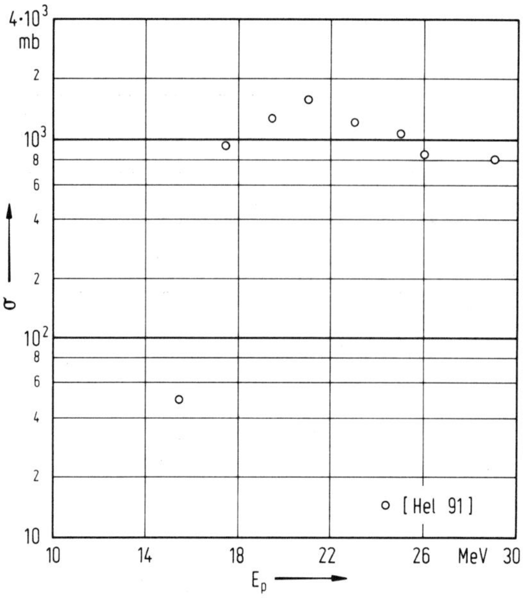

Fig. 343. Reaction $^{78}_{36}Kr \rightarrow ^{77}_{36}Kr$

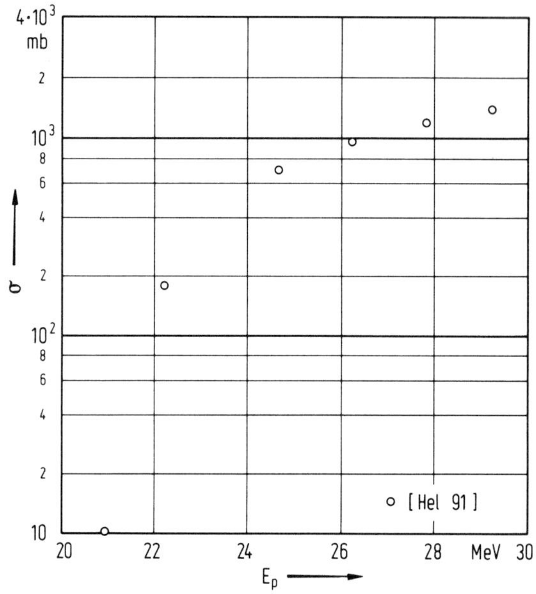

Fig. 344. Reaction $^{78}_{36}Kr \rightarrow ^{77}_{37}Rb$

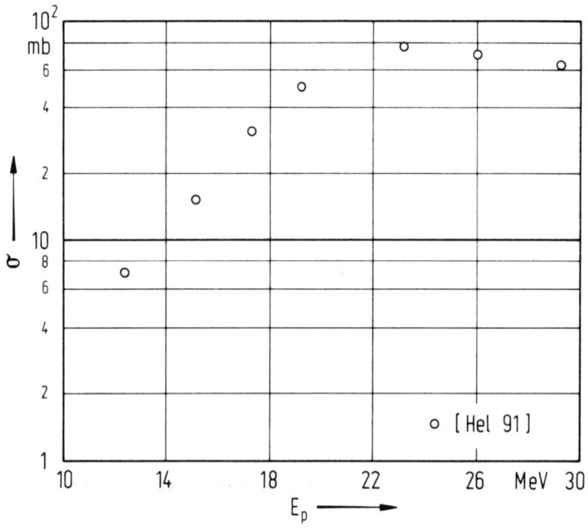

Fig. 345. Reaction $^{80}_{36}Kr \rightarrow ^{77}_{35}Br$

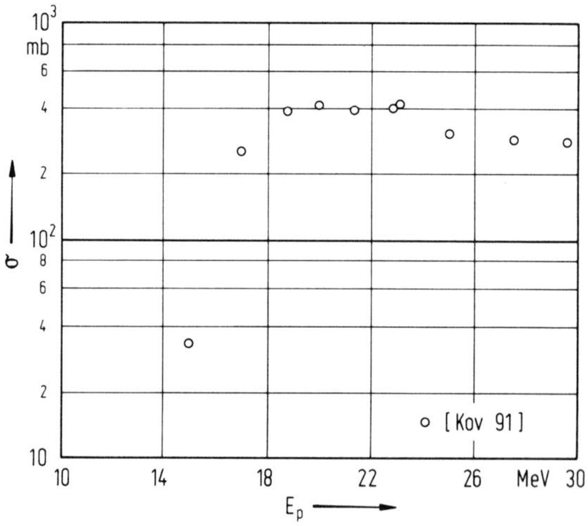

Fig. 346. Reaction $^{82}_{36}Kr \rightarrow ^{81}_{37}Rb$

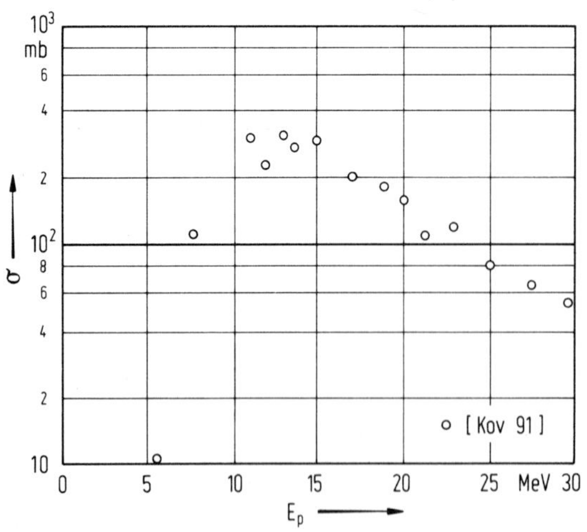

◀ Fig. 347. Reaction $^{82}_{36}Kr \rightarrow ^{82m}_{37}Rb$

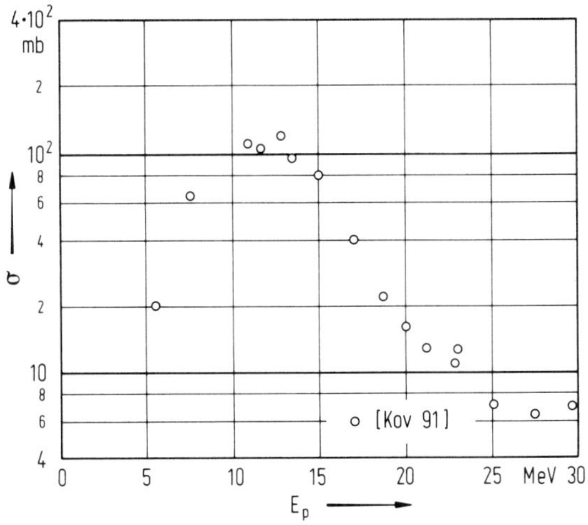

Fig. 348. Reaction $^{82}_{36}Kr \rightarrow {}^{83}_{37}Rb$

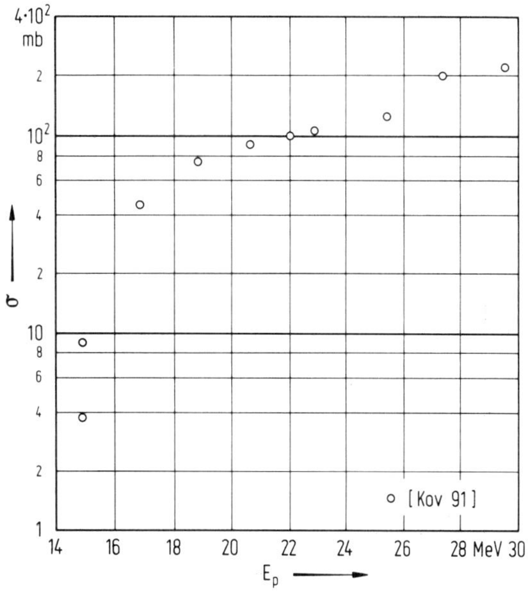

Fig. 349. Reaction $^{83}_{36}Kr \rightarrow {}^{81}_{37}Rb$

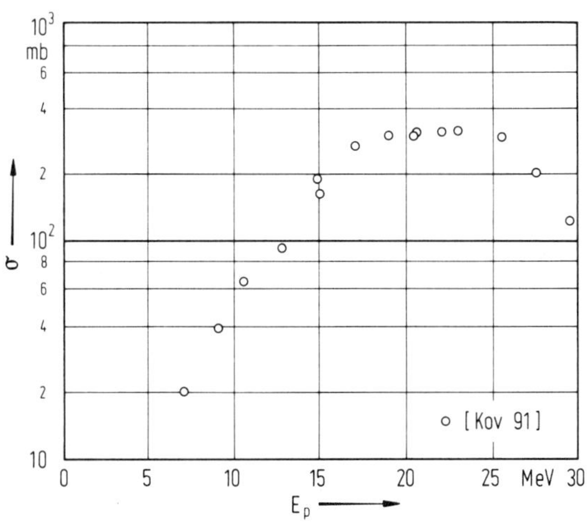

Fig. 350. Reaction $^{83}_{36}Kr \rightarrow {}^{82m}_{37}Rb$

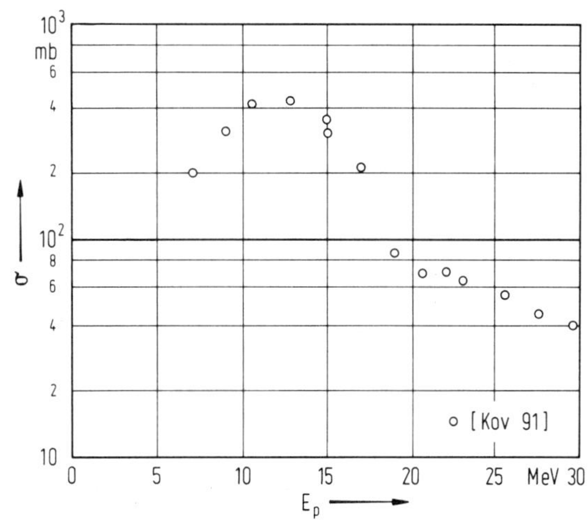

Fig. 351. Reaction $^{83}_{36}Kr \rightarrow {}^{83}_{37}Rb$

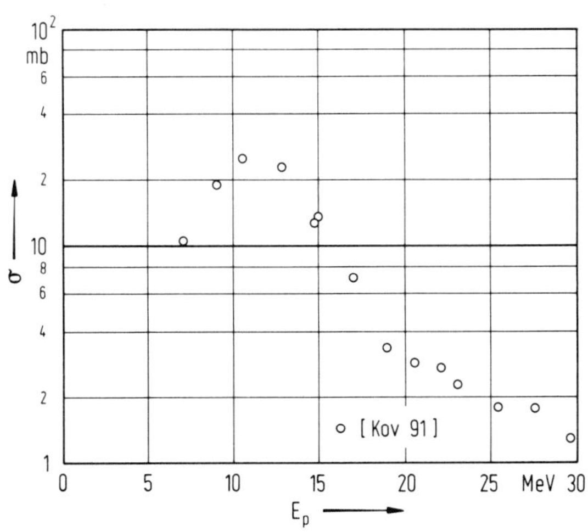

◀ Fig. 352. Reaction $^{83}_{36}Kr \rightarrow {}^{84}_{37}Rb$

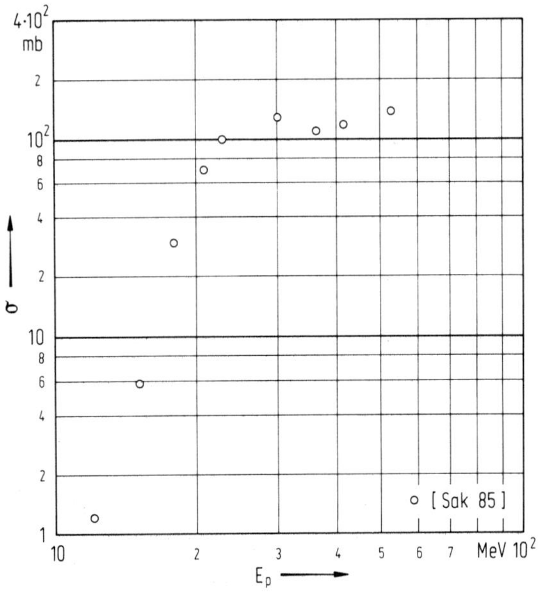

Fig. 353. Reaction $_{37}Rb \rightarrow {}^{84}{}_{37}Rb$

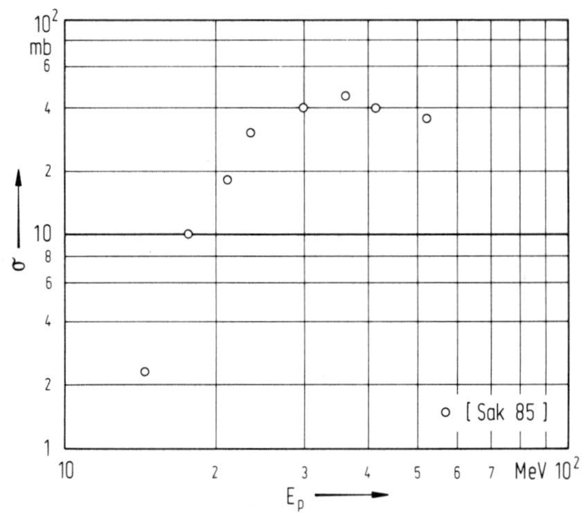

Fig. 354. Reaction $_{37}Rb \rightarrow {}^{86}{}_{37}Rb$

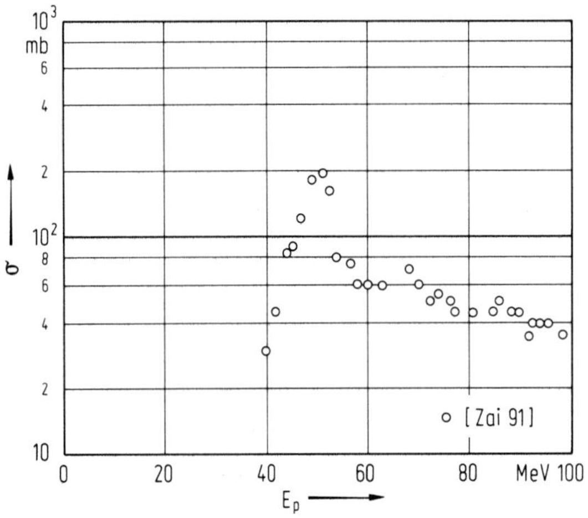

Fig. 355. Reaction $_{37}Rb \rightarrow {}^{82}{}_{38}Sr$

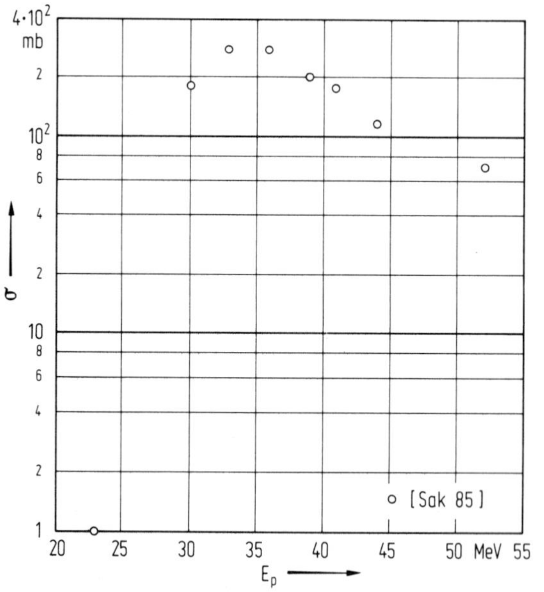

Fig. 356. Reaction $_{37}Rb \rightarrow {}^{83m+g}{}_{38}Sr$

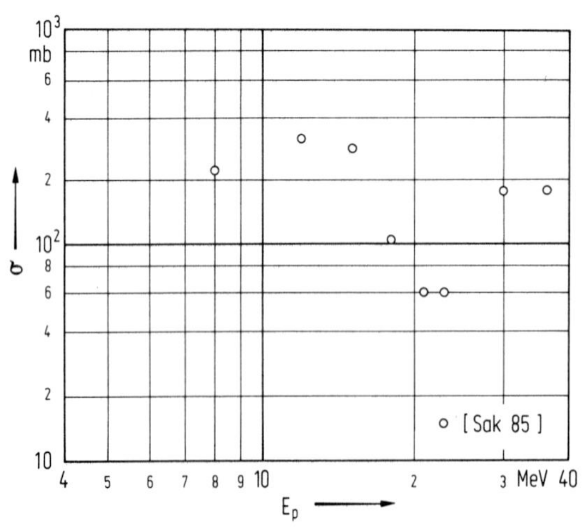

◀ Fig. 357. Reaction $_{37}Rb \rightarrow {}^{85m+g}{}_{38}Sr$

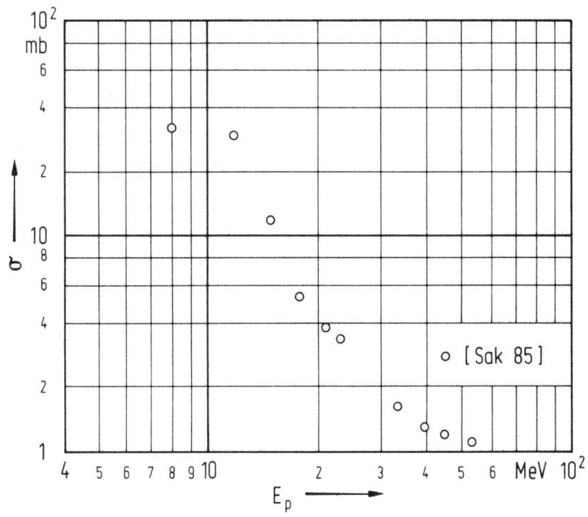

Fig. 358. Reaction $_{37}Rb \rightarrow {}^{87m}_{38}Sr$

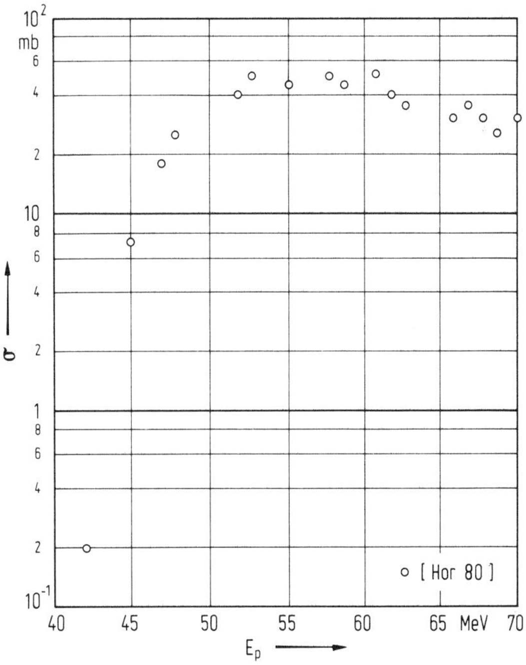

Fig. 359. Reaction $^{85}_{37}Rb \rightarrow {}^{79}_{36}Kr$

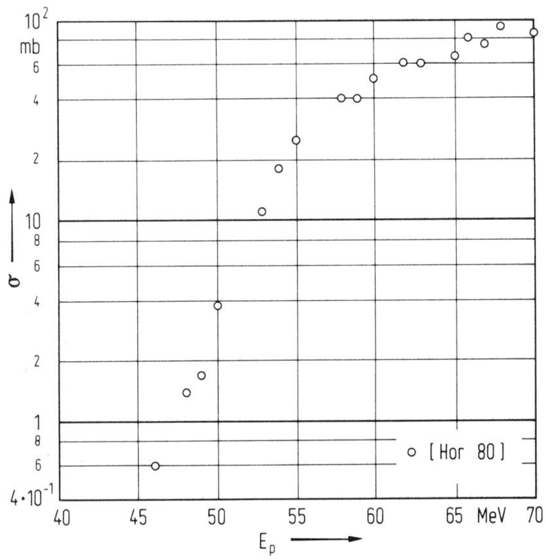

◀ Fig. 360. Reaction $^{85}_{37}Rb \rightarrow {}^{81}_{37}Rb$

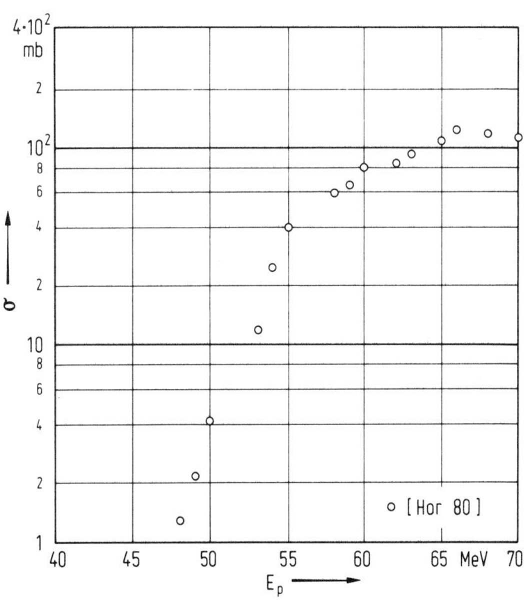

Fig. 361. Reaction $^{85}_{37}Rb \rightarrow {}^{81m}_{37}Rb$

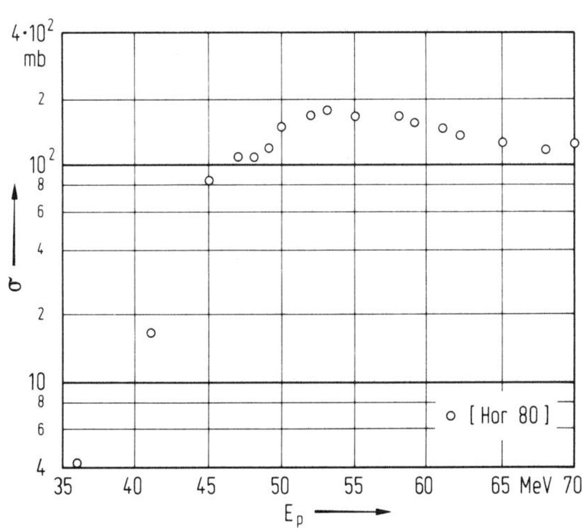

◀ Fig. 362. Reaction $^{85}_{37}Rb \rightarrow {}^{82m}_{37}Rb$

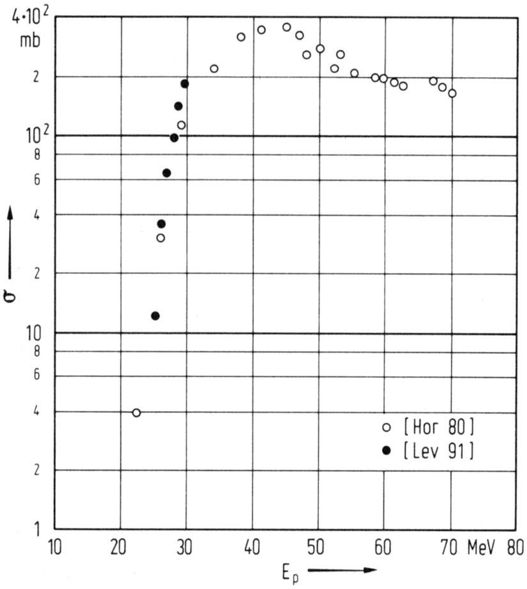

Fig. 363. Reaction $^{85}_{37}Rb \rightarrow {}^{83}_{37}Rb$

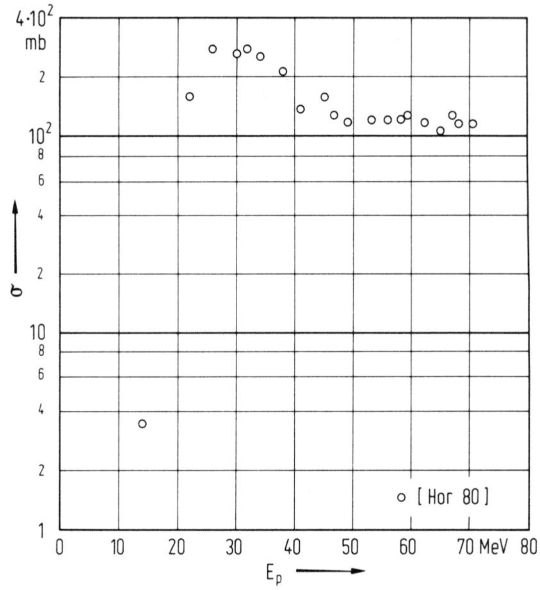

Fig. 364. Reaction $^{85}_{37}Rb \rightarrow {}^{84}_{37}Rb$

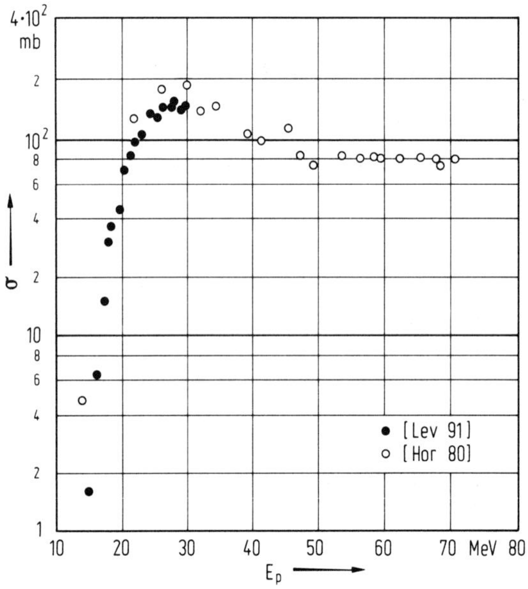

Fig. 365. Reaction $^{85}_{37}Rb \rightarrow {}^{84m}_{37}Rb$

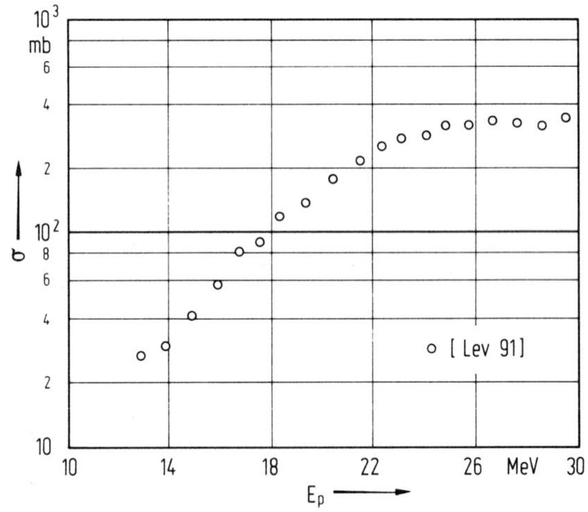

Fig. 366. Reaction $^{85}_{37}Rb \rightarrow {}^{84m+g}_{37}Rb$

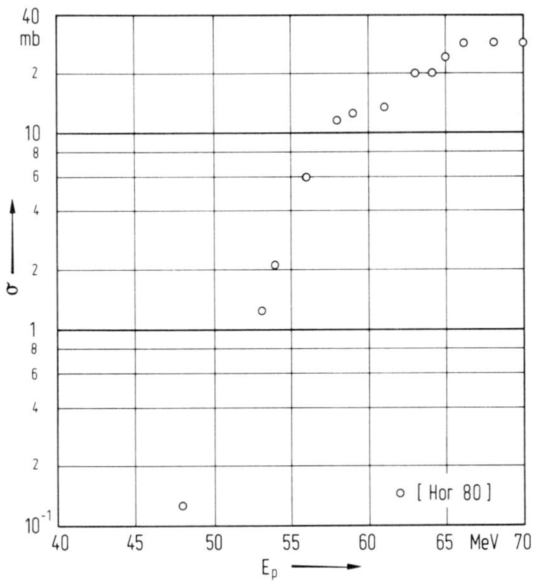

Fig. 367. Reaction $^{85}_{37}Rb \rightarrow {}^{81}_{38}Sr$

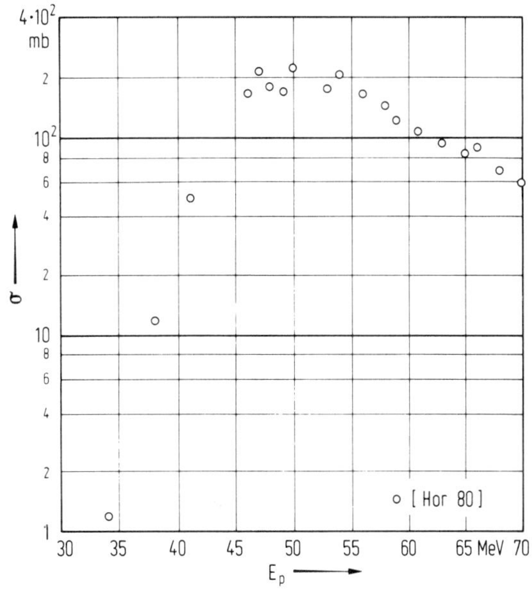

Fig. 368. Reaction $^{85}_{37}Rb \rightarrow {}^{82}_{38}Sr$

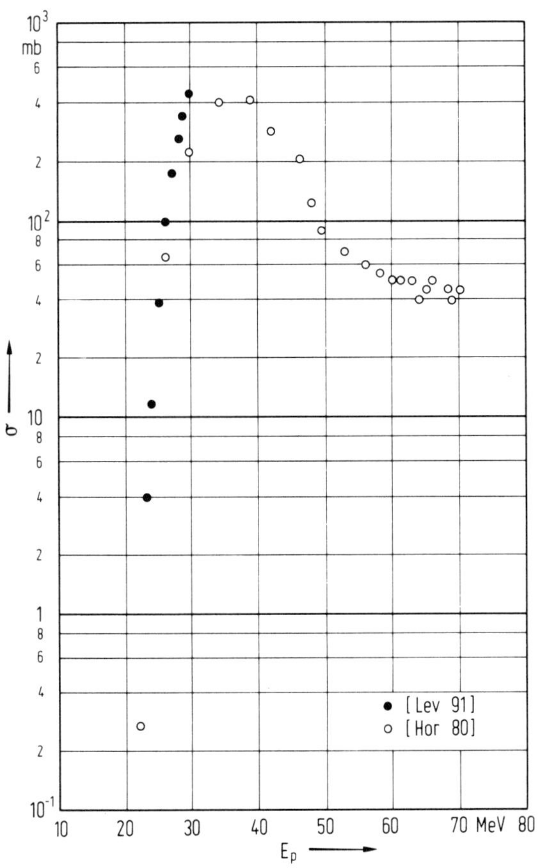

◀ Fig. 369. Reaction $^{85}_{37}Rb \rightarrow {}^{83}_{38}Sr$

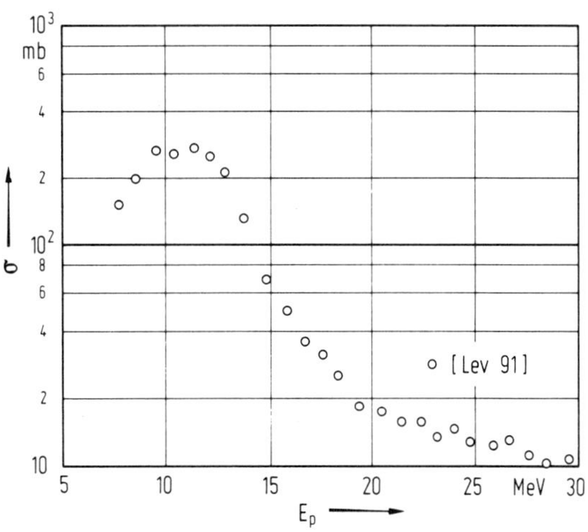

Fig. 370. Reaction $^{85}_{37}Rb \rightarrow {}^{85m}_{38}Sr$

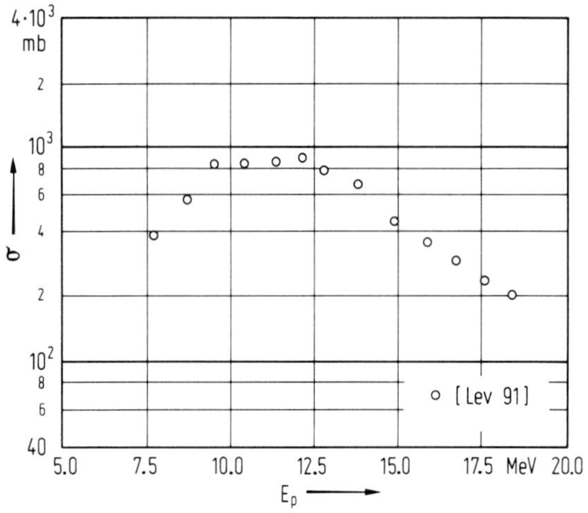

Fig. 371. Reaction $^{85}{}_{37}Rb \rightarrow {}^{85m+g}{}_{38}Sr$

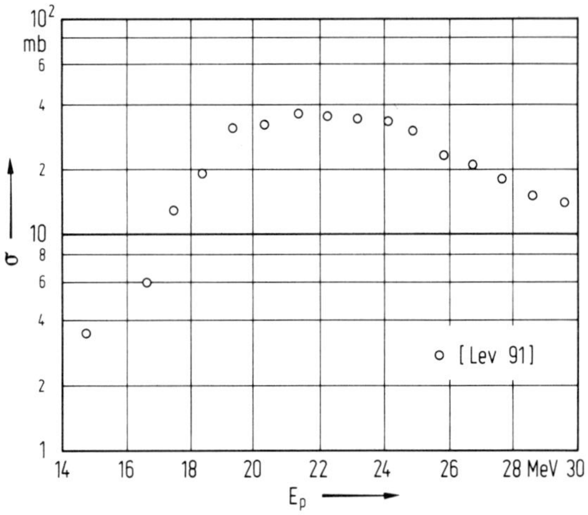

Fig. 372. Reaction $^{86}{}_{38}Sr \rightarrow {}^{83}{}_{37}Rb$

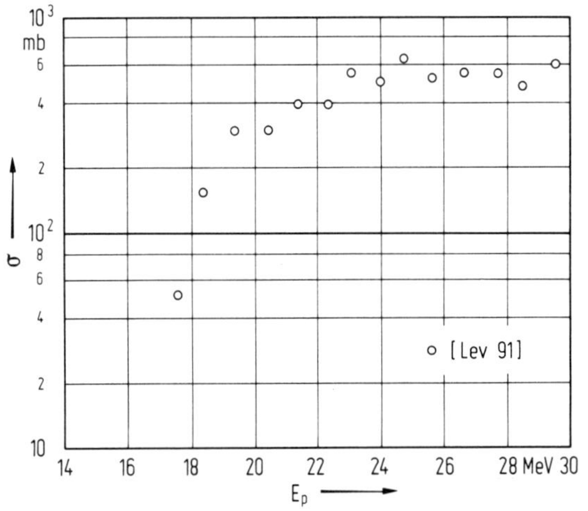

Fig. 373. Reaction $^{86}{}_{38}Sr \rightarrow {}^{85m+g}{}_{38}Sr$

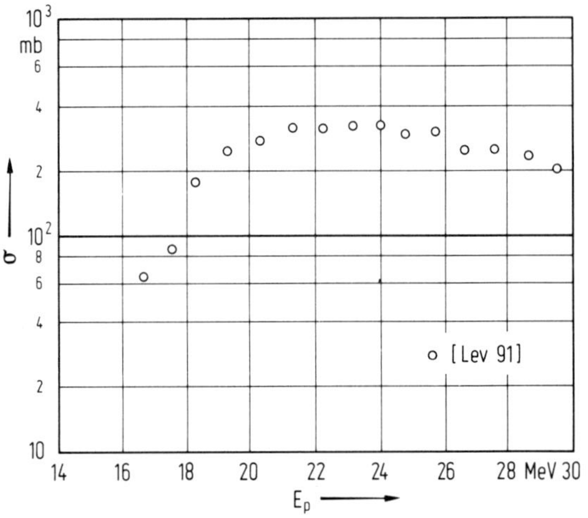

Fig. 374. Reaction $^{86}{}_{38}Sr \rightarrow {}^{85}{}_{39}Y$

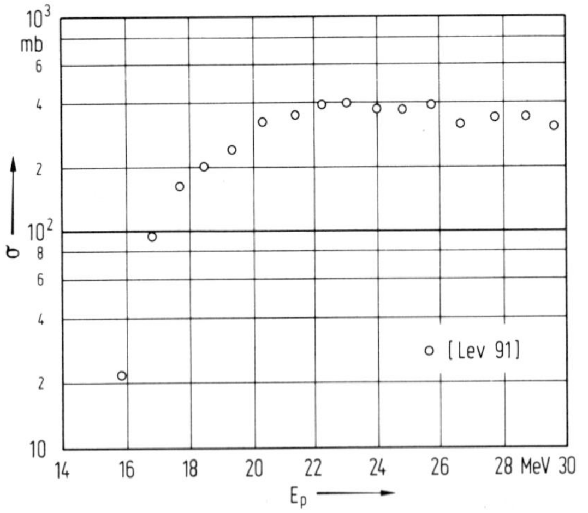

Fig. 375. Reaction $^{86}{}_{38}Sr \rightarrow {}^{85m}{}_{39}Y$

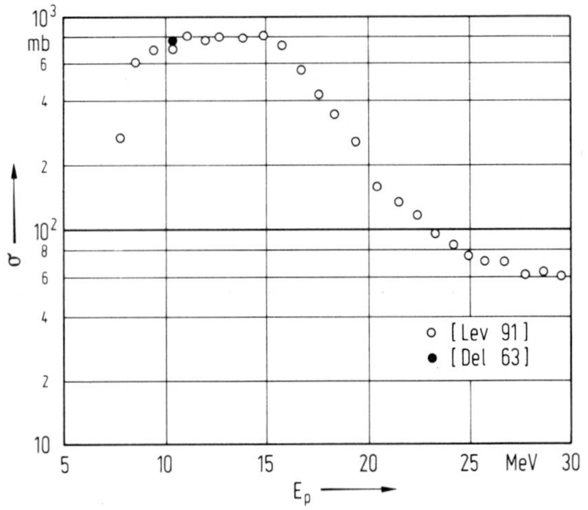

Fig. 376. Reaction $^{86}{}_{38}Sr \rightarrow {}^{86}{}_{39}Y$

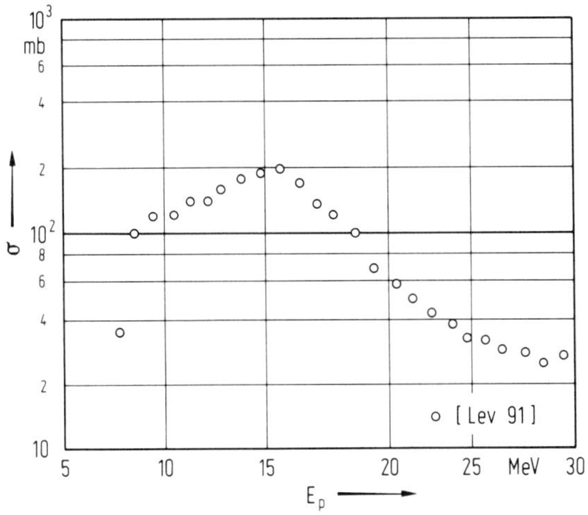

Fig. 377. Reaction ${}^{86}_{38}Sr \rightarrow {}^{86m}_{39}Y$

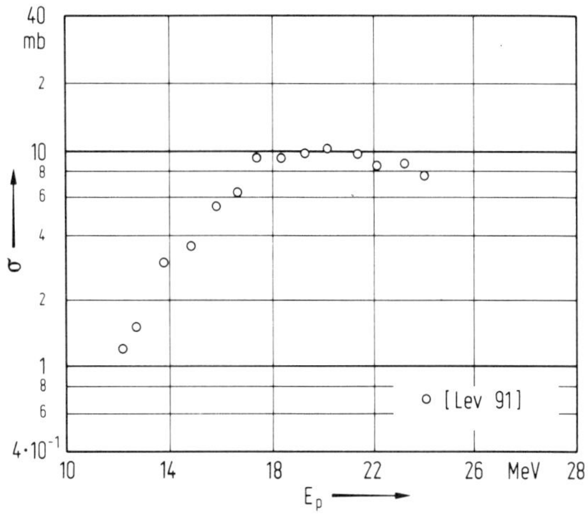

Fig. 378. Reaction ${}^{87}_{38}Sr \rightarrow {}^{84m}_{37}Rb$

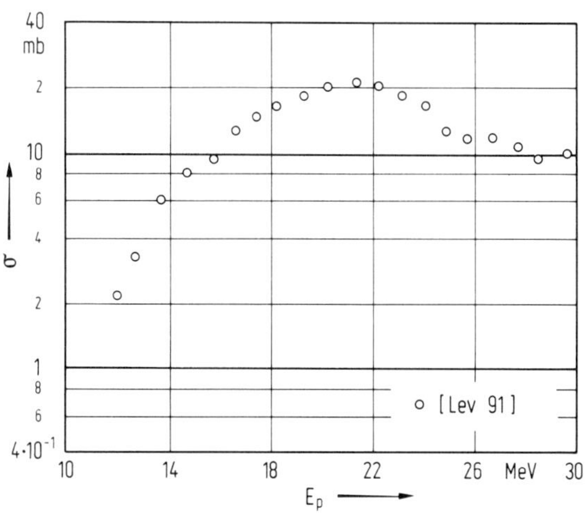

Fig. 379. Reaction ${}^{87}_{38}Sr \rightarrow {}^{84m+g}_{37}Rb$

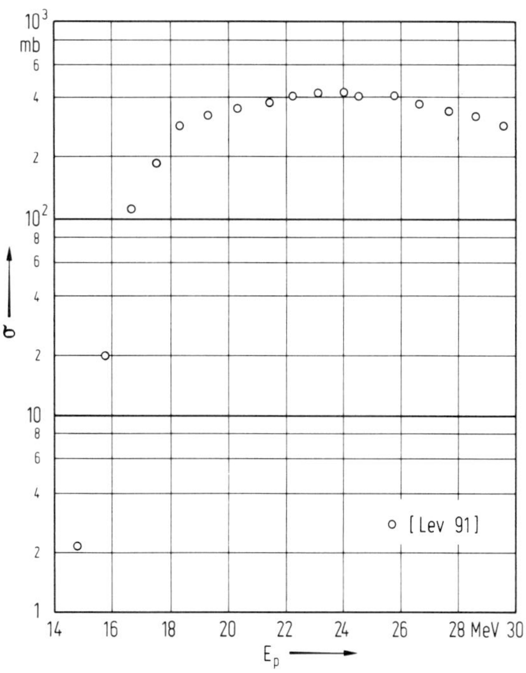

Fig. 381. Reaction ${}^{87}_{38}Sr \rightarrow {}^{86m}_{39}Y$

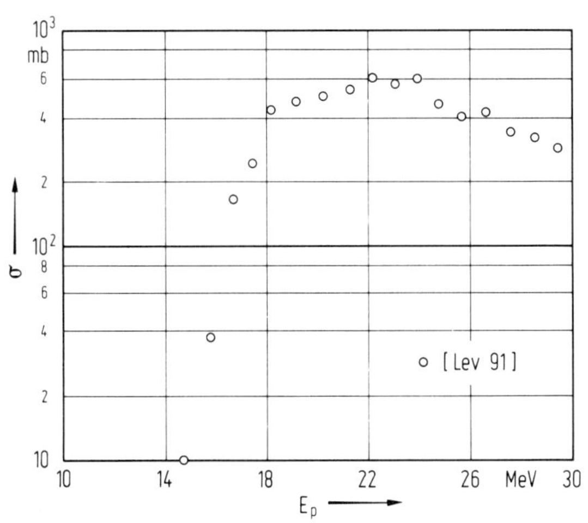

◀ Fig. 380. Reaction ${}^{87}_{38}Sr \rightarrow {}^{86}_{39}Y$

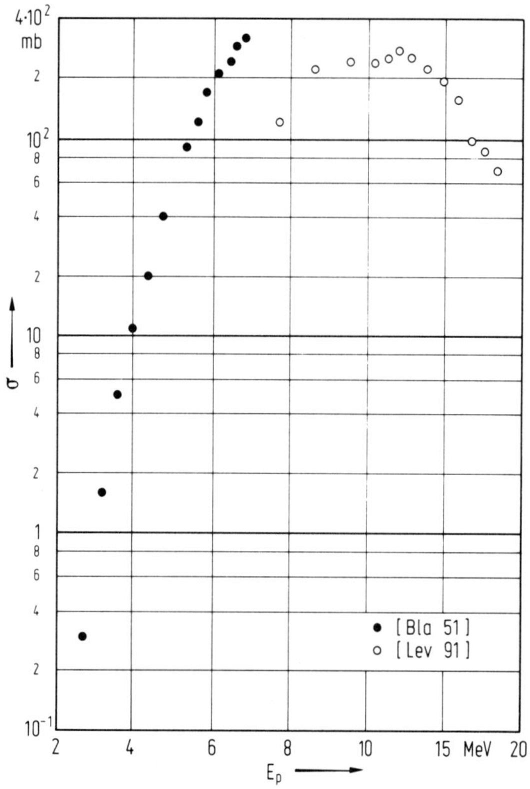

Fig. 382. Reaction $^{87}_{38}Sr \rightarrow ^{87}_{39}Y$

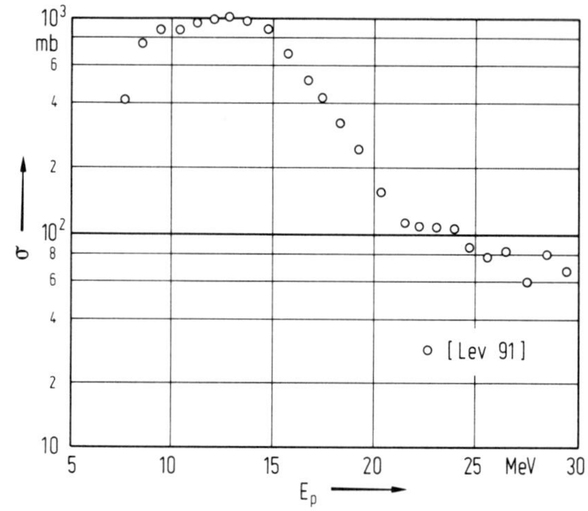

Fig. 383. Reaction $^{87}_{38}Sr \rightarrow ^{87m}_{39}Y$

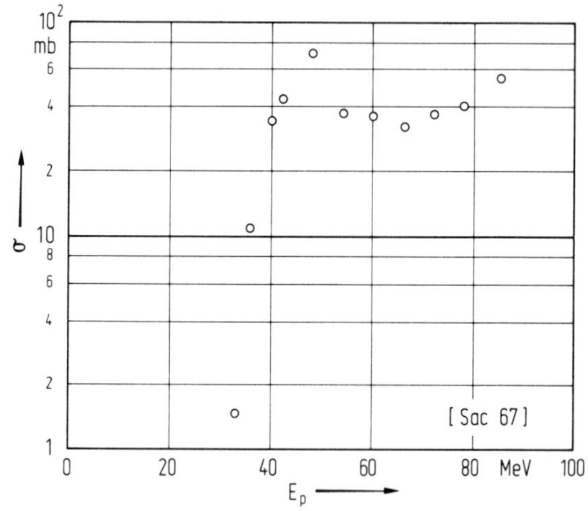

Fig. 384. Reaction $^{88}_{38}Sr \rightarrow ^{83}_{37}Rb$

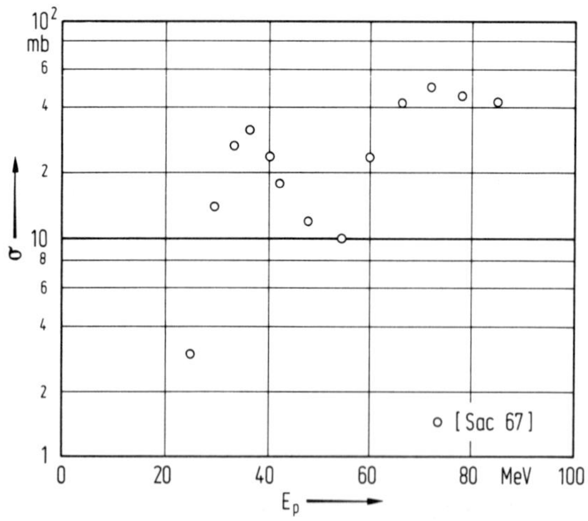

Fig. 385. Reaction $^{88}_{38}Sr \rightarrow ^{84}_{37}Rb$

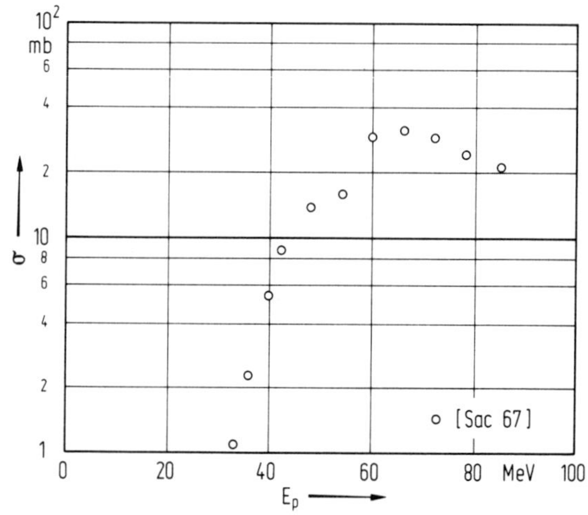

Fig. 386. Reaction $^{88}_{38}Sr \rightarrow ^{86}_{37}Rb$

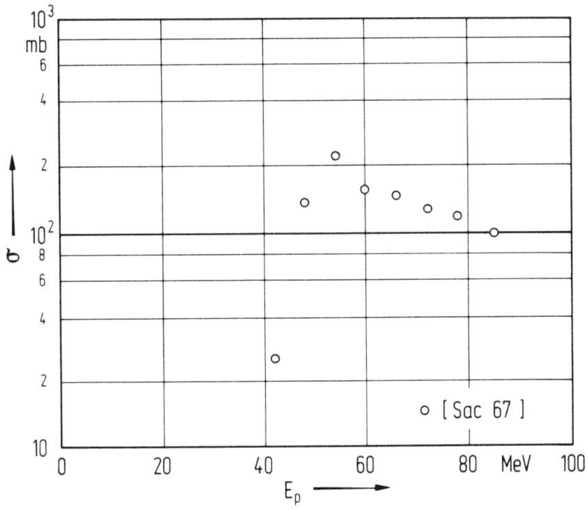

Fig. 387. Reaction $^{88}_{38}Sr \rightarrow ^{85}_{38}Sr$

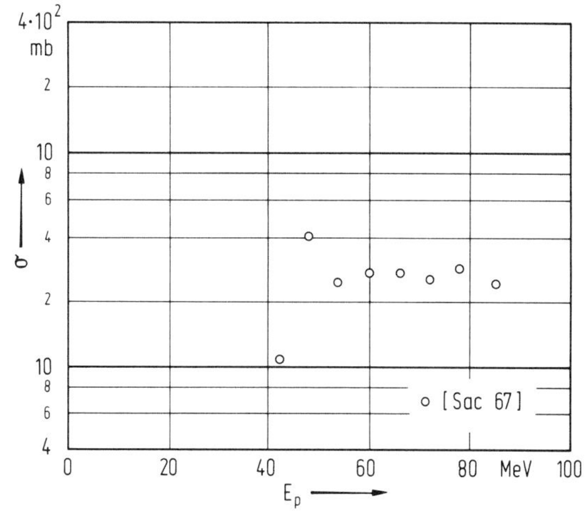

Fig. 388. Reaction $^{88}_{38}Sr \rightarrow ^{85m}_{38}Sr$

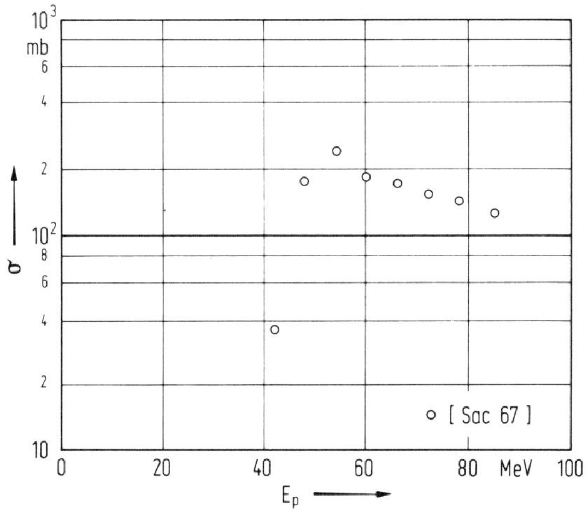

Fig. 389. Reaction $^{88}_{38}Sr \rightarrow ^{85m+g}_{38}Sr$

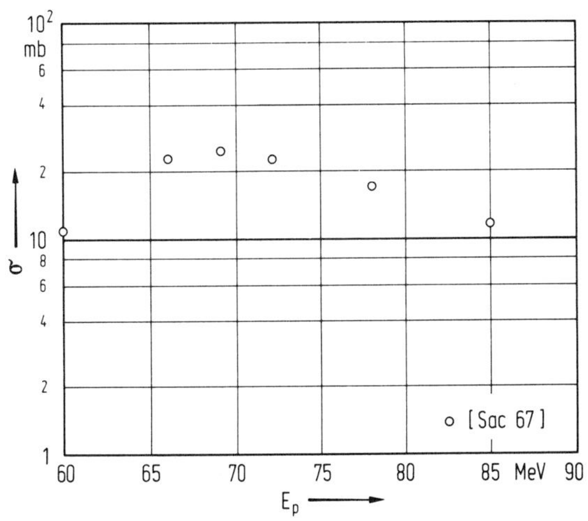

Fig. 390. Reaction $^{88}_{38}Sr \rightarrow ^{84}_{39}Y$

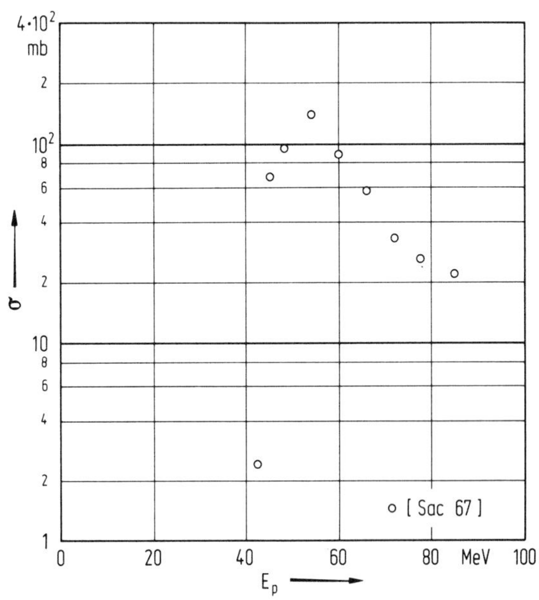

Fig. 391. Reaction $^{88}_{38}Sr \rightarrow ^{85}_{39}Y$

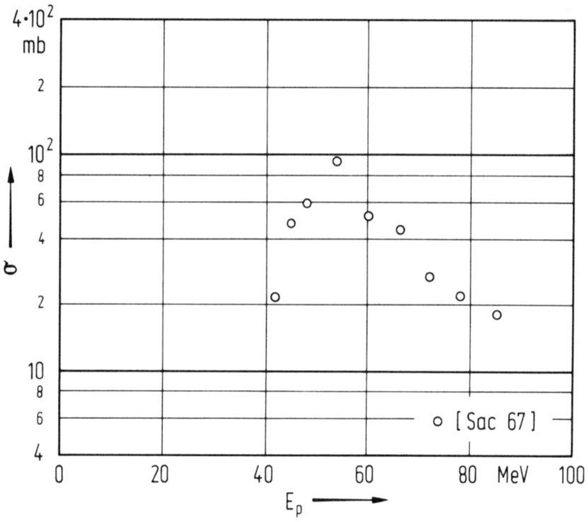

Fig. 392. Reaction $^{88}_{38}Sr \rightarrow {}^{85m}_{39}Y$

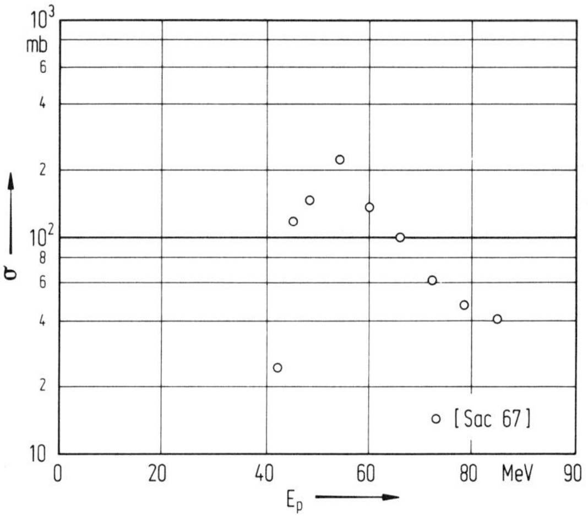

Fig. 393. Reaction $^{88}_{38}Sr \rightarrow {}^{85m+g}_{39}Y$

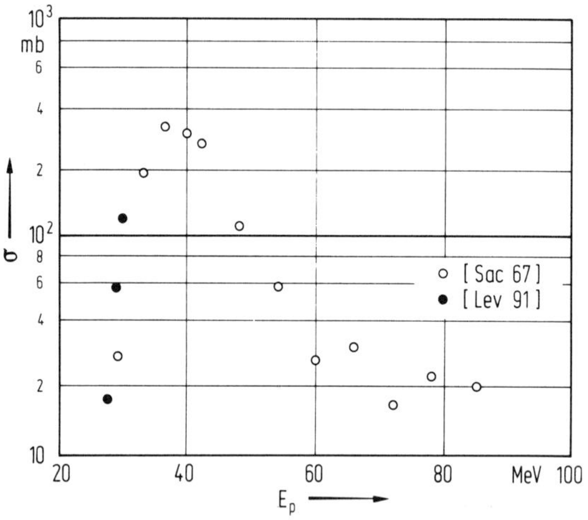

Fig. 394. Reaction $^{88}_{38}Sr \rightarrow {}^{86}_{39}Y$

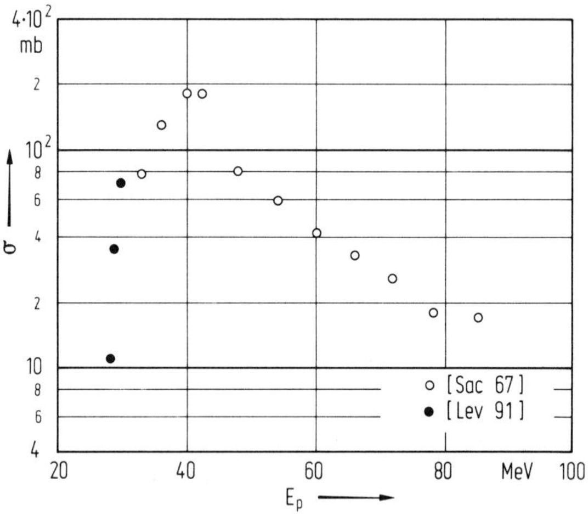

Fig. 395. Reaction $^{88}_{38}Sr \rightarrow {}^{86m}_{39}Y$

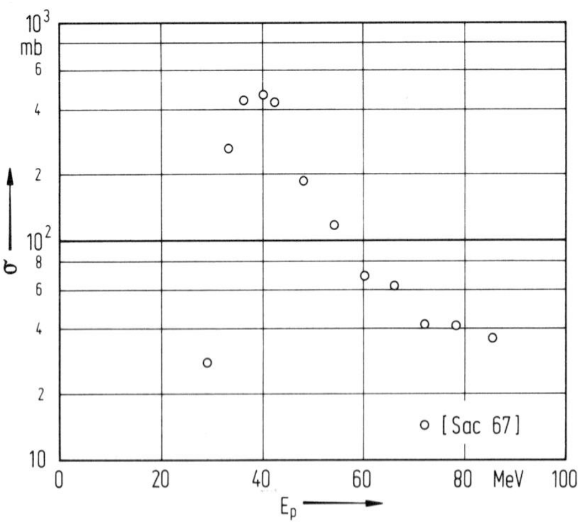

Fig. 396. Reaction $^{88}_{38}Sr \rightarrow {}^{86m+g}_{39}Y$

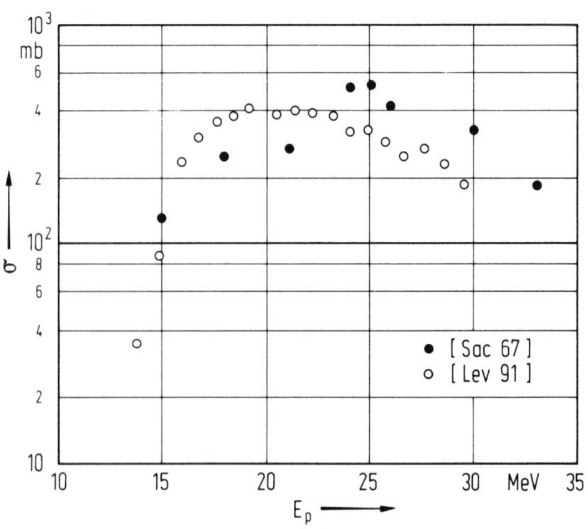

Fig. 397. Reaction $^{88}_{38}Sr \rightarrow {}^{87}_{39}Y$

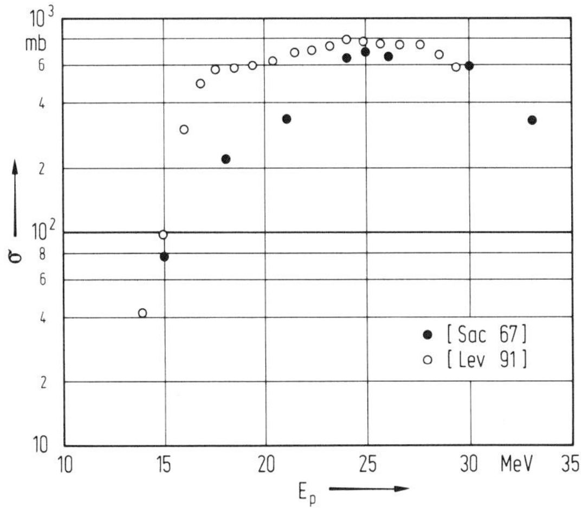

Fig. 398. Reaction $^{88}{}_{38}Sr \rightarrow {}^{87m}{}_{39}Y$

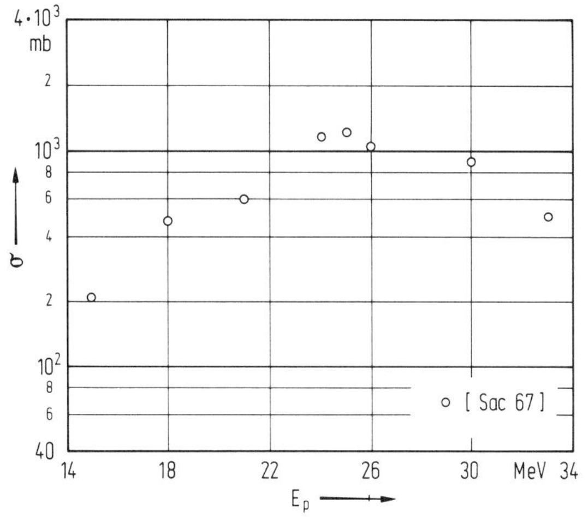

Fig. 399. Reaction $^{88}{}_{38}Sr \rightarrow {}^{87m+g}{}_{39}Y$

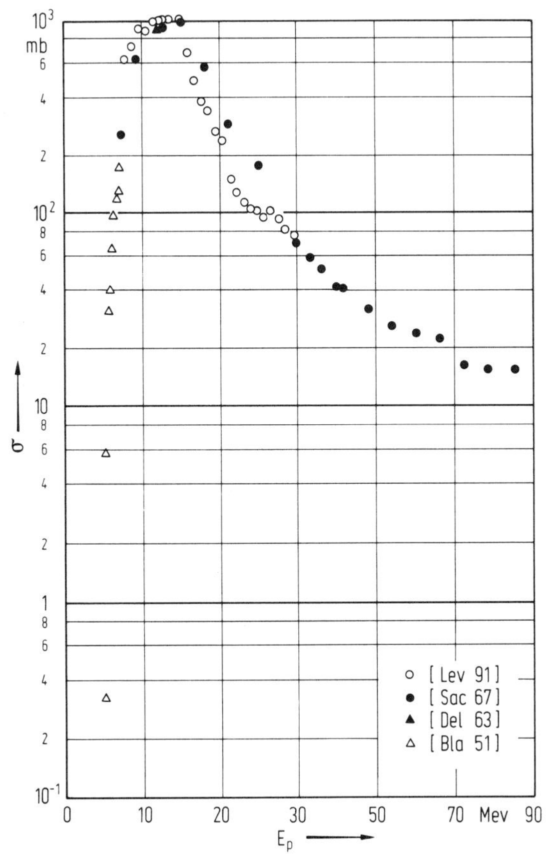

Fig. 400. Reaction $^{88}{}_{38}Sr \rightarrow {}^{88}{}_{39}Y$

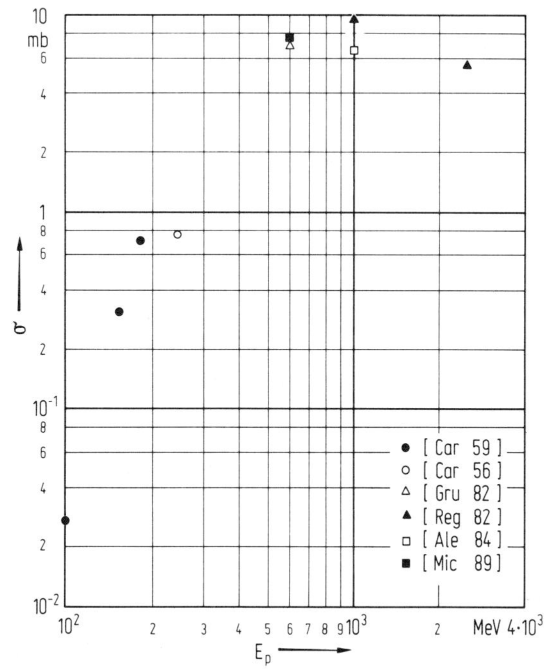

Fig. 401. Reaction $_{39}Y \rightarrow {}^{74}{}_{33}As$ (cumulative)

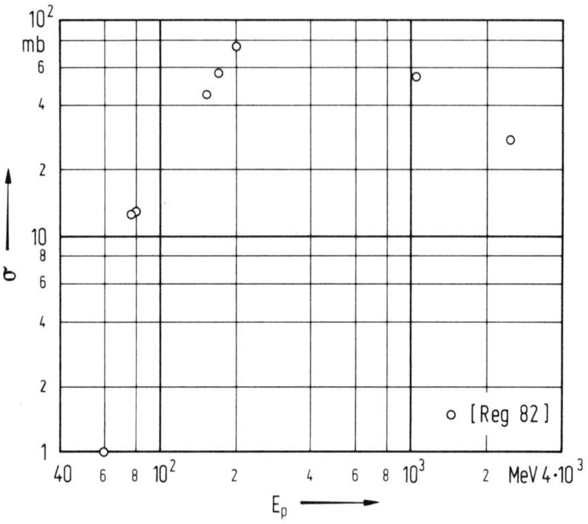

Fig. 402. Reaction $_{39}Y \rightarrow {}^{81}_{36}Kr$

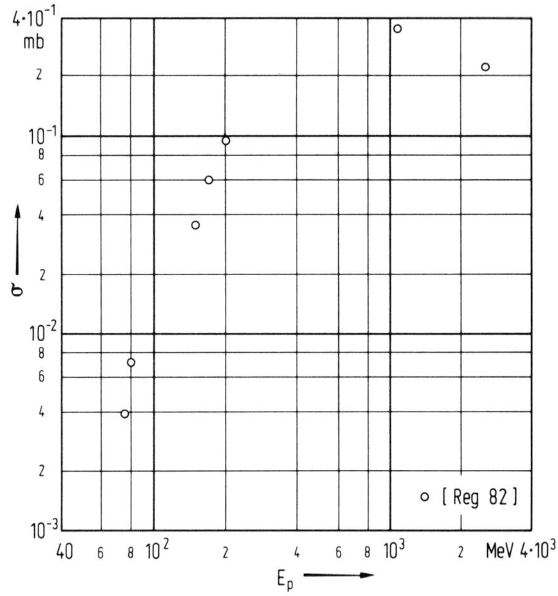

Fig. 403. Reaction $_{39}Y \rightarrow {}^{85}_{36}Kr$

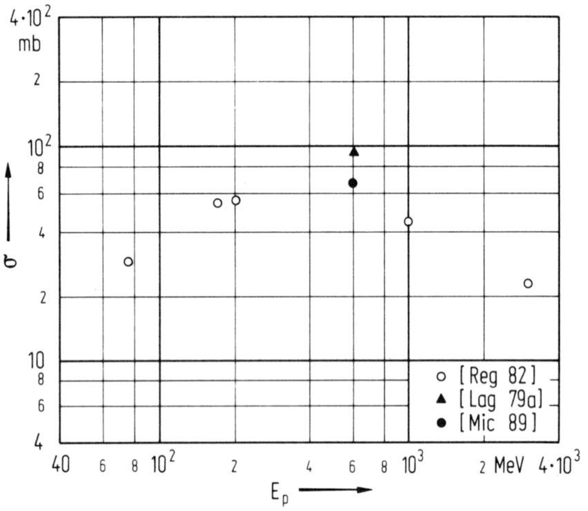

Fig. 404. Reaction $_{39}Y \rightarrow {}^{83}_{37}Rb$ (cumulative)

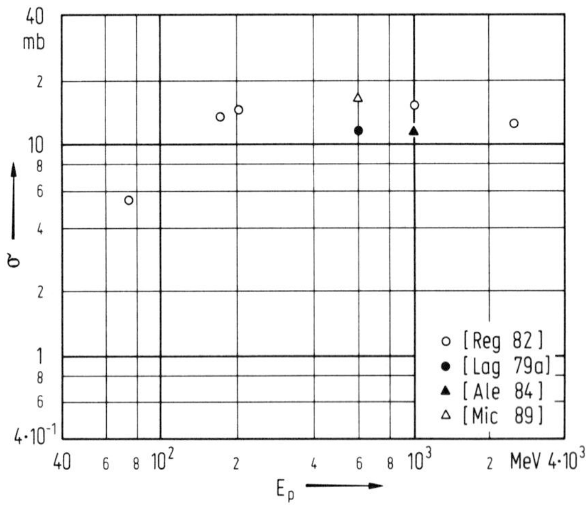

Fig. 405. Reaction $_{39}Y \rightarrow {}^{84}_{37}Rb$

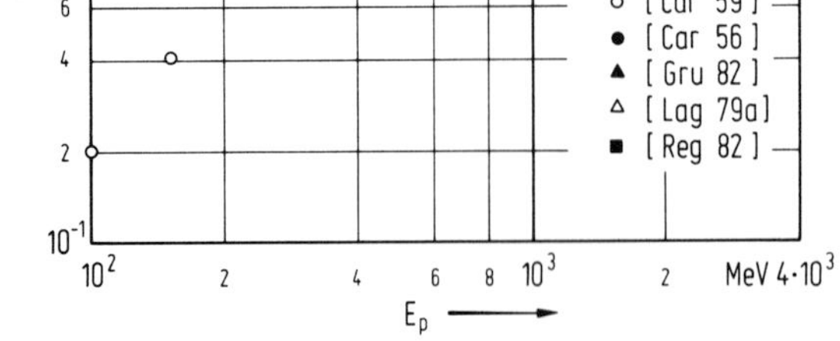

◀ Fig. 406. Reaction $_{39}Y \rightarrow {}^{86}_{37}Rb$

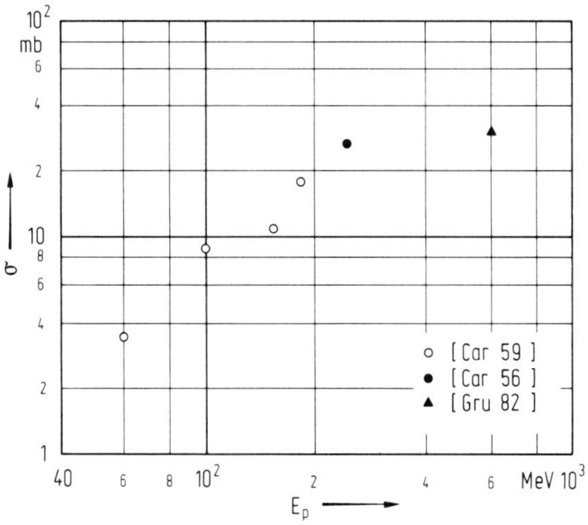

Fig. 407. Reaction $_{39}Y \rightarrow {}^{83}_{38}Sr$

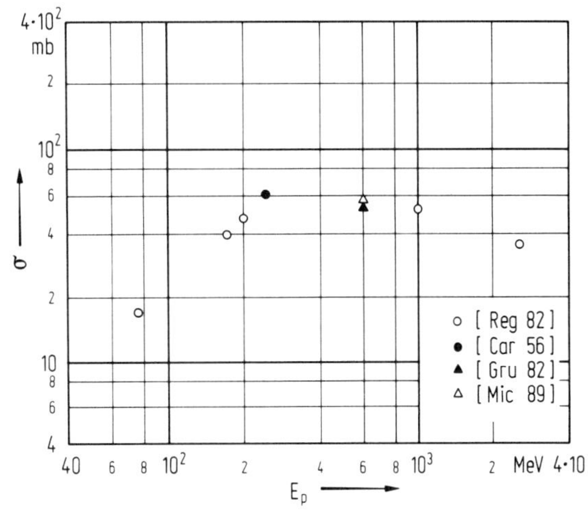

Fig. 408. Reaction $_{39}Y \rightarrow {}^{85}_{38}Sr$

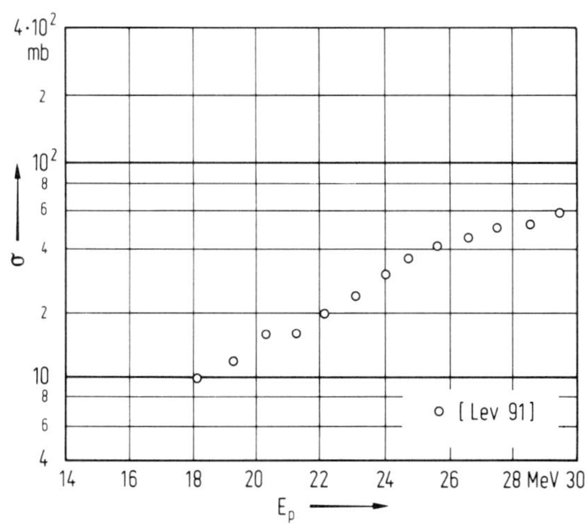

Fig. 409. Reaction $_{39}Y \rightarrow {}^{85m+g}_{38}Sr$

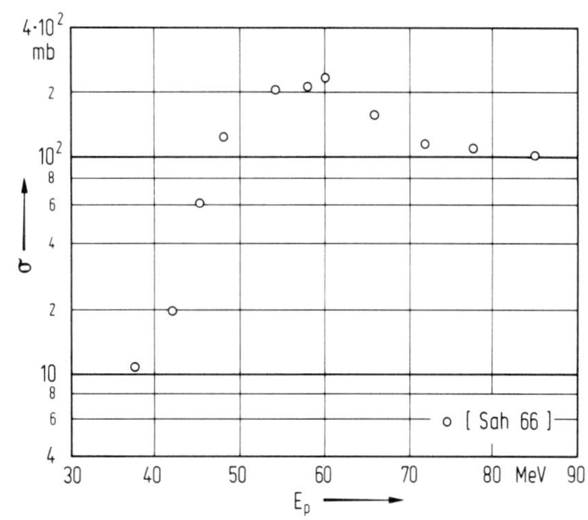

Fig. 410. Reaction $_{39}Y \rightarrow {}^{86m+g}_{39}Y$

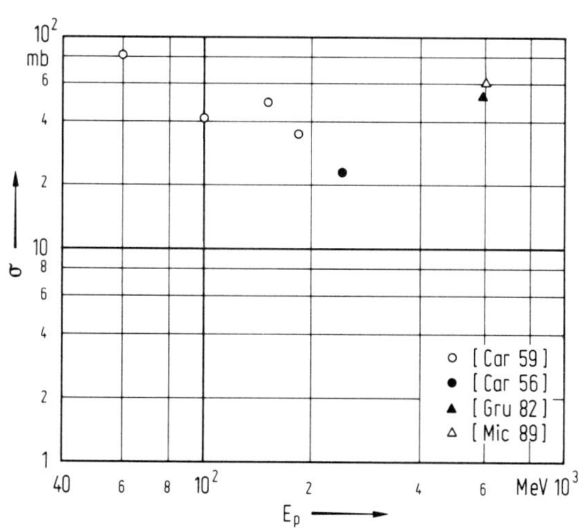

◀ Fig. 411. Reaction $_{39}Y \rightarrow {}^{87}_{39}Y$

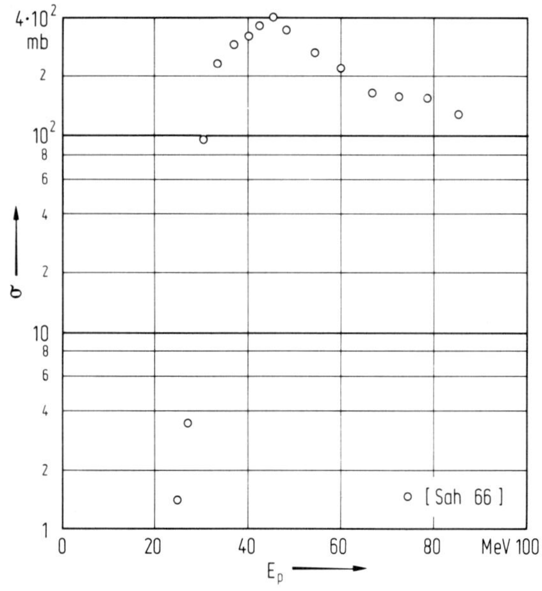

Fig. 412. Reaction $_{39}Y \rightarrow {}^{87m+g}_{39}Y$

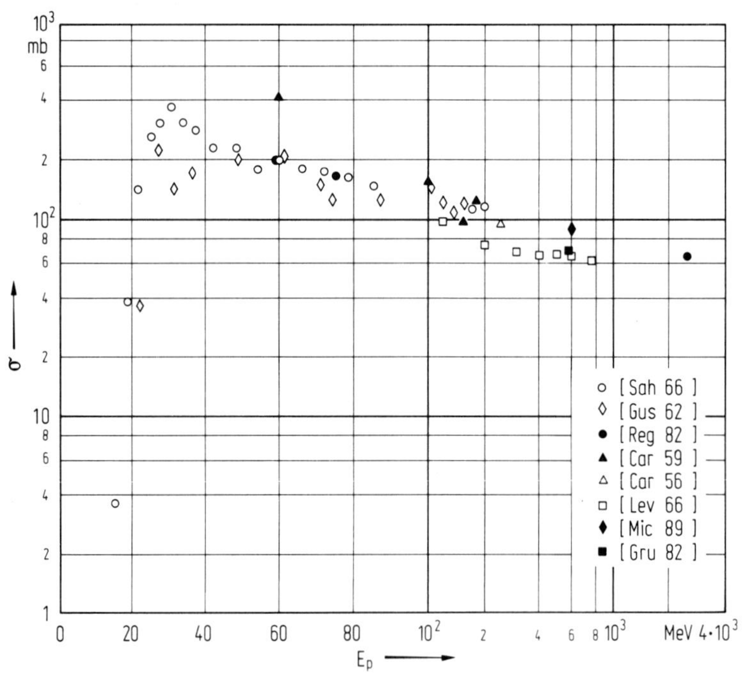

Fig. 413. Reaction $_{39}Y \rightarrow {}^{88}_{39}Y$

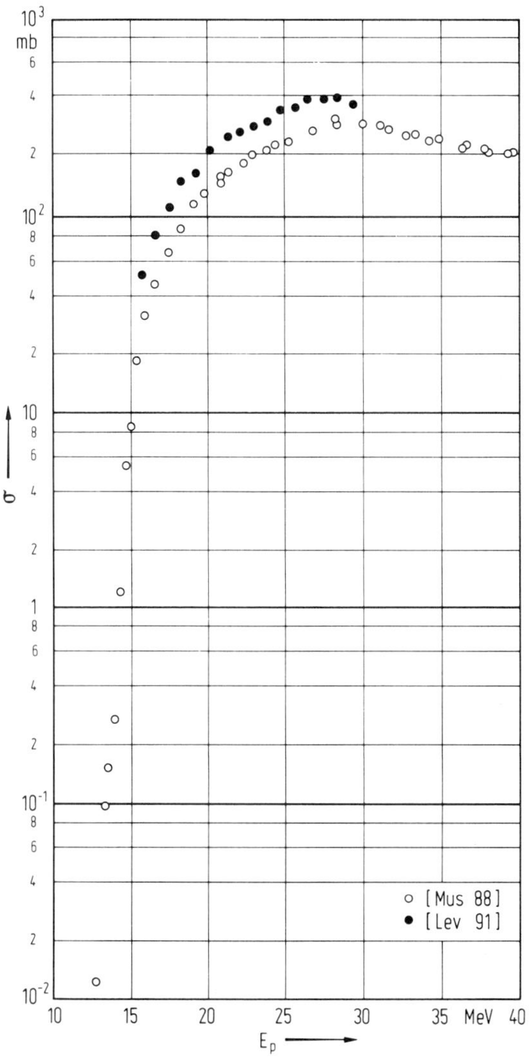

Fig. 414. Reaction $_{39}Y \rightarrow {}^{88}{}_{39}Y$

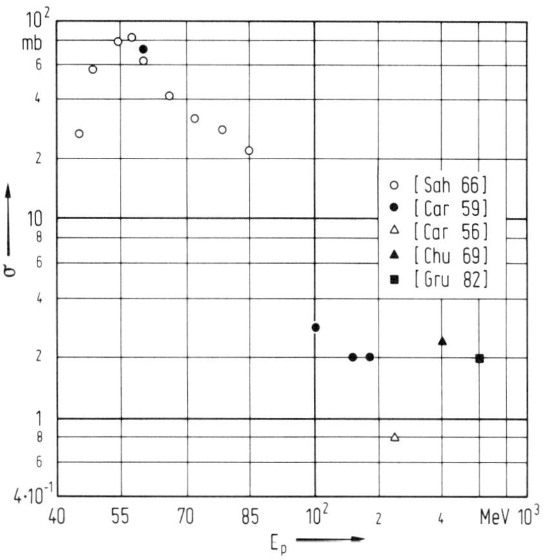

Fig. 415. Reaction $_{39}Y \rightarrow {}^{86}{}_{40}Zr$

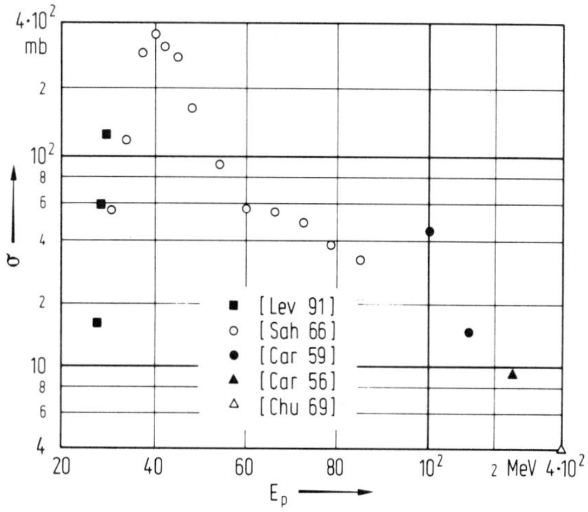

Fig. 416. Reaction $_{39}Y \rightarrow {}^{87}{}_{40}Zr$

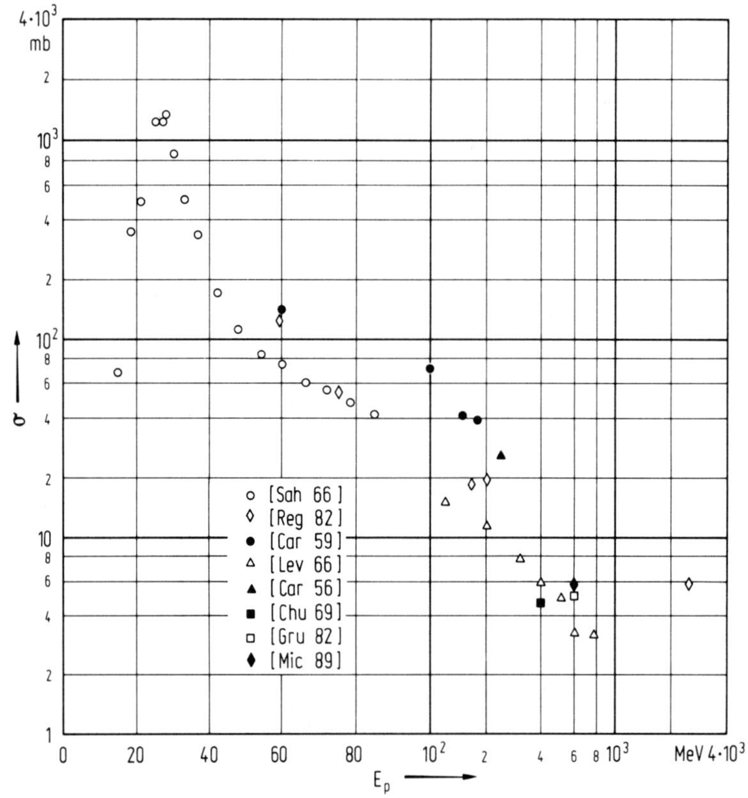

Fig. 417. Reaction ${}_{39}Y \rightarrow {}^{88}{}_{40}Zr$

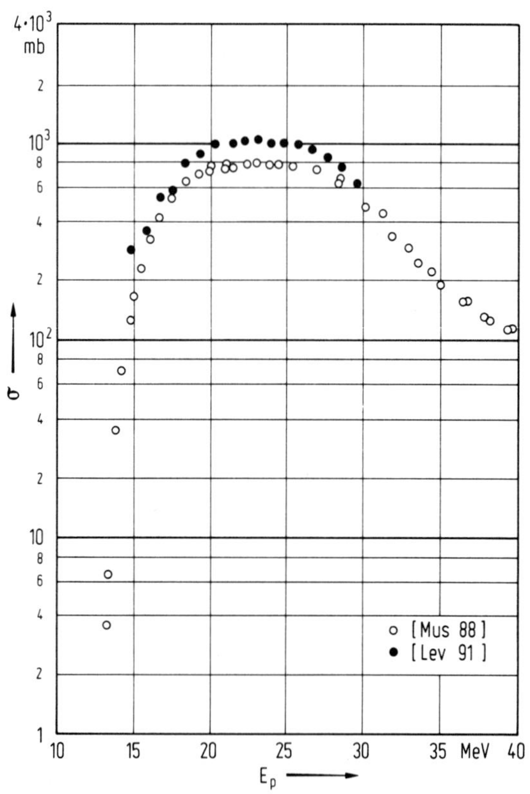

◀ Fig. 418. Reaction ${}_{39}Y \rightarrow {}^{88}{}_{40}Zr$

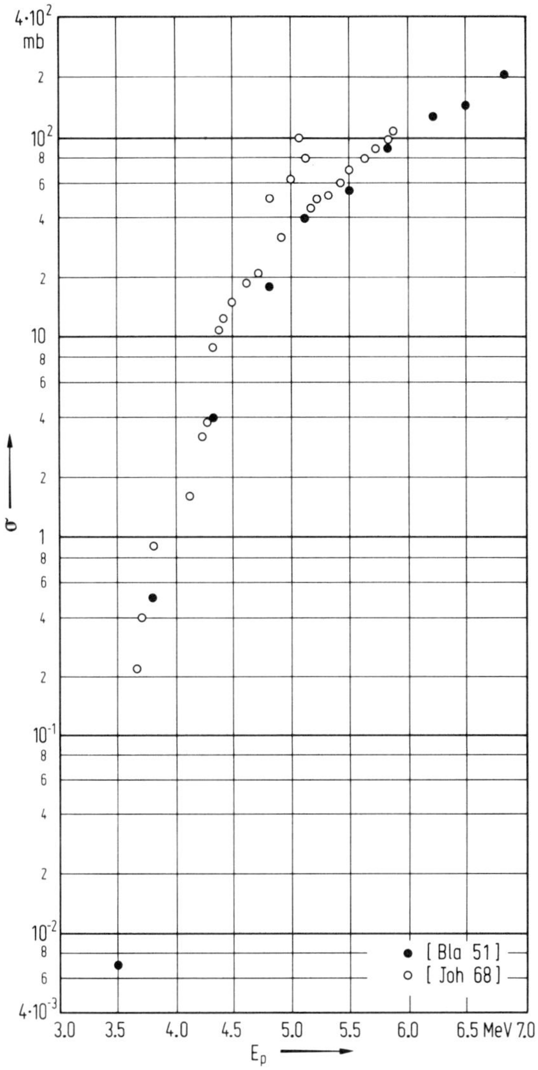

Fig. 419. Reaction $_{39}Y \rightarrow {}^{89}_{40}Zr$

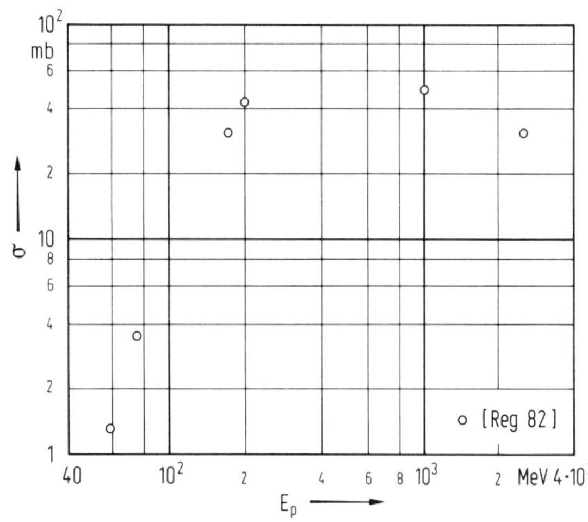

Fig. 422. Reaction $_{40}Zr \rightarrow {}^{81}_{36}Kr$

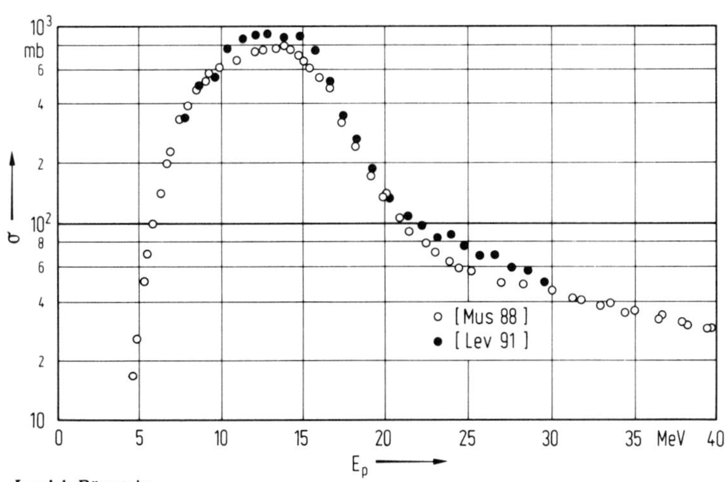

Fig. 420, see next page

Fig. 421. Reaction $_{39}Y \rightarrow {}^{89}_{40}Zr$

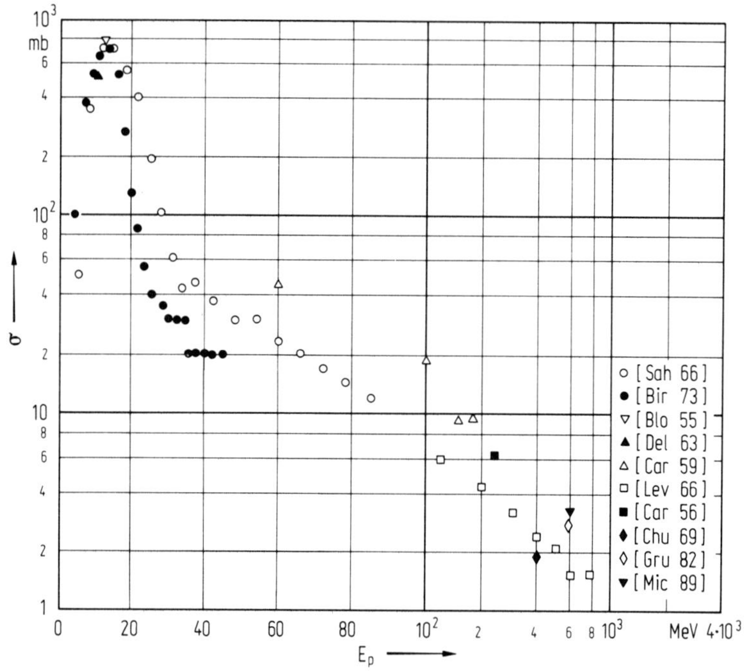

Fig. 420. Reaction $_{39}Y \rightarrow {}^{89}{}_{40}Zr$

Fig. 424. Reaction $_{40}Zr \rightarrow {}^{83}{}_{37}Rb$ ▶

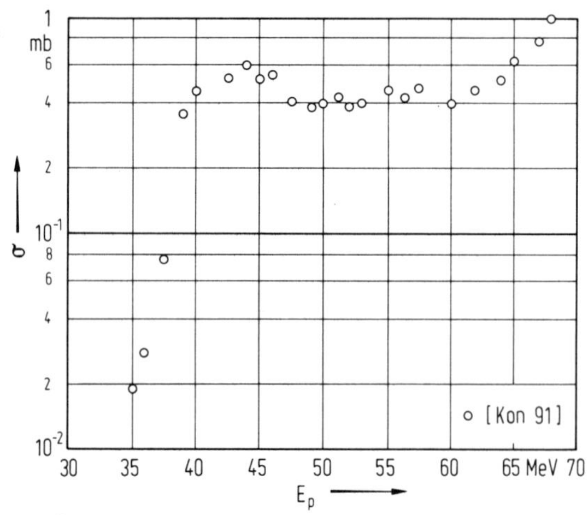

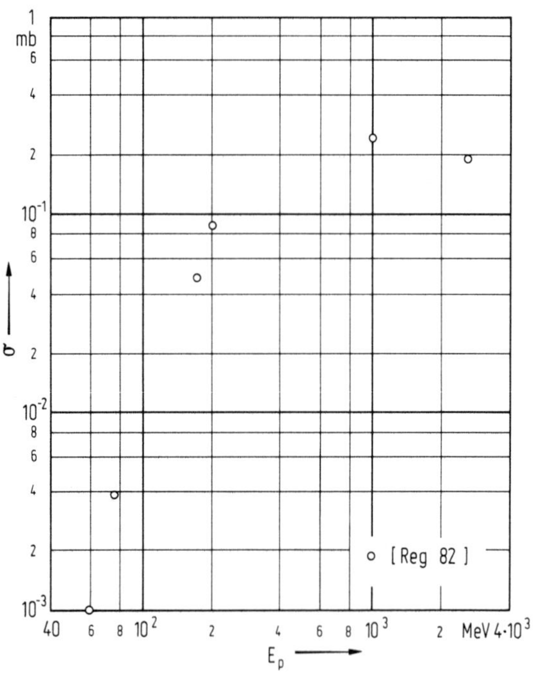

Fig. 423. Reaction $_{40}Zr \rightarrow {}^{85}{}_{36}Kr$

Fig. 425. Reaction $_{40}Zr \rightarrow {}^{83}{}_{37}Rb$ (cumulative) ▶

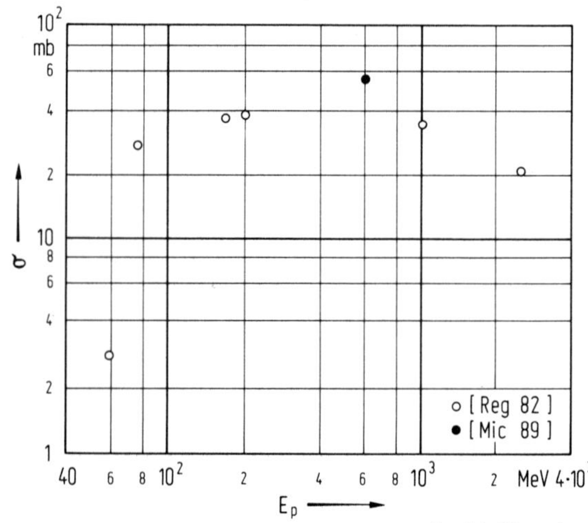

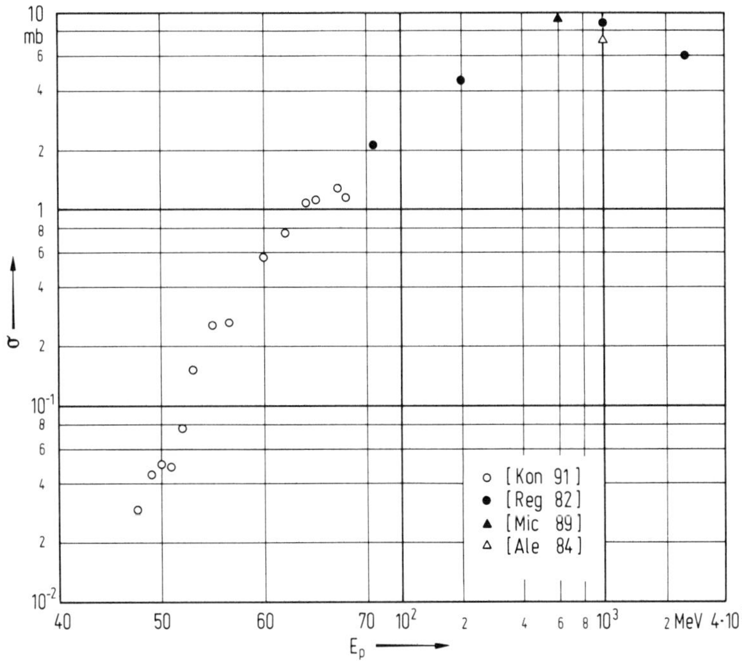

Fig. 426. Reaction $_{40}Zr \rightarrow {}^{84}{}_{37}Rb$

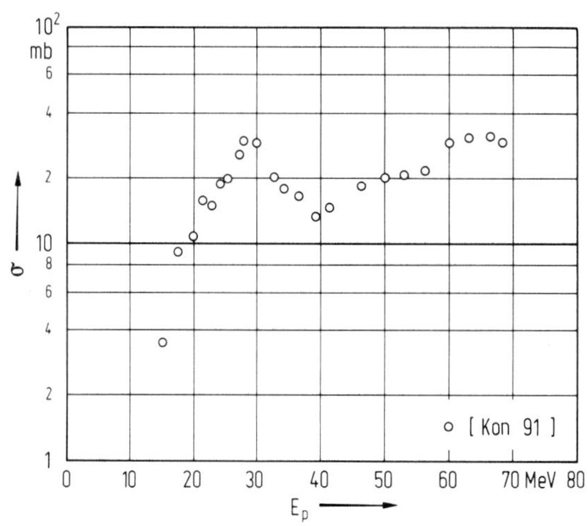

◀ Fig. 427. Reaction $_{40}Zr \rightarrow {}^{86}{}_{39}Y$

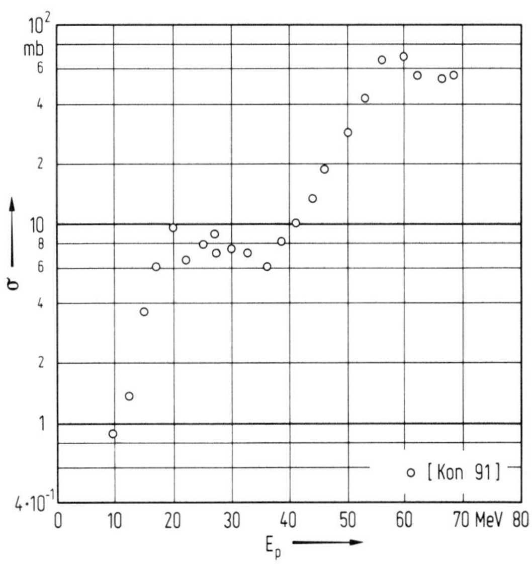

Fig. 428. Reaction $_{40}Zr \rightarrow {}^{87}{}_{39}Y$

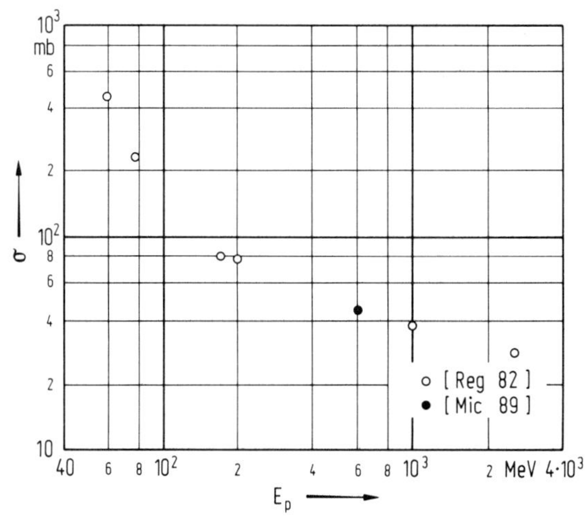

◀ Fig. 429. Reaction $_{40}Zr \rightarrow {}^{88}{}_{40}Zr$ (cumulative)

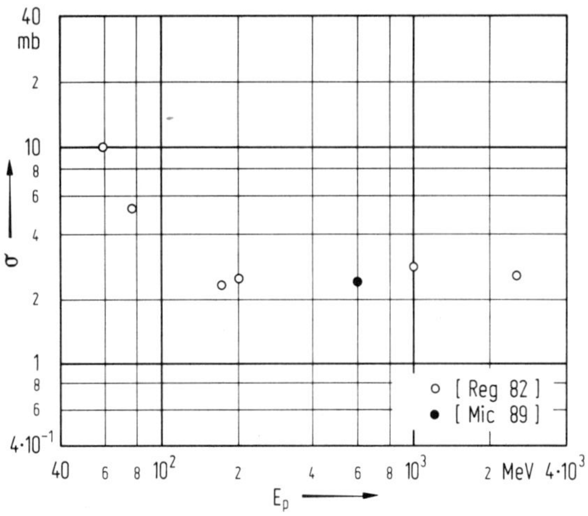

Fig. 430. Reaction $_{40}Zr \rightarrow {}^{95}_{40}Zr$ (cumulative)

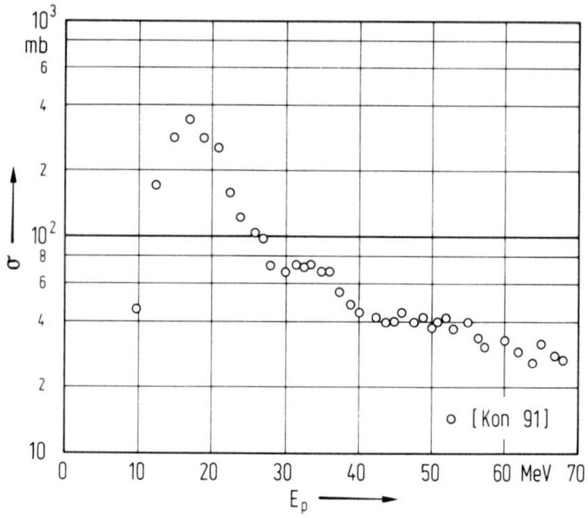

Fig. 431. Reaction $_{40}Zr \rightarrow {}^{90}_{41}Nb$

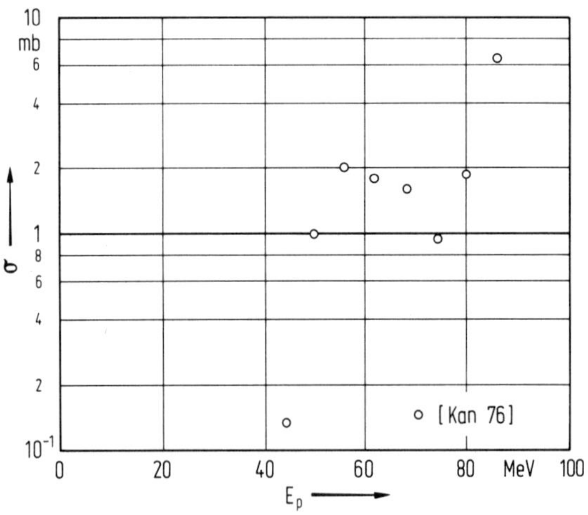

◀ Fig. 432. Reaction ${}^{90}_{40}Zr \rightarrow {}^{82m}_{37}Rb$

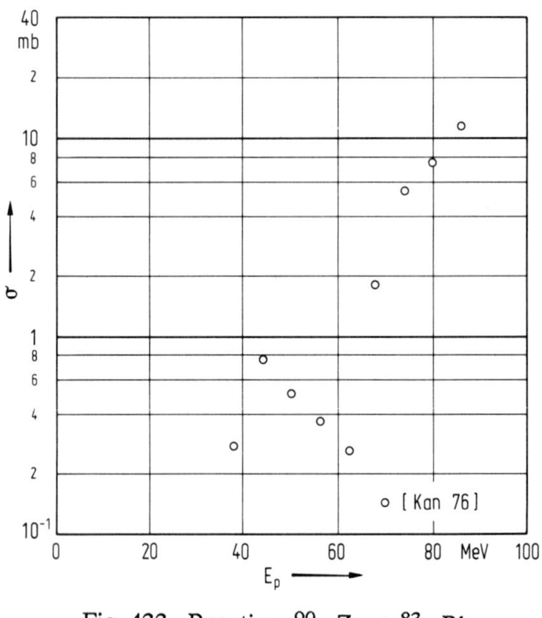

Fig. 433. Reaction ${}^{90}_{40}Zr \rightarrow {}^{83}_{37}Rb$

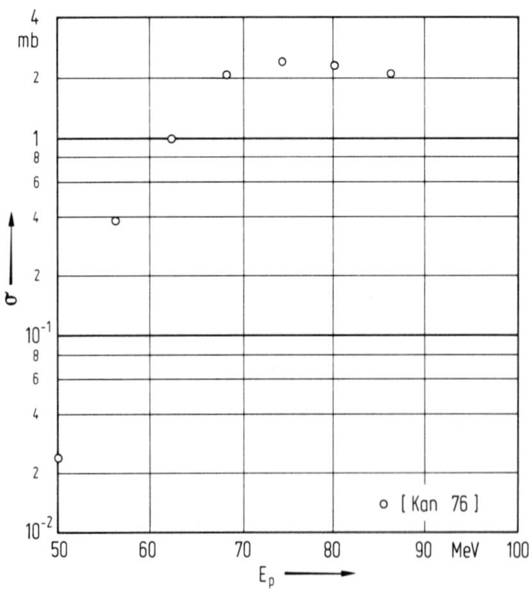

◀ Fig. 434. Reaction ${}^{90}_{40}Zr \rightarrow {}^{84}_{37}Rb$

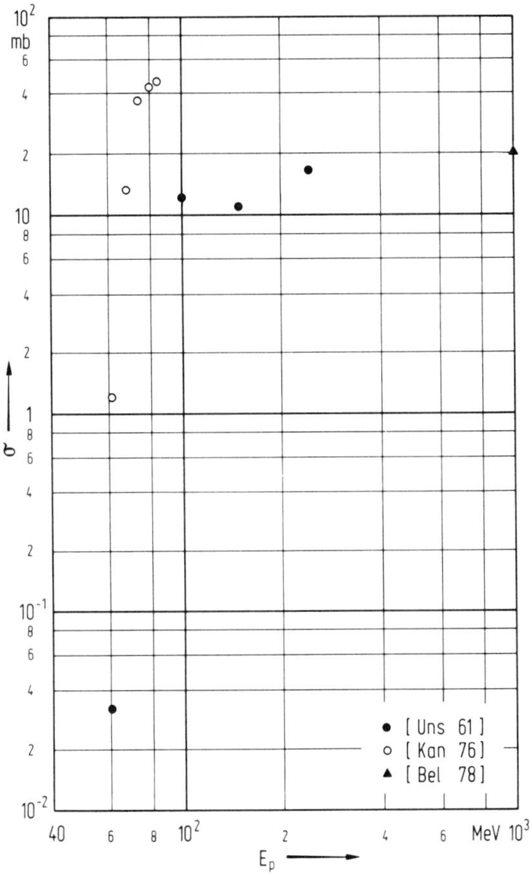

Fig. 435. Reaction ${}^{90}_{40}Zr \rightarrow {}^{83}_{38}Sr$ (cumulative)

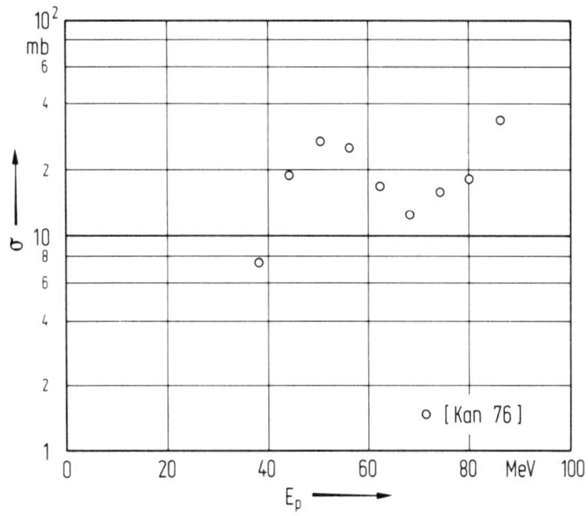

Fig. 436. Reaction ${}^{90}_{40}Zr \rightarrow {}^{85}_{38}Sr$

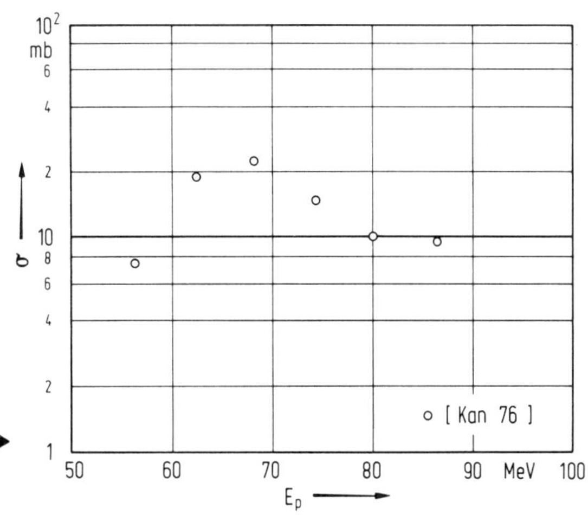

Fig. 437. Reaction ${}^{90}_{40}Zr \rightarrow {}^{84}_{39}Y$ ▶

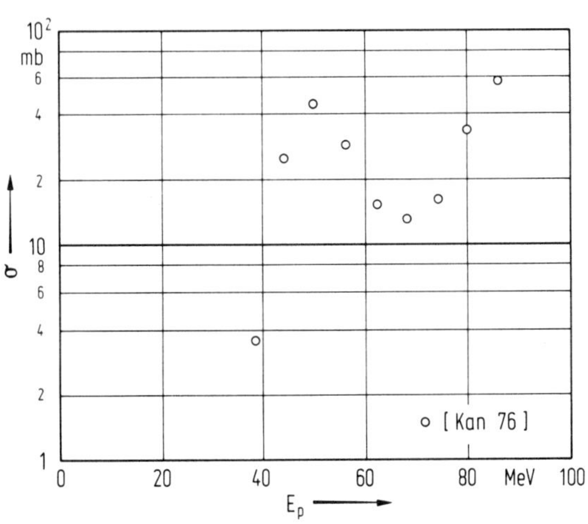

Fig. 438. Reaction ${}^{90}_{40}Zr \rightarrow {}^{85}_{39}Y$

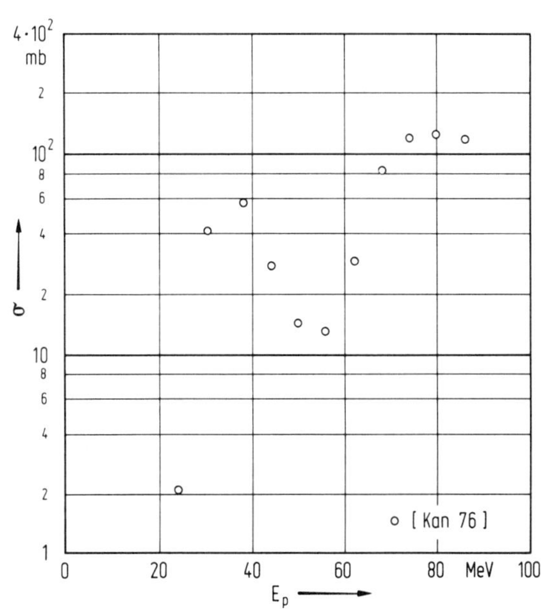

Fig. 439. Reaction ${}^{90}_{40}Zr \rightarrow {}^{86}_{39}Y$ ▶

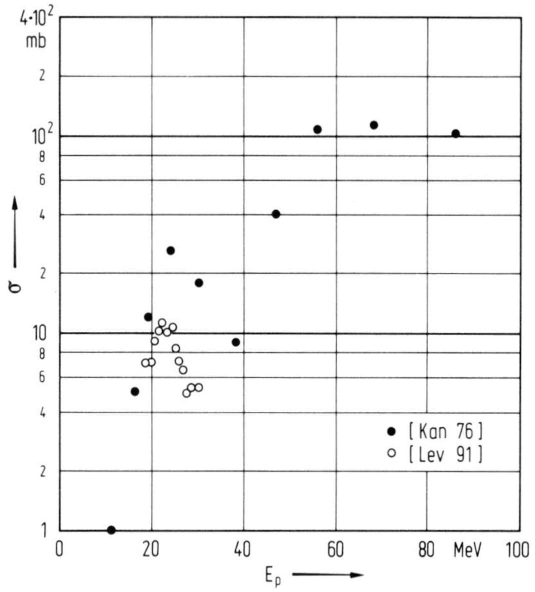

Fig. 440. Reaction $^{90}_{40}Zr \rightarrow ^{87}_{39}Y$

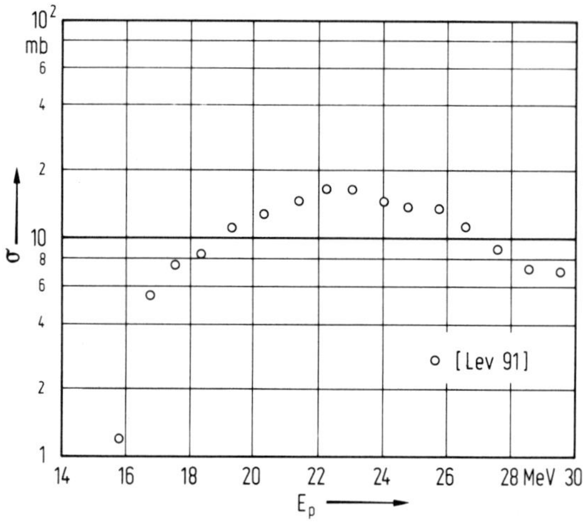

Fig. 441. Reaction $^{90}_{40}Zr \rightarrow ^{87m}_{39}Y$

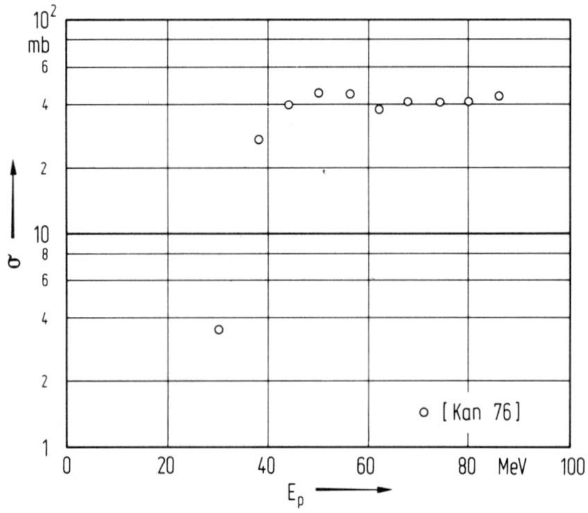

Fig. 442. Reaction $^{90}_{40}Zr \rightarrow ^{88}_{39}Y$

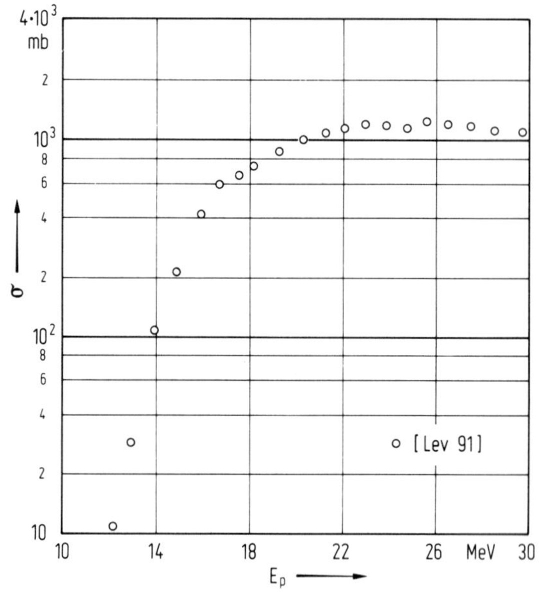

Fig. 443. Reaction $^{90}_{40}Zr \rightarrow ^{89}_{40}Zr$

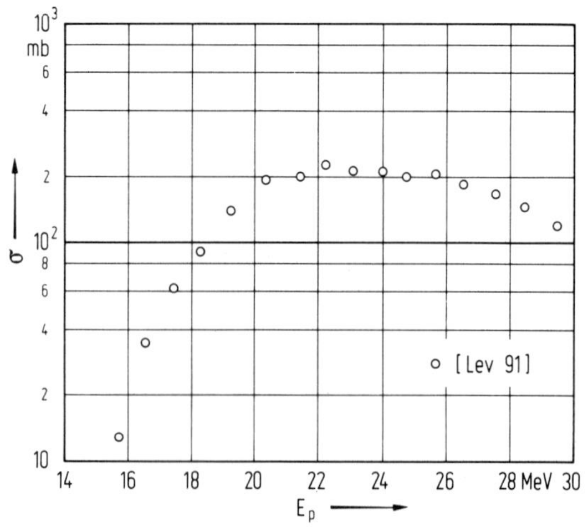

◀ Fig. 444. Reaction $^{90}_{40}Zr \rightarrow ^{89m}_{41}Nb$

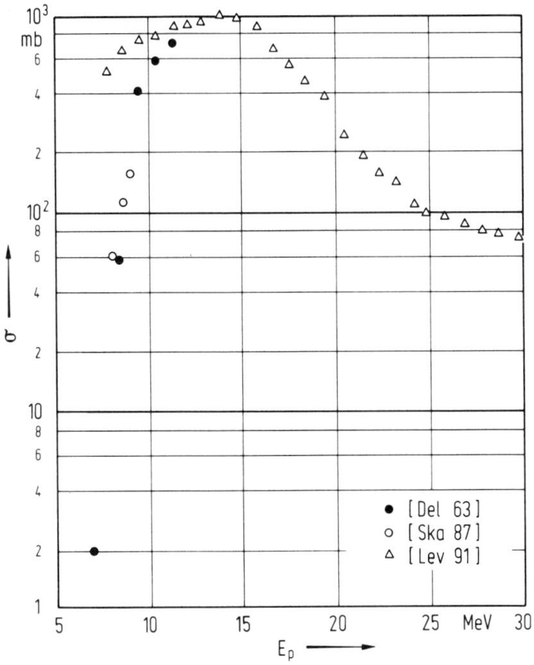

◀ Fig. 445. Reaction ${}^{90}_{40}Zr \rightarrow {}^{90}_{41}Nb$

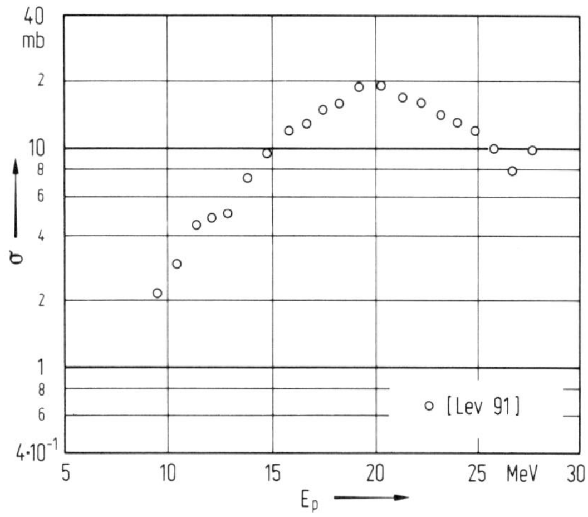

Fig. 446. Reaction ${}^{91}_{40}Zr \rightarrow {}^{88}_{39}Y$

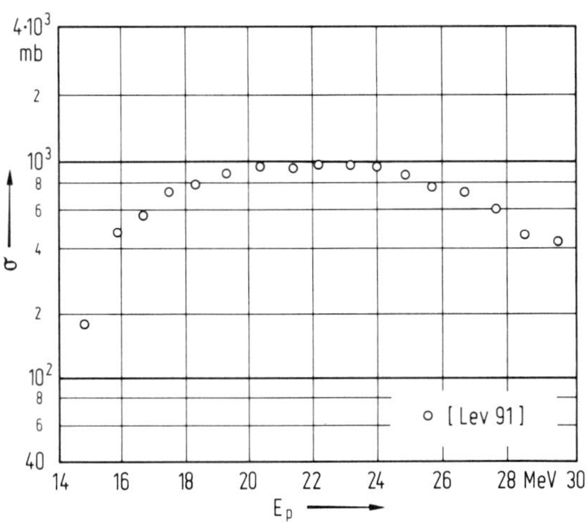

Fig. 447. Reaction ${}^{91}_{40}Zr \rightarrow {}^{90}_{41}Nb$

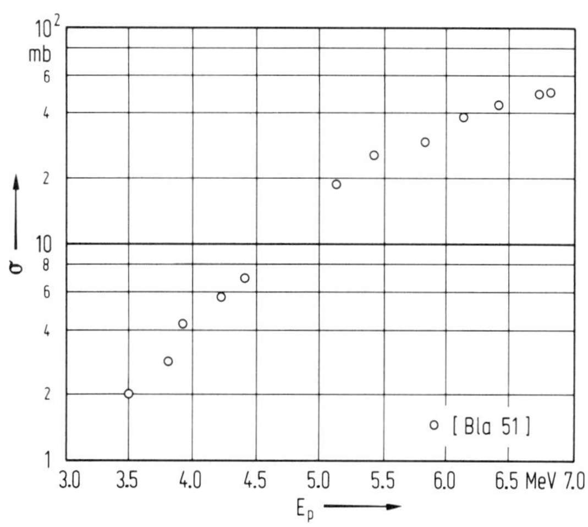

Fig. 448. Reaction ${}^{91}_{40}Zr \rightarrow {}^{91}_{41}Nb$

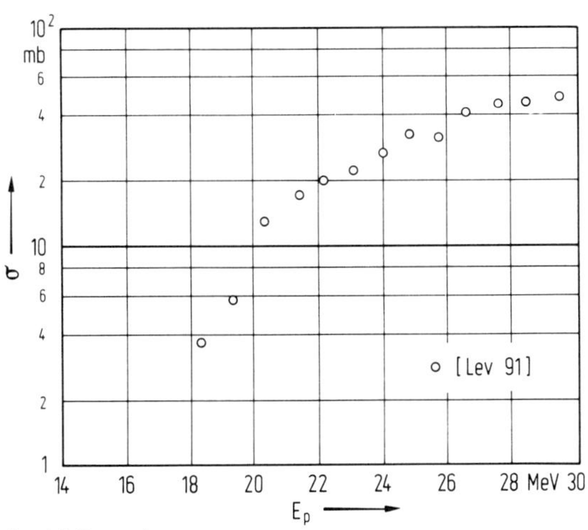

◀ Fig. 449. Reaction ${}^{92}_{40}Zr \rightarrow {}^{88}_{39}Y$

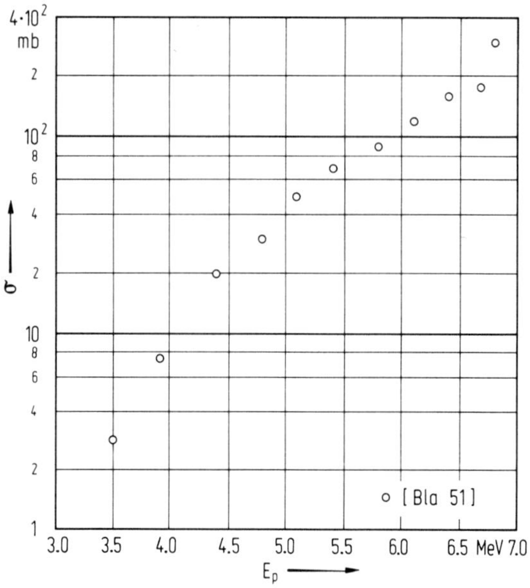

Fig. 450. Reaction $^{92}_{40}Zr \rightarrow {}^{92}_{41}Nb$

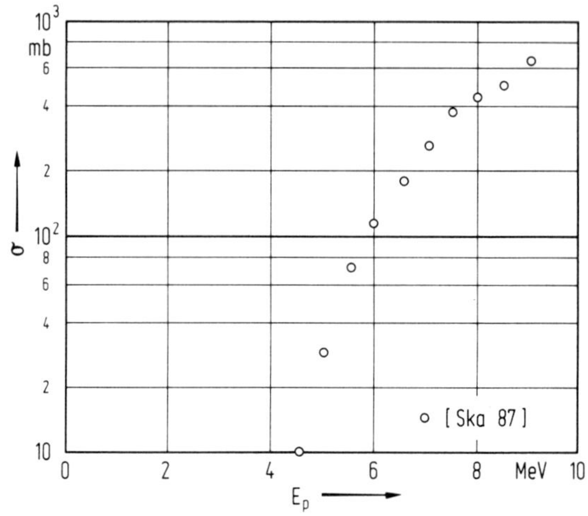

Fig. 451. Reaction $^{92}_{40}Zr \rightarrow {}^{92m}_{41}Nb$

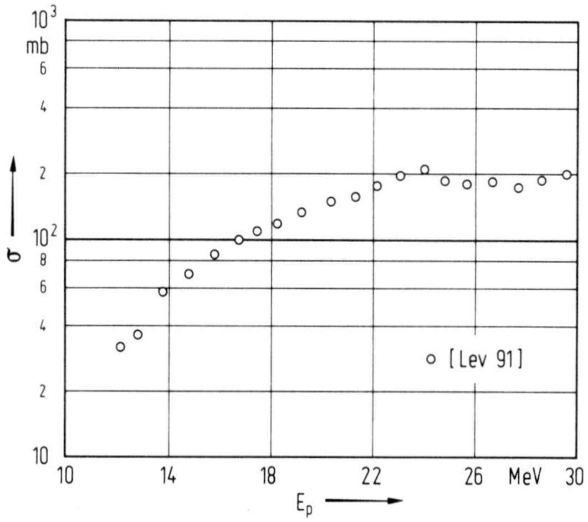

Fig. 452. Reaction $^{96}_{40}Zr \rightarrow {}^{95}_{40}Zr$

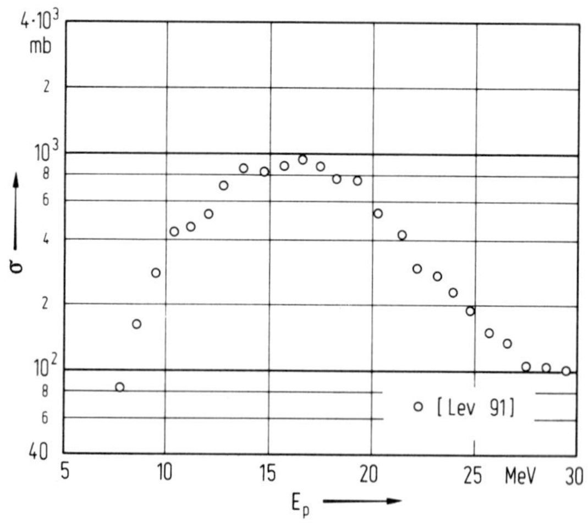

Fig. 453. Reaction $^{96}_{40}Zr \rightarrow {}^{95}_{41}Nb$

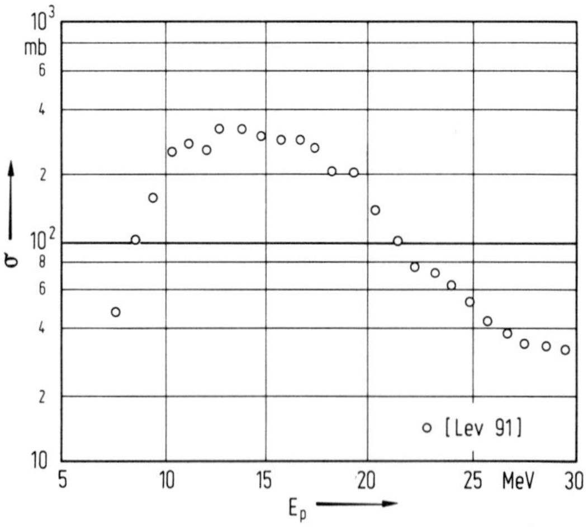

◀ Fig. 454. Reaction $^{96}_{40}Zr \rightarrow {}^{95m}_{41}Nb$

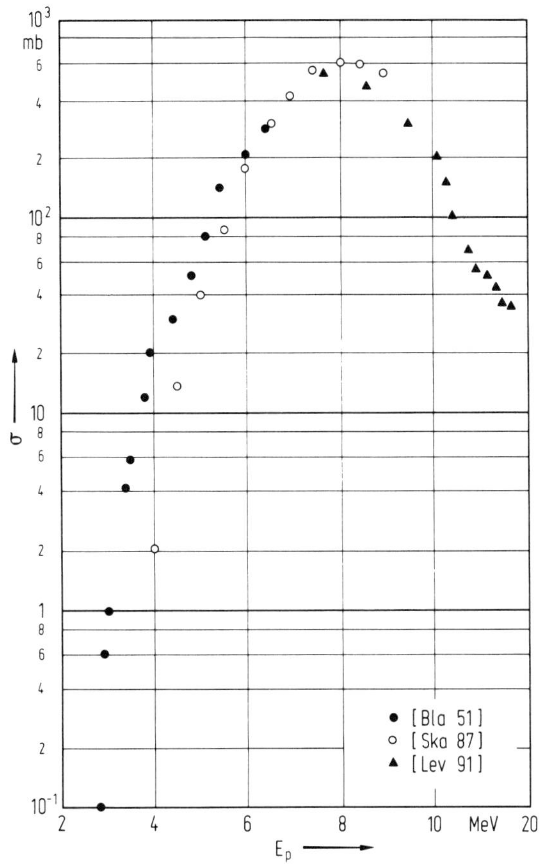

Fig. 455. Reaction $^{96}_{40}Zr \rightarrow ^{96}_{41}Nb$

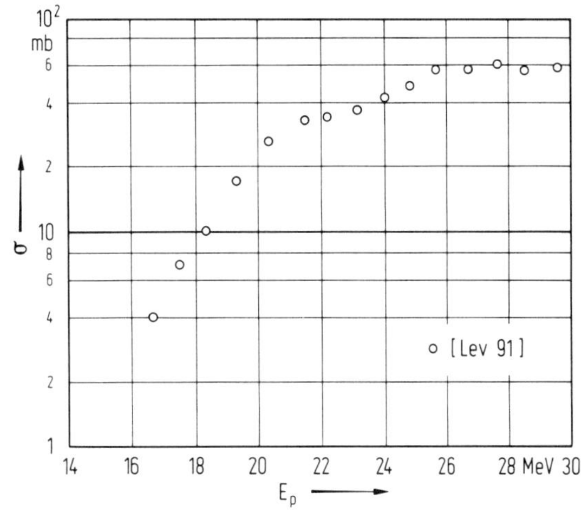

Fig. 456. Reaction $_{41}Nb \rightarrow ^{89}_{40}Zr$

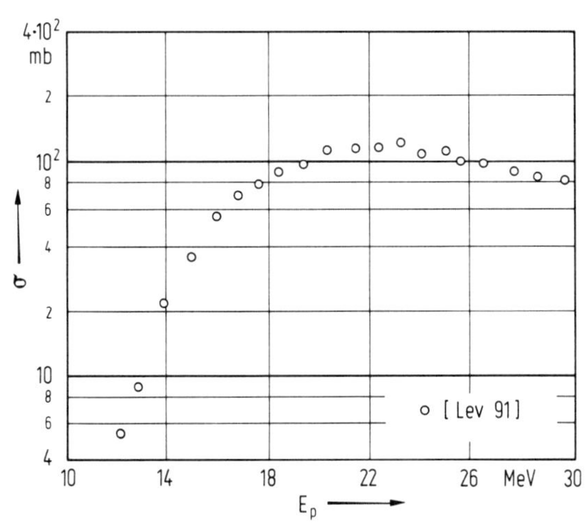

Fig. 457. Reaction $_{41}Nb \rightarrow ^{92m}_{41}Nb$

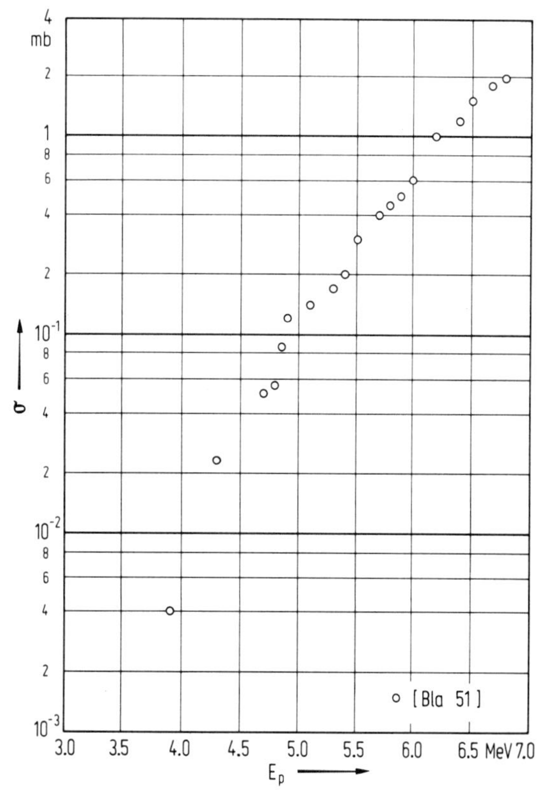

Fig. 458. Reaction $_{41}Nb \rightarrow ^{93}_{42}Mo$

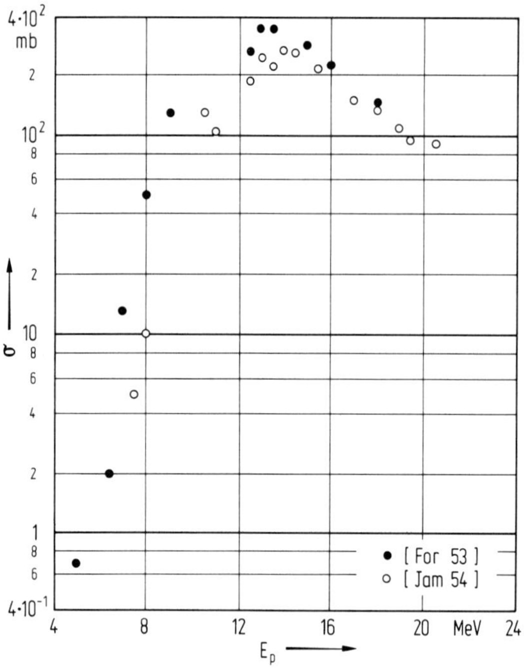

◀ Fig. 459. Reaction $_{41}Nb \rightarrow {}^{93m}{}_{42}Mo$

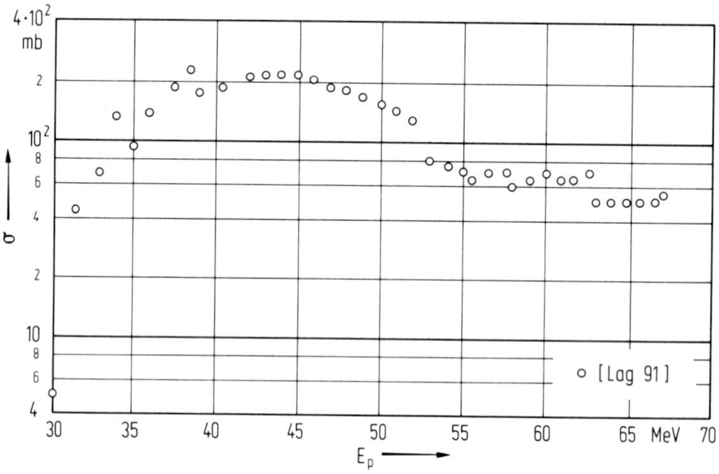

Fig. 460. Reaction $_{42}Mo \rightarrow {}^{90}{}_{42}Mo$

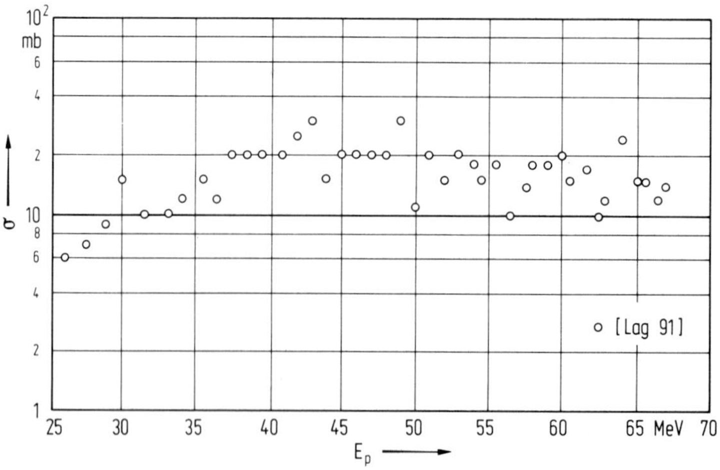

Fig. 461. Reaction $_{42}Mo \rightarrow {}^{93m}{}_{42}Mo$

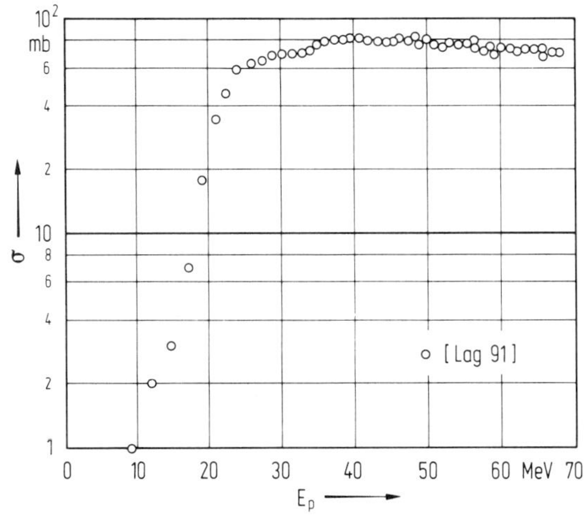

Fig. 462. Reaction $_{42}Mo \rightarrow {}^{99}_{42}Mo$

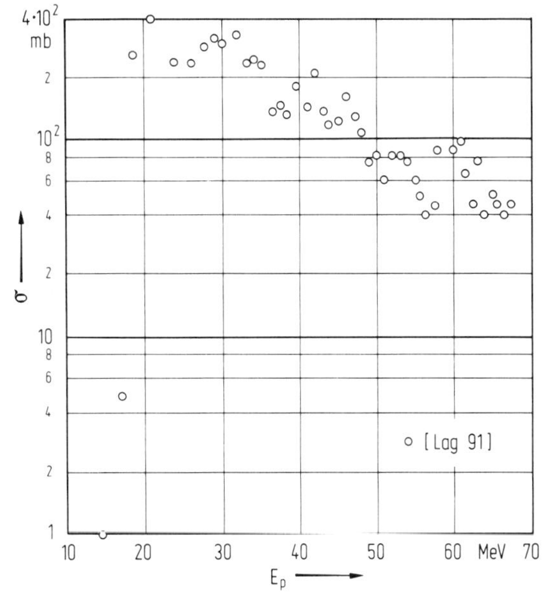

Fig. 463. Reaction $_{42}Mo \rightarrow {}^{93}_{43}Tc$

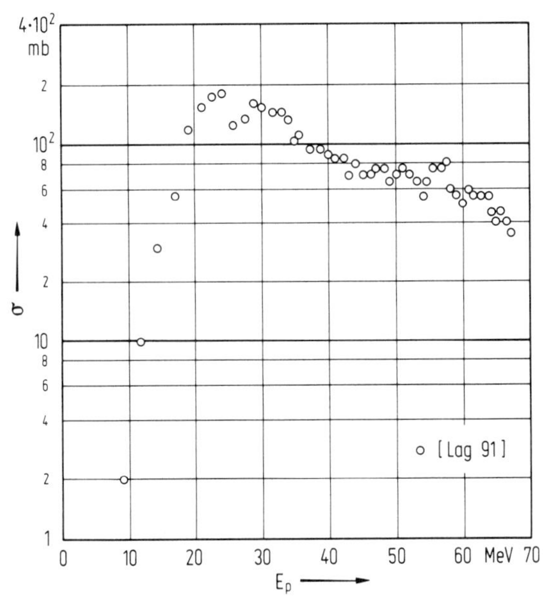

Fig. 464. Reaction $_{42}Mo \rightarrow {}^{94}_{43}Tc$

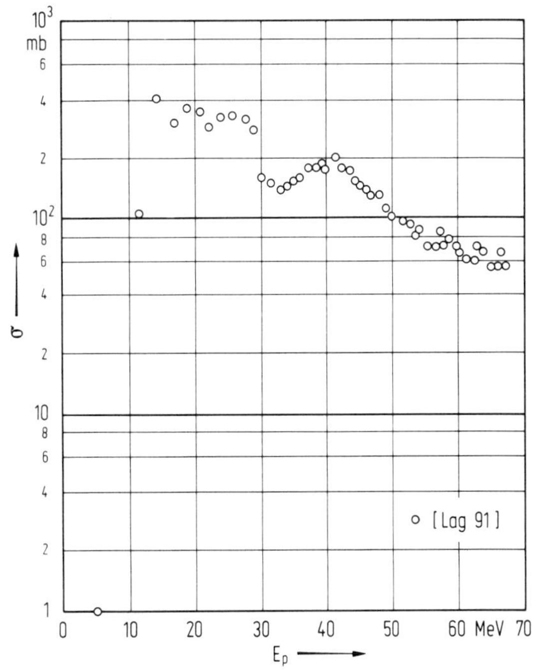

Fig. 465. Reaction $_{42}Mo \rightarrow {}^{95}_{43}Tc$

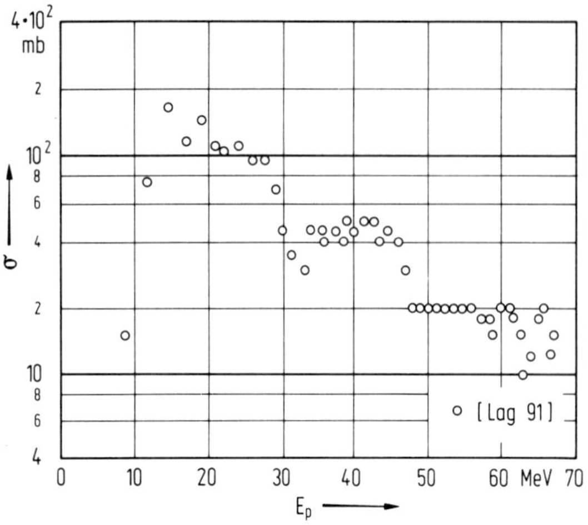

Fig. 466. Reaction ${}_{42}Mo \rightarrow {}^{95m}{}_{43}Tc$

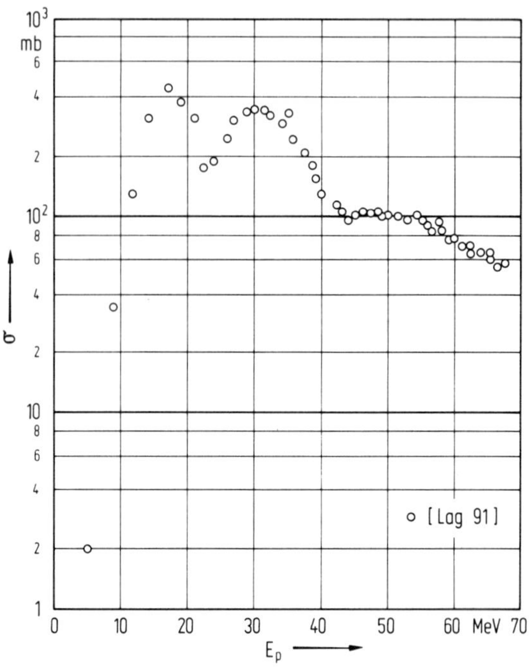

Fig. 467. Reaction ${}_{42}Mo \rightarrow {}^{96}{}_{43}Tc$

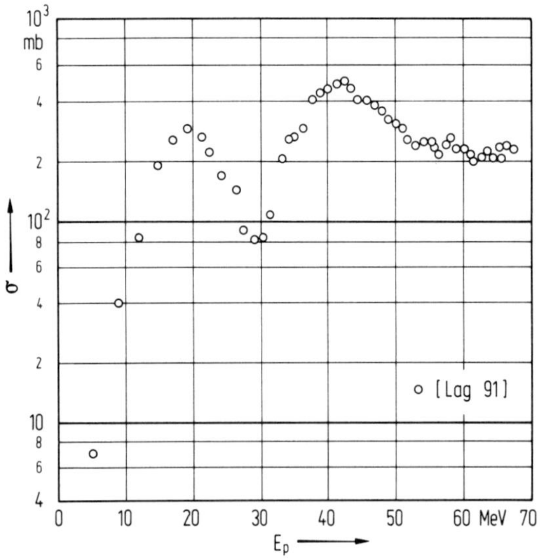

Fig. 468. Reaction ${}_{42}Mo \rightarrow {}^{99m}{}_{43}Tc$

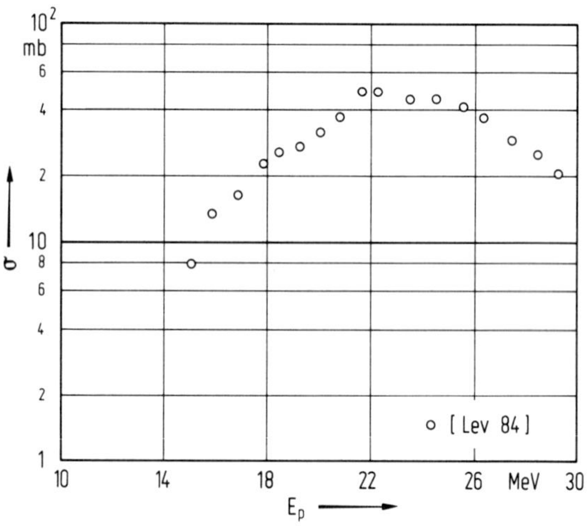

Fig. 469. Reaction ${}^{92}{}_{42}Mo \rightarrow {}^{89}{}_{41}Nb$

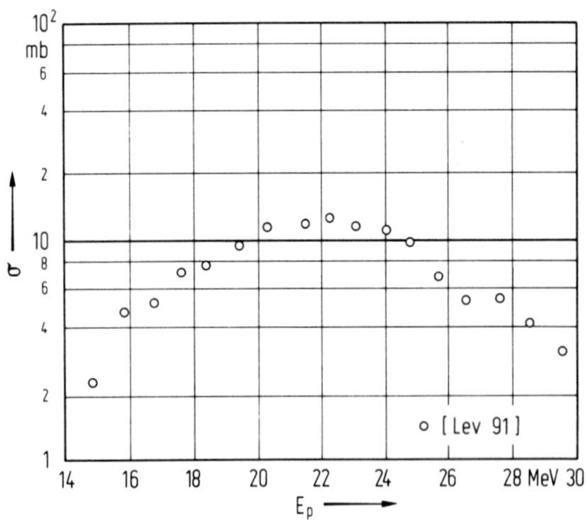

◀ Fig. 470. Reaction ${}^{92}{}_{42}Mo \rightarrow {}^{89m}{}_{41}Nb$

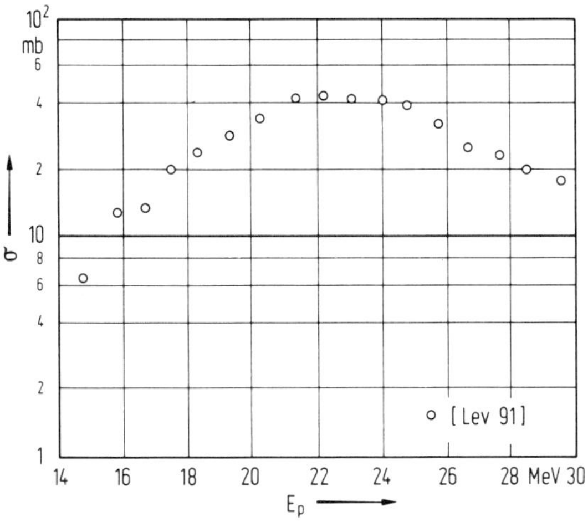

Fig. 471. Reaction ${}^{92}_{42}Mo \rightarrow {}^{89m+g}_{41}Nb$

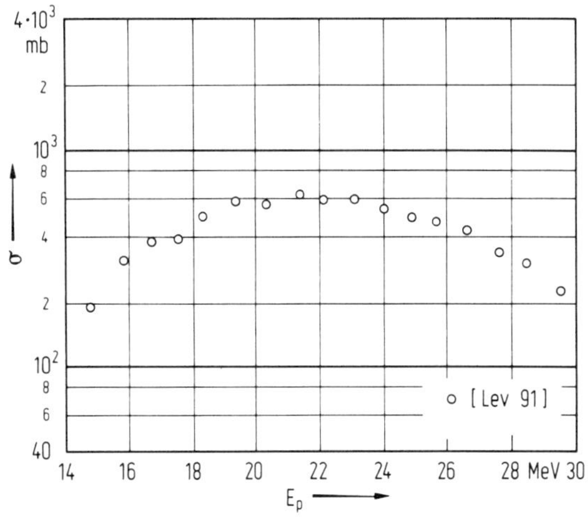

Fig. 472. Reaction ${}^{94}_{42}Mo \rightarrow {}^{90}_{41}Nb$

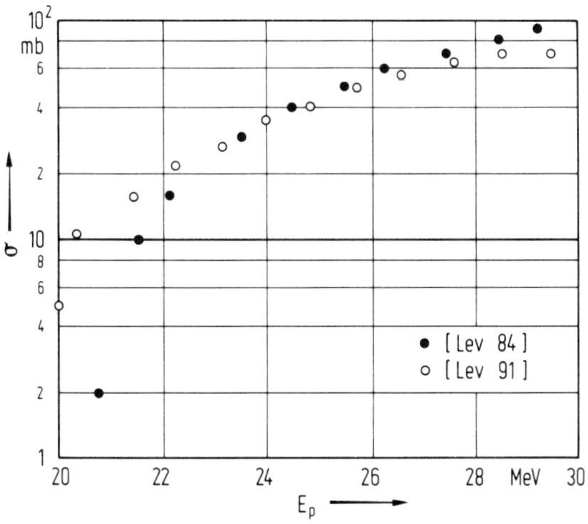

Fig. 473. Reaction ${}^{94}_{42}Mo \rightarrow {}^{93}_{43}Tc$

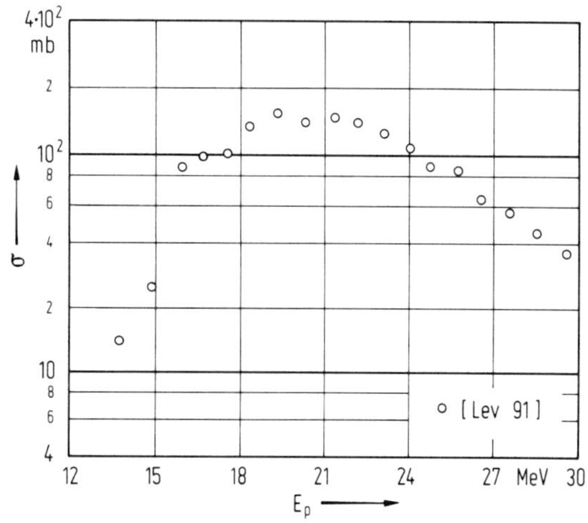

Fig. 474. Reaction ${}^{94}_{42}Mo \rightarrow {}^{93m}_{43}Tc$

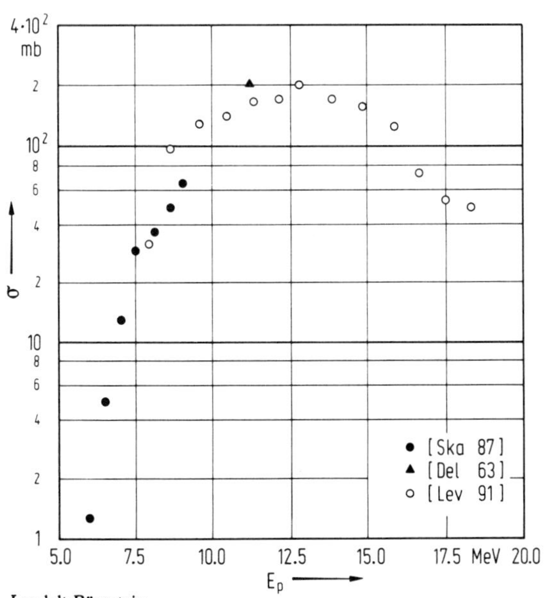

◀ Fig. 475. Reaction ${}^{94}_{42}Mo \rightarrow {}^{94}_{43}Tc$

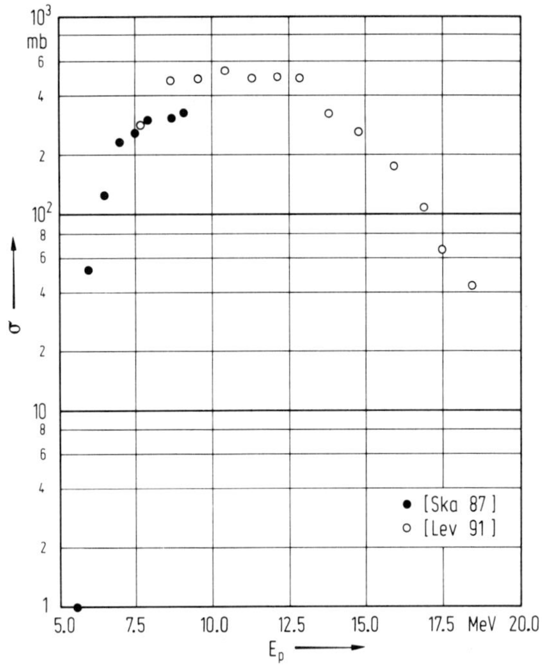

Fig. 476. Reaction $^{94}{}_{42}Mo \rightarrow {}^{94m}{}_{43}Tc$

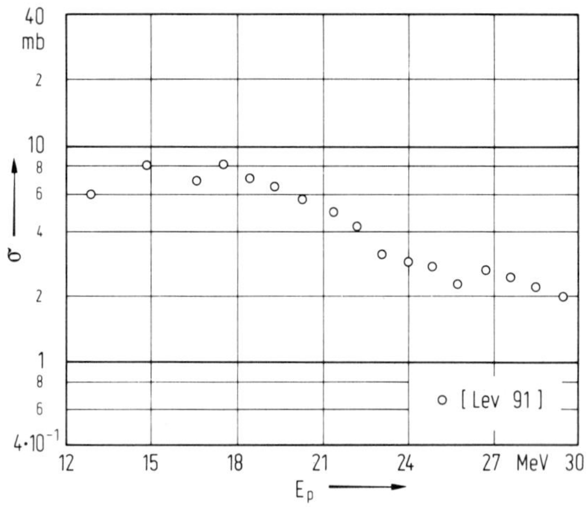

Fig. 477. Reaction $^{95}{}_{42}Mo \rightarrow {}^{92m}{}_{41}Nb$

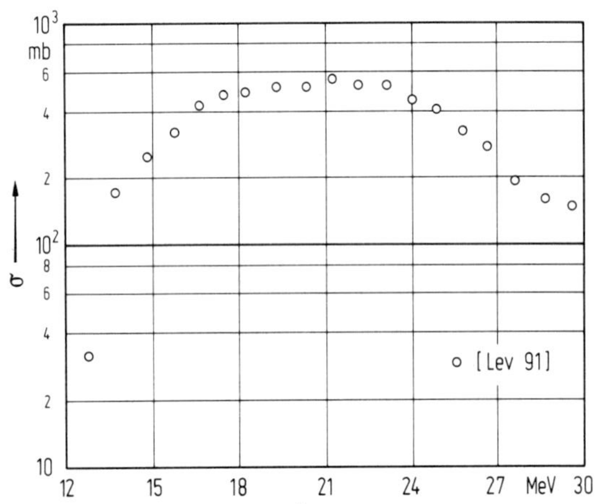

Fig. 478. Reaction $^{95}{}_{42}Mo \rightarrow {}^{94}{}_{43}Tc$

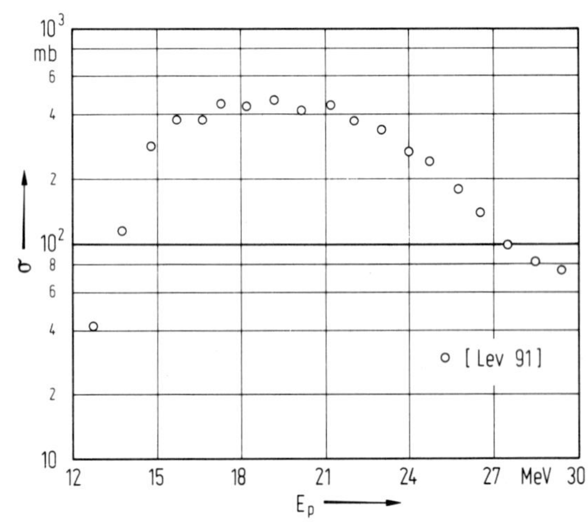

Fig. 479. Reaction $^{95}{}_{42}Mo \rightarrow {}^{94m}{}_{43}Tc$

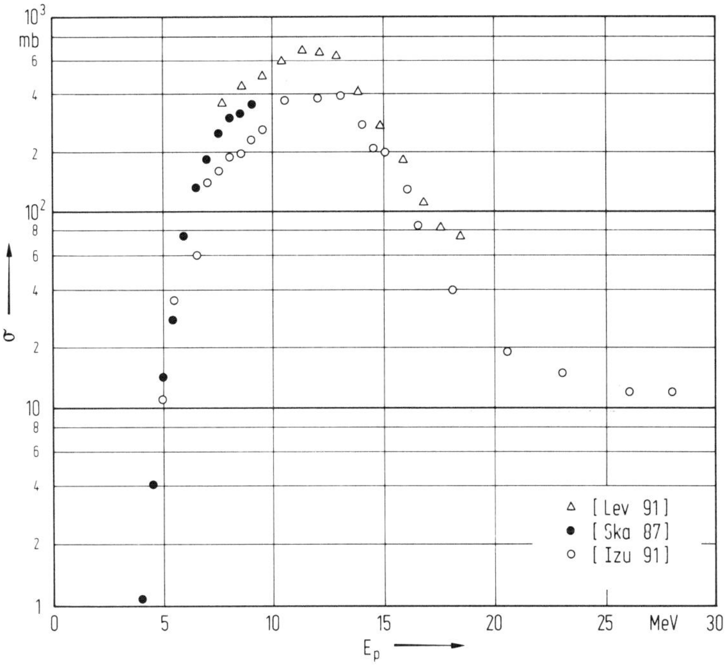

Fig. 480. Reaction $^{95}_{42}Mo \rightarrow ^{95}_{43}Tc$

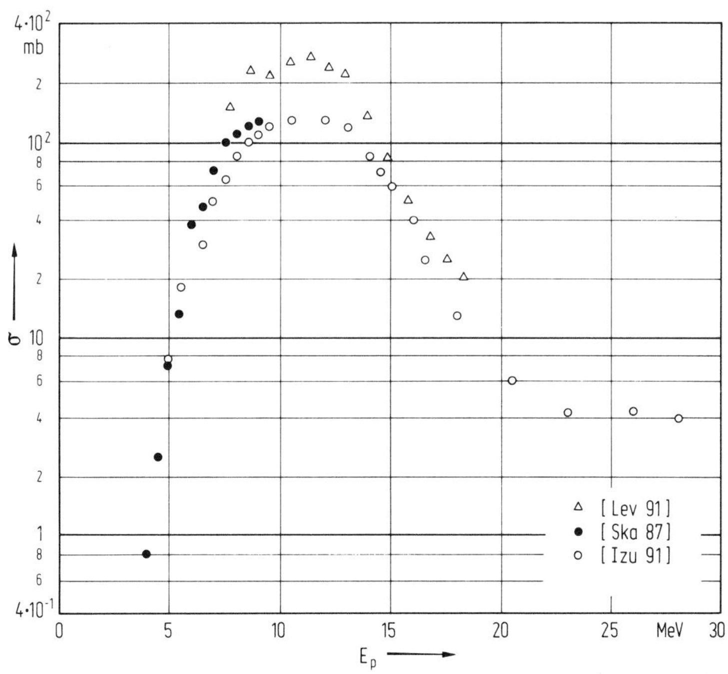

Fig. 481. Reaction $^{95}_{42}Mo \rightarrow ^{95m}_{43}Tc$

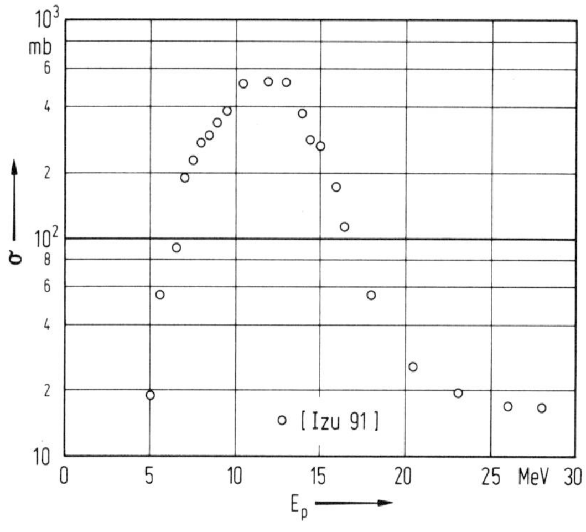

◀ Fig. 482. Reaction $^{95}{}_{42}Mo \rightarrow {}^{95m+g}{}_{43}Tc$

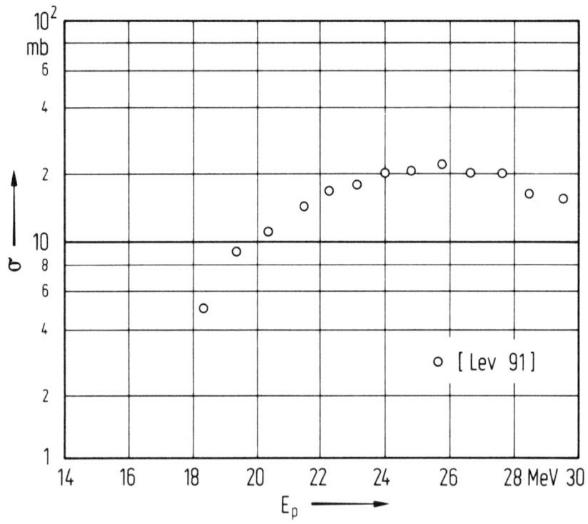

Fig. 483. Reaction $^{96}{}_{42}Mo \rightarrow {}^{92m}{}_{41}Nb$

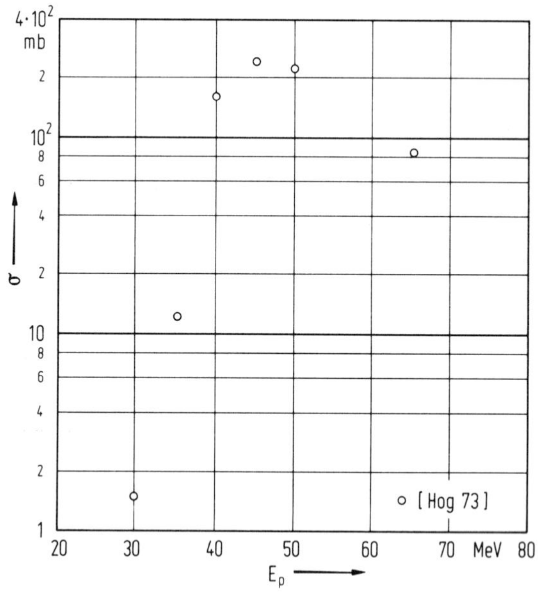

Fig. 484. Reaction $^{96}{}_{42}Mo \rightarrow {}^{93}{}_{43}Tc$

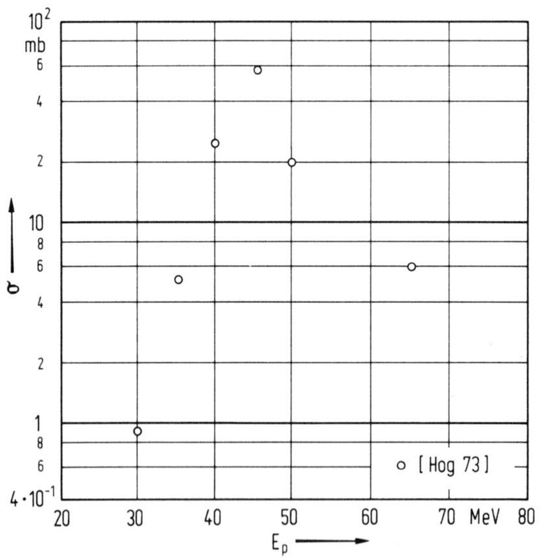

◀ Fig. 485. Reaction $^{96}{}_{42}Mo \rightarrow {}^{93m}{}_{43}Tc$

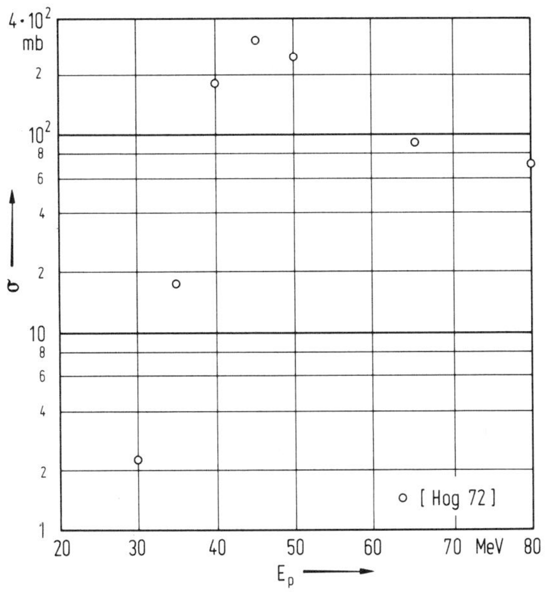

Fig. 486. Reaction $^{96}_{42}Mo \rightarrow {}^{93m+g}_{43}Tc$

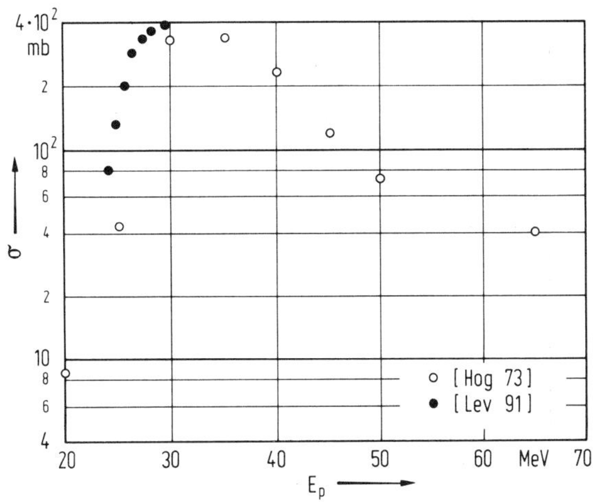

Fig. 487. Reaction $^{96}_{42}Mo \rightarrow {}^{94}_{43}Tc$

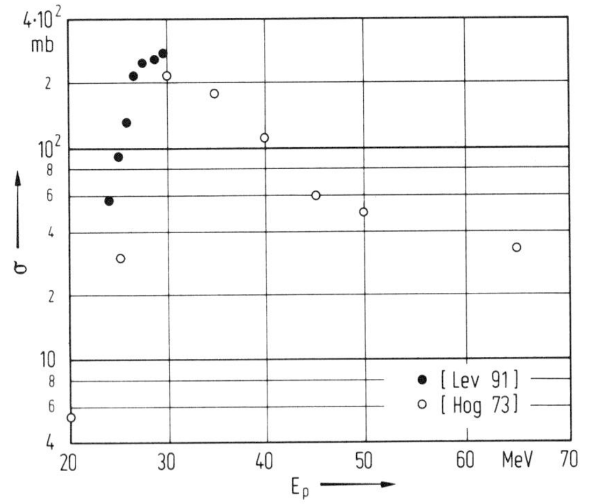

◀ Fig. 488. Reaction $^{96}_{42}Mo \rightarrow {}^{94m}_{43}Tc$

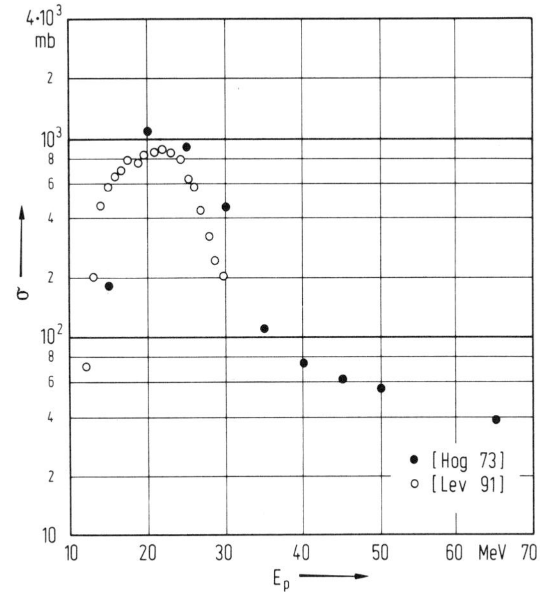

Fig. 490. Reaction $^{96}_{42}Mo \rightarrow {}^{95}_{43}Tc$

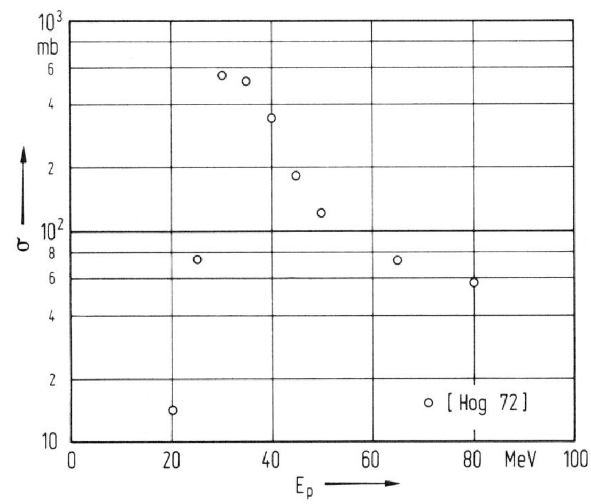

◀ Fig. 489. Reaction $^{96}_{42}Mo \rightarrow {}^{94m+g}_{43}Tc$

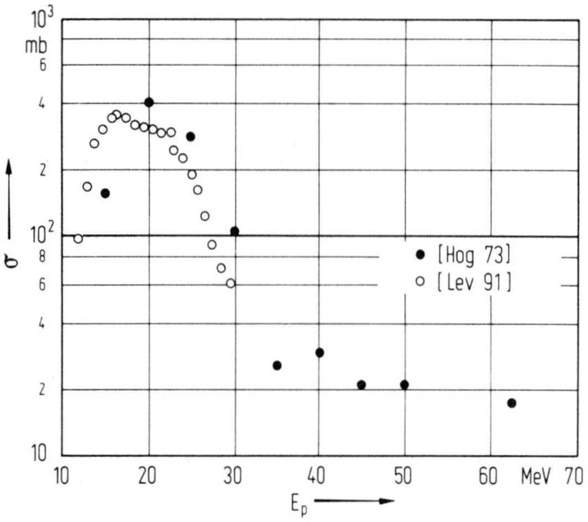

Fig. 491. Reaction $^{96}_{42}Mo \rightarrow {}^{95m}_{43}Tc$

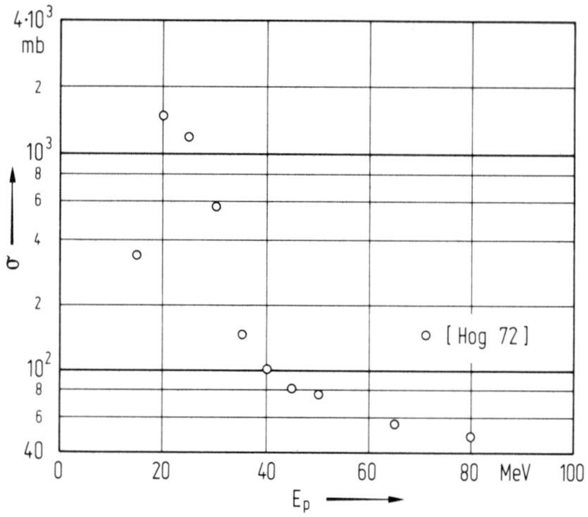

Fig. 492. Reaction $^{96}_{42}Mo \rightarrow {}^{95m+g}_{43}Tc$

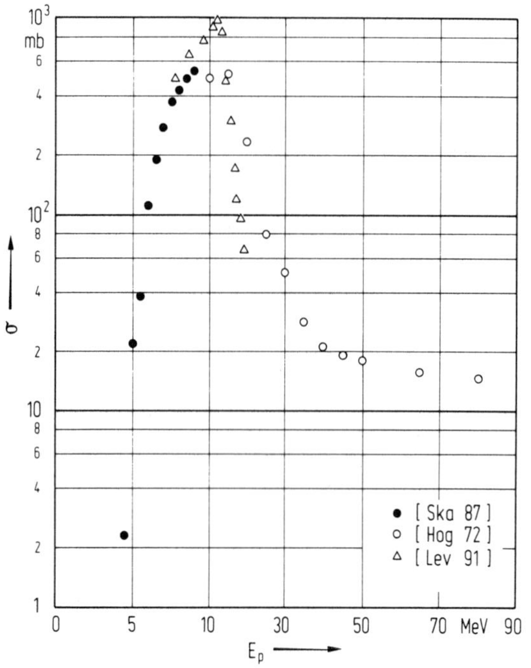

Fig. 493. Reaction $^{96}_{42}Mo \rightarrow {}^{96}_{43}Tc$

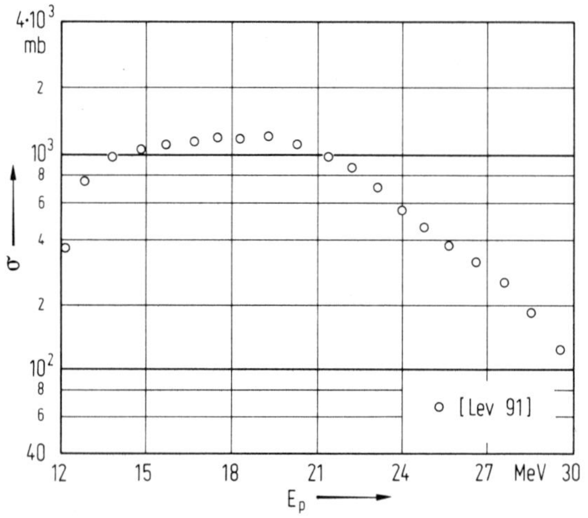

Fig. 494. Reaction $^{97}_{42}Mo \rightarrow {}^{96}_{43}Tc$

Fig. 495. Reaction $^{98}_{42}Mo \rightarrow {}^{95}_{41}Nb$ ▶

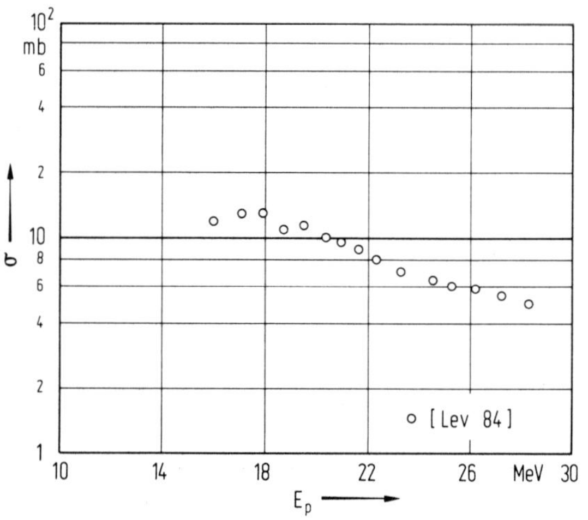

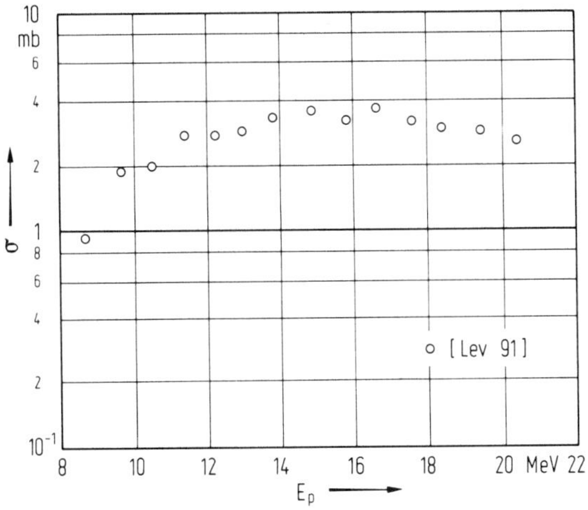

Fig. 496. Reaction $^{98}{}_{42}Mo \rightarrow {}^{95m}{}_{41}Nb$

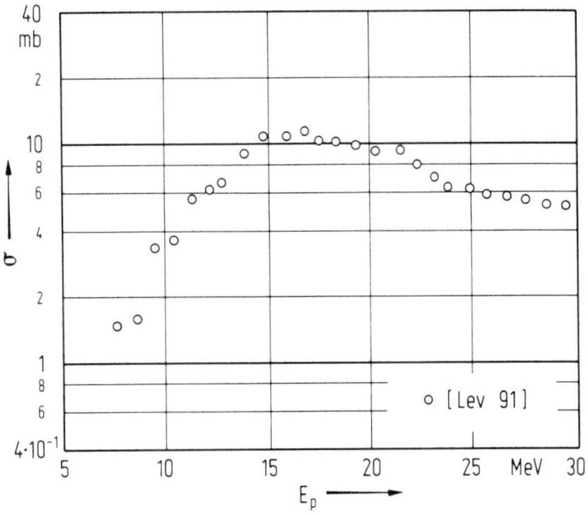

Fig. 497. Reaction $^{98}{}_{42}Mo \rightarrow {}^{95m+g}{}_{41}Nb$

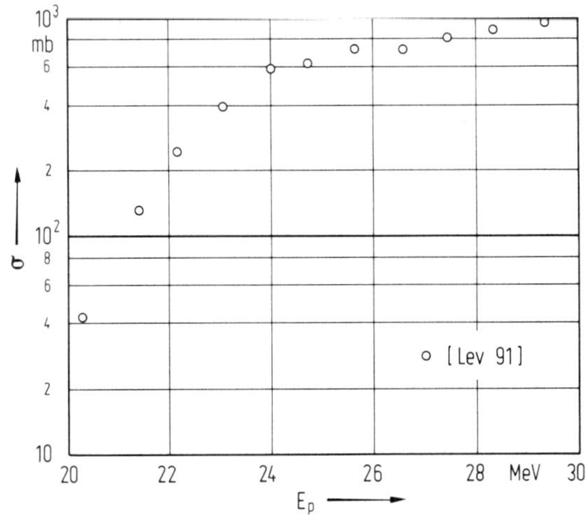

Fig. 498. Reaction $^{98}{}_{42}Mo \rightarrow {}^{96}{}_{43}Tc$

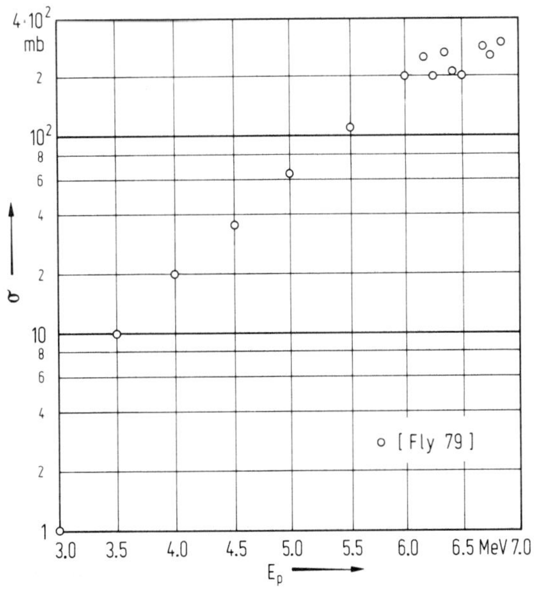

Fig. 499. Reaction $^{98}{}_{42}Mo \rightarrow {}^{98}{}_{43}Tc$

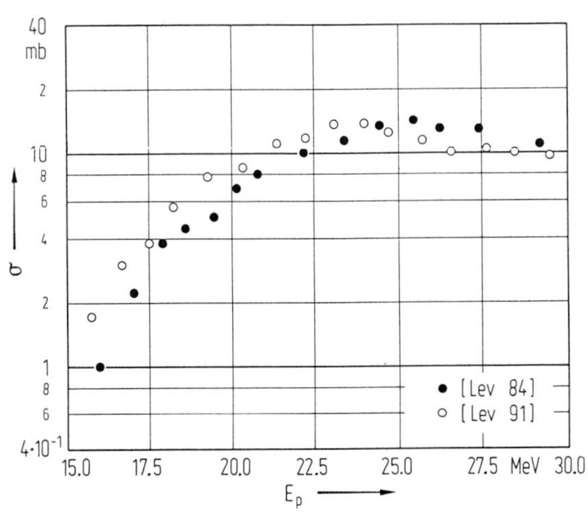

◀ Fig. 500. Reaction $^{100}{}_{42}Mo \rightarrow {}^{96}{}_{41}Nb$

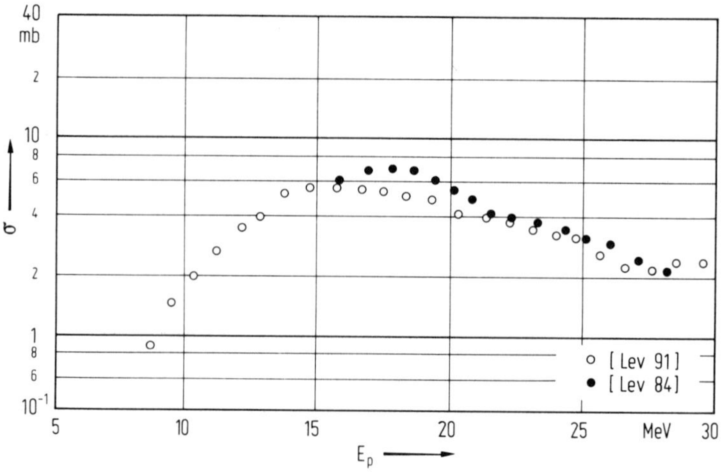

Fig. 501. Reaction $^{100}_{42}Mo \rightarrow ^{97}_{41}Nb$

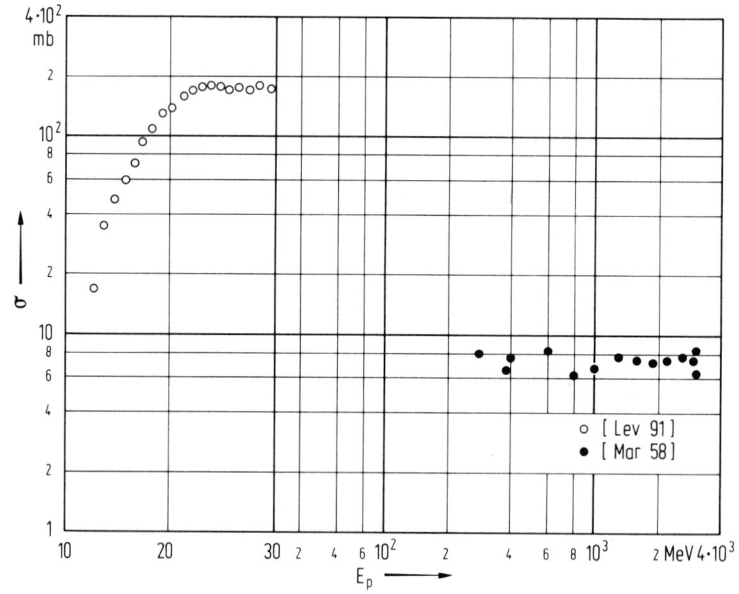

Fig. 502. Reaction $^{100}_{42}Mo \rightarrow ^{99}_{42}Mo$

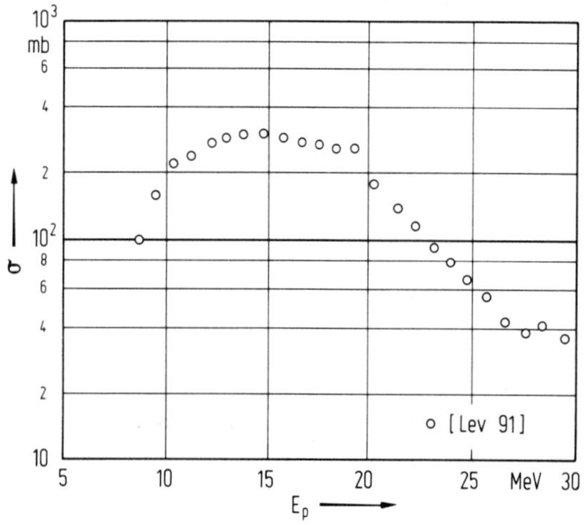

◀ Fig. 503. Reaction $^{100}_{42}Mo \rightarrow ^{99m}_{43}Tc$

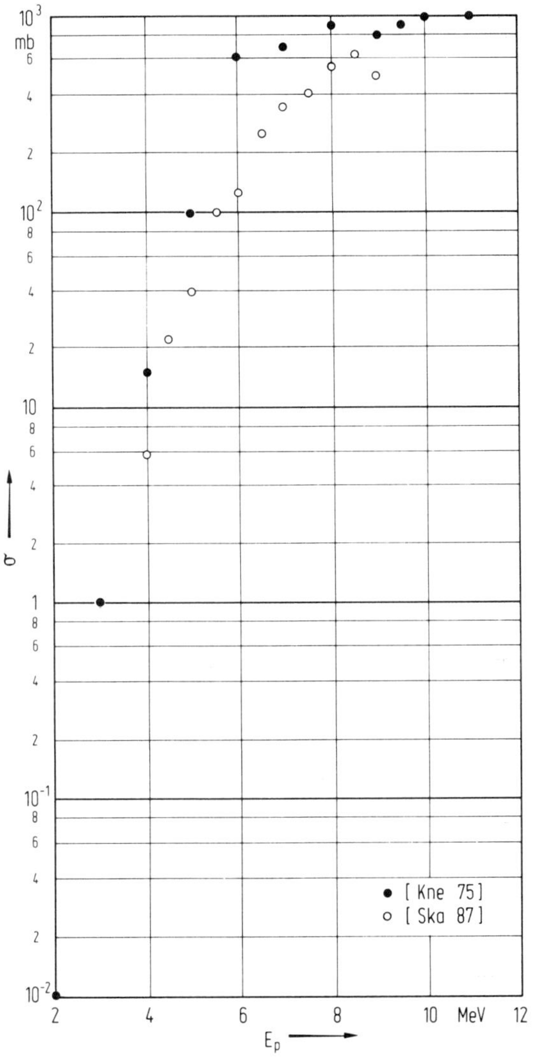

Fig. 504. Reaction $^{100}{}_{42}Mo \rightarrow {}^{100}{}_{43}Tc$

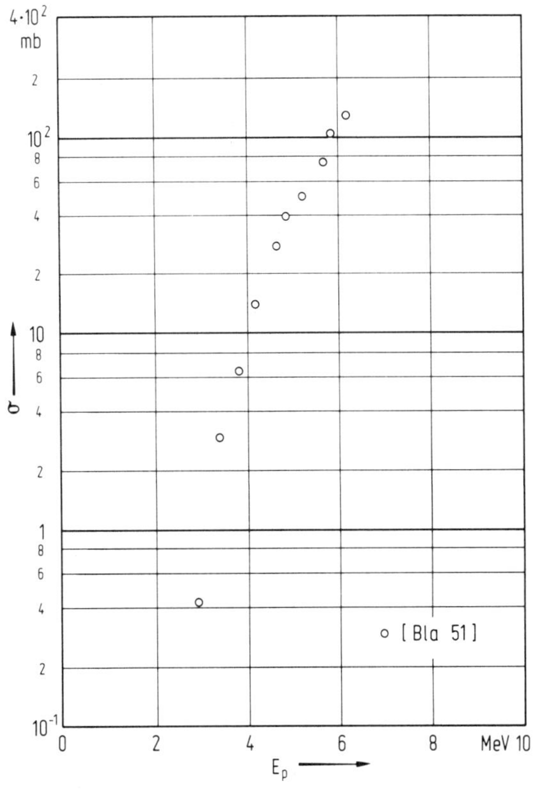

Fig. 505. Reaction $^{101}{}_{44}Ru \rightarrow {}^{101}{}_{45}Rh$

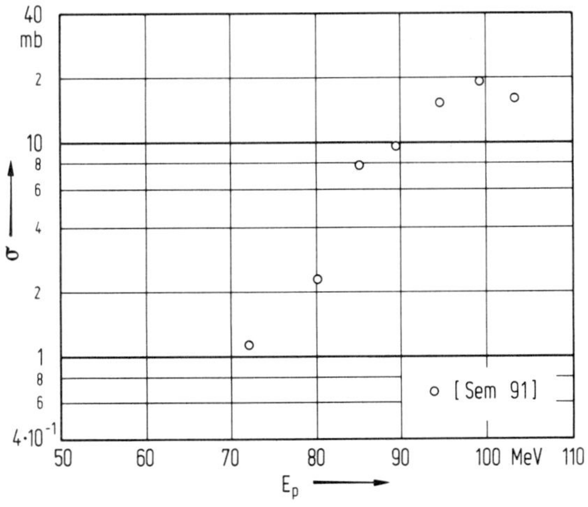

Fig. 506. Reaction $_{45}Rh \rightarrow {}^{94}{}_{43}Tc$

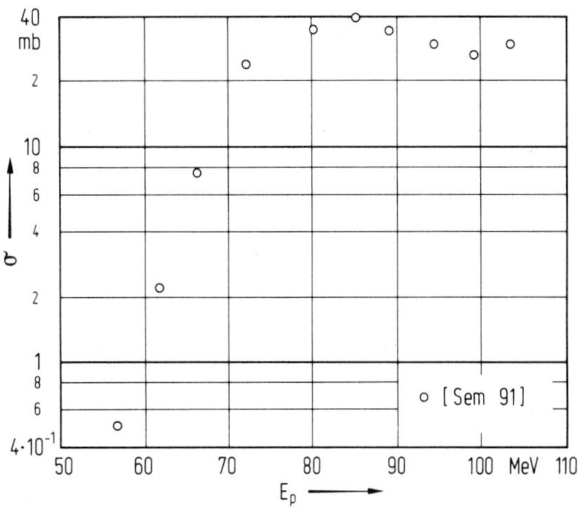

Fig. 507. Reaction $_{45}Rh \rightarrow {}^{95}{}_{43}Tc$

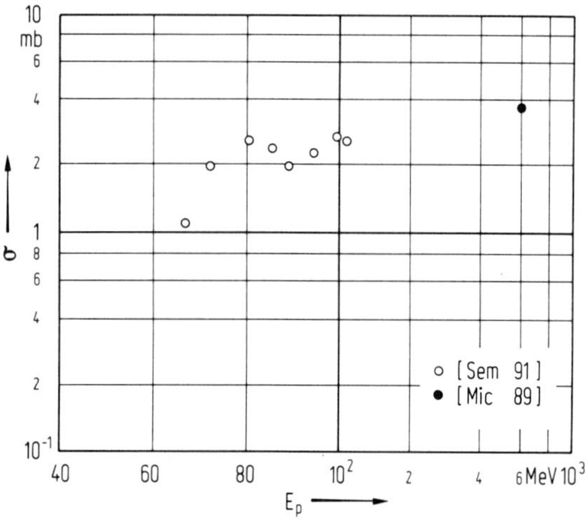

Fig. 508. Reaction $_{45}Rh \rightarrow {}^{95m}{}_{43}Tc$

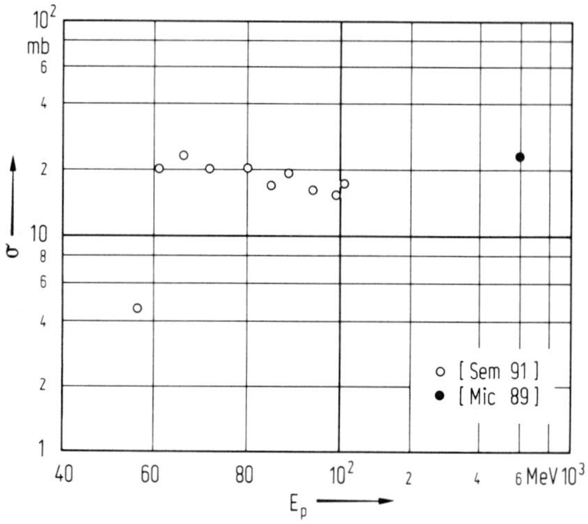

Fig. 509. Reaction $_{45}Rh \rightarrow {}^{96}{}_{43}Tc$

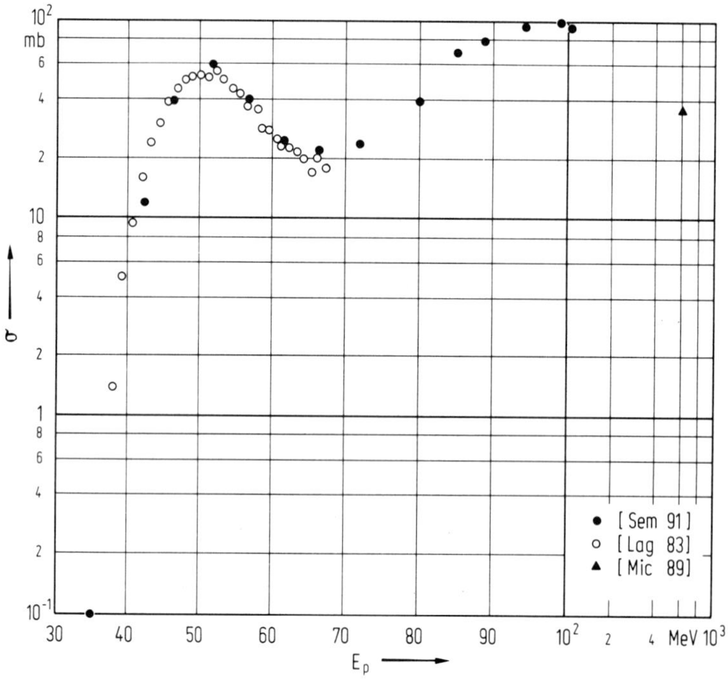

▲ Fig. 510. Reaction $_{45}Rh \rightarrow {}^{97}{}_{44}Ru$ (cumulative)

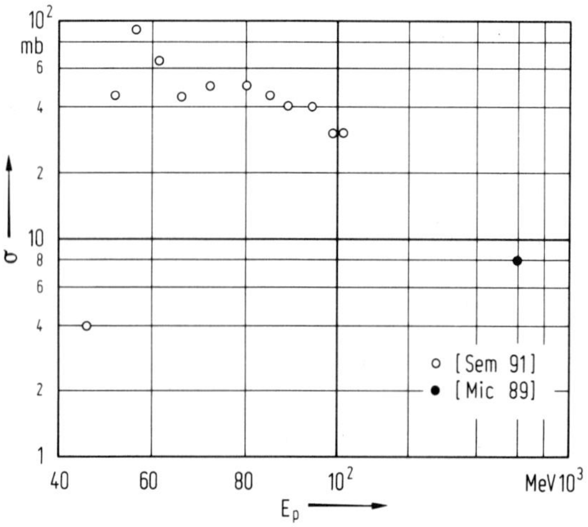

◀ Fig. 511. Reaction $_{45}Rh \rightarrow {}^{99}{}_{45}Rh$

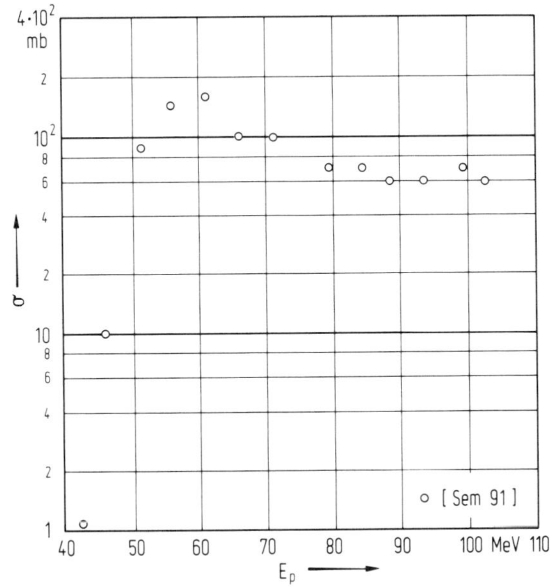

◀ Fig. 512. Reaction $_{45}Rh \rightarrow {}^{99m}{}_{45}Rh$

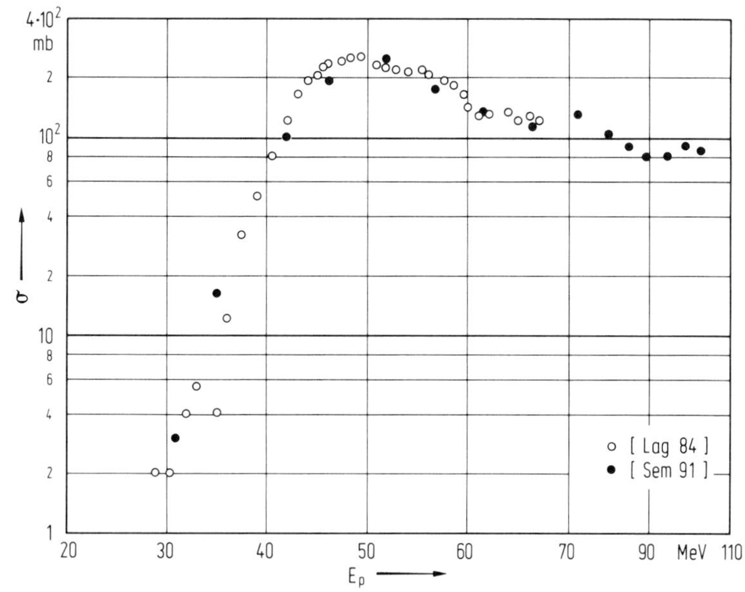

◀ Fig. 513. Reaction $_{45}Rh \rightarrow {}^{100}{}_{45}Rh$

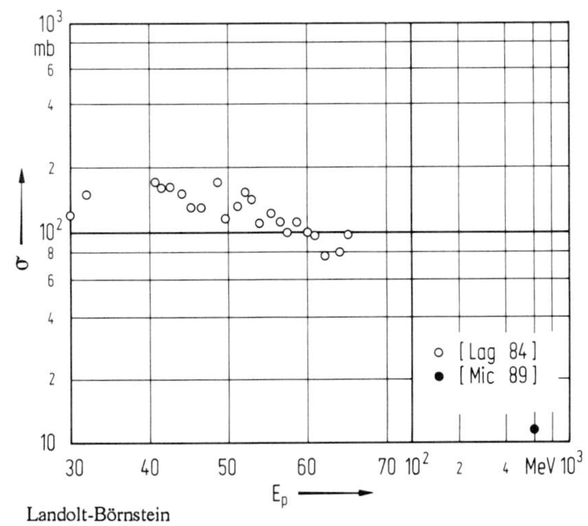

◀ Fig. 514. Reaction $_{45}Rh \rightarrow {}^{101}{}_{45}Rh$

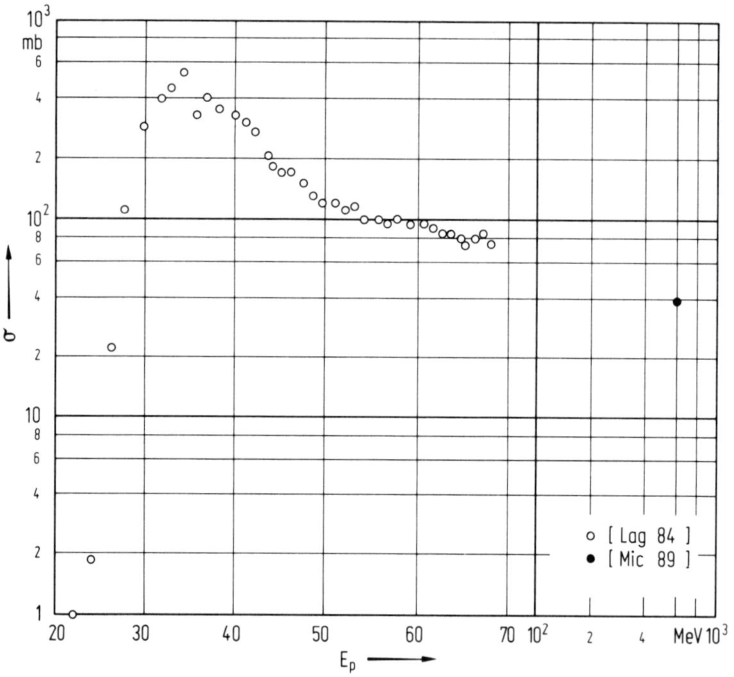

Fig. 515. Reaction $_{45}Rh \rightarrow {}^{101m}{}_{45}Rh$

Fig. 516, see next page

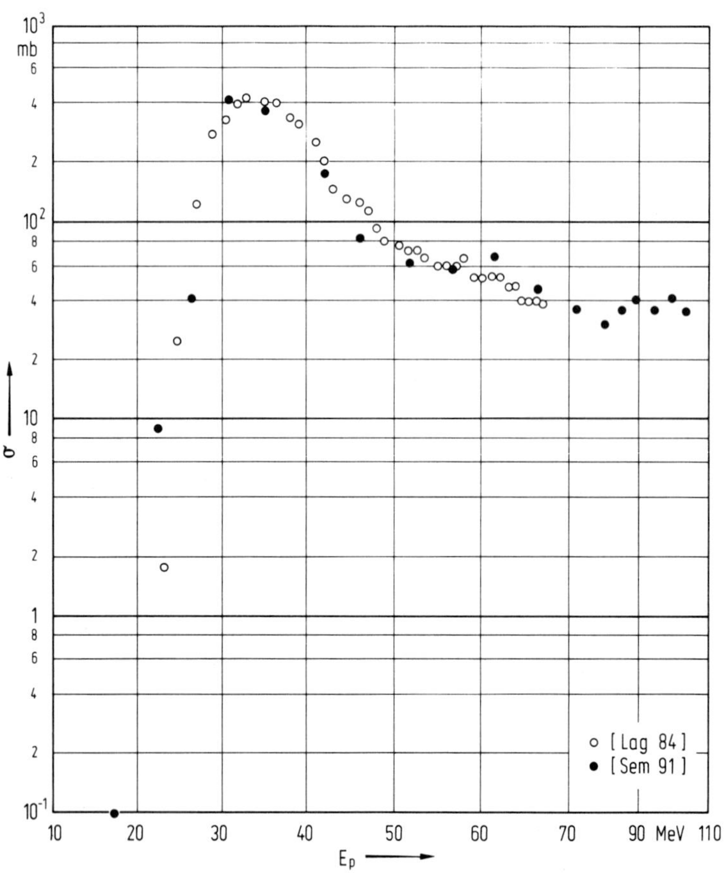

Fig. 517. Reaction $_{45}Rh \rightarrow {}^{101}{}_{46}Pd$

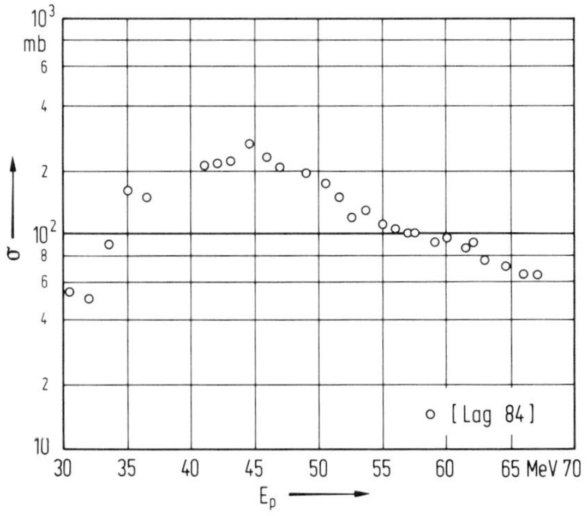

Fig. 516. Reaction $_{45}Rh \rightarrow {}^{100}_{46}Pd$

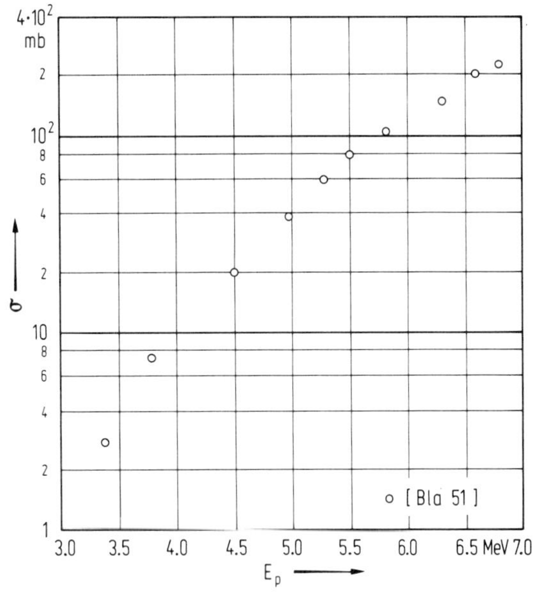

Fig. 518. Reaction $_{45}Rh \rightarrow {}^{103}_{46}Pd$

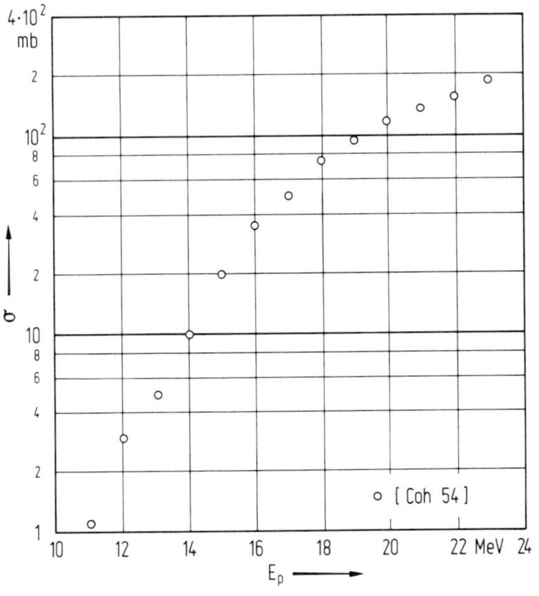

Fig. 519. Reaction $_{46}Pd \rightarrow {}^{109}_{46}Pd$

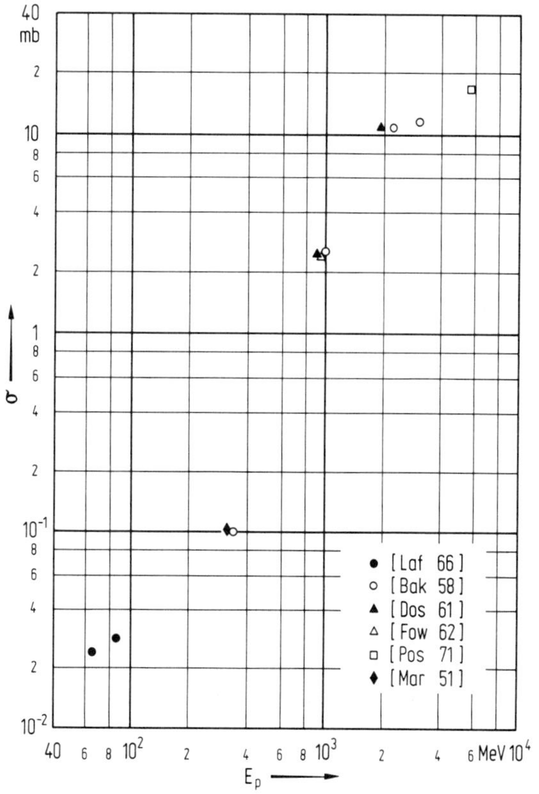

Fig. 520. Reaction $_{47}Ag \rightarrow {}^{7}_{4}Be$

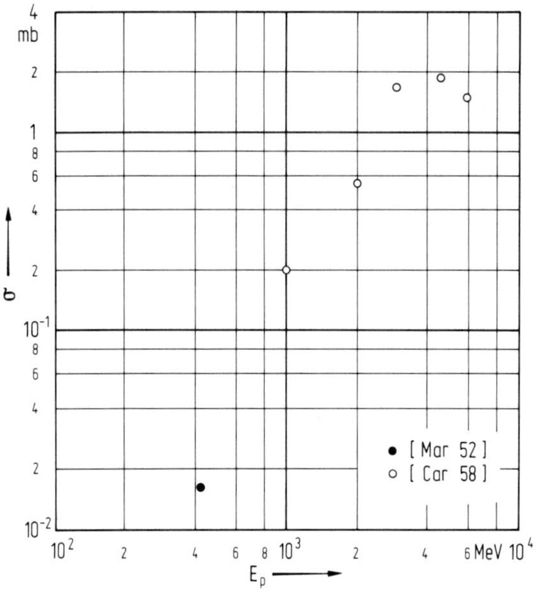

Fig. 521. Reaction ${}_{47}Ag \rightarrow {}^{18}{}_{9}F$

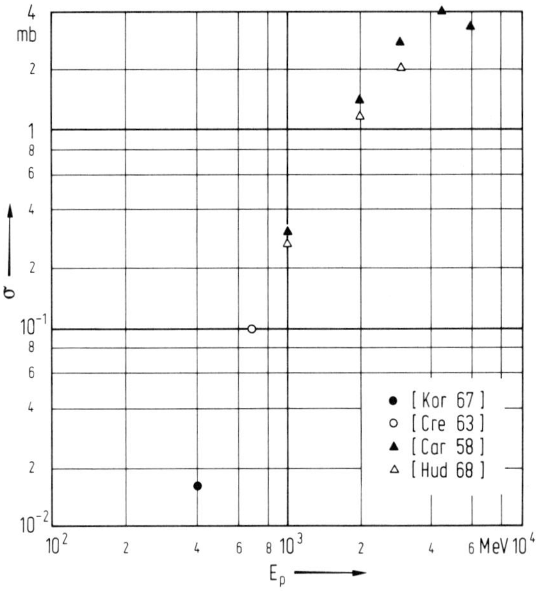

Fig. 522. Reaction ${}_{47}Ag \rightarrow {}^{24}{}_{11}Na$

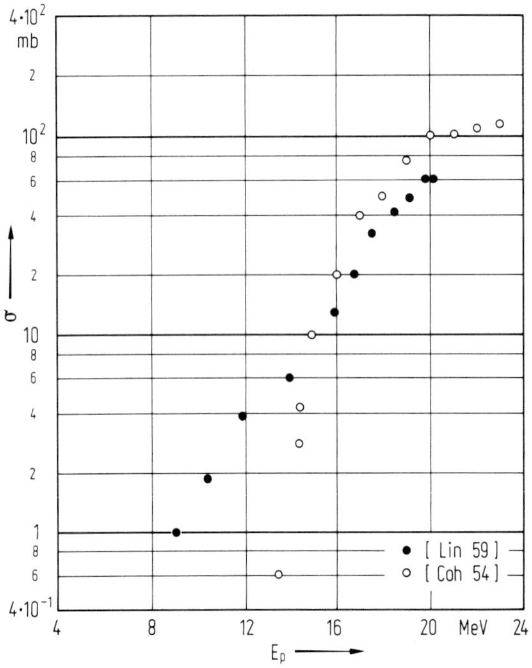

Fig. 523. Reaction ${}_{47}Ag \rightarrow {}^{106}{}_{47}Ag$

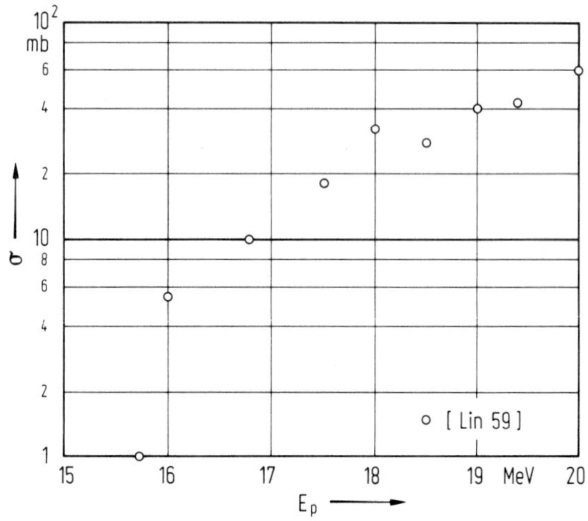

Fig. 524. Reaction ${}_{47}Ag \rightarrow {}^{106m}{}_{47}Ag$

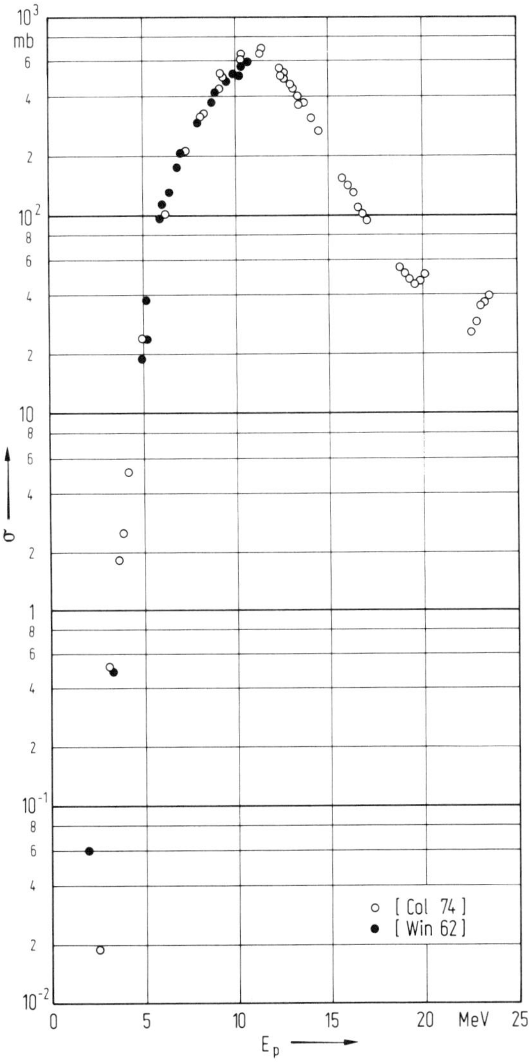

Fig. 525. Reaction $_{47}Ag \rightarrow {}^{107}{}_{48}Cd$

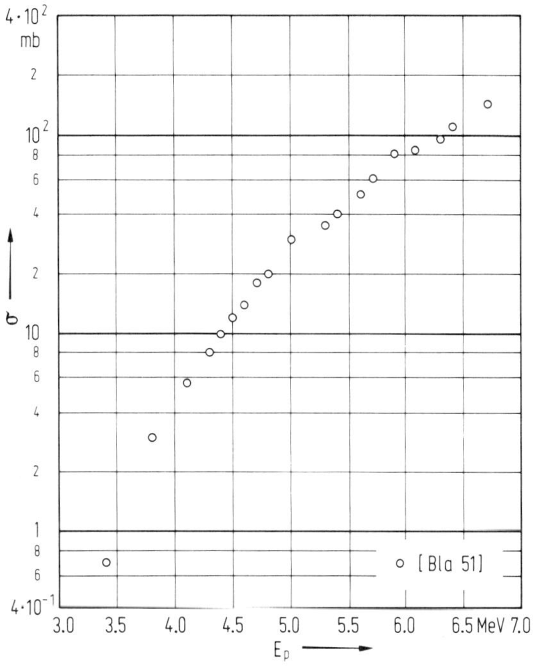

Fig. 526. Reaction $^{107}{}_{47}Ag \rightarrow {}^{107}{}_{48}Cd$

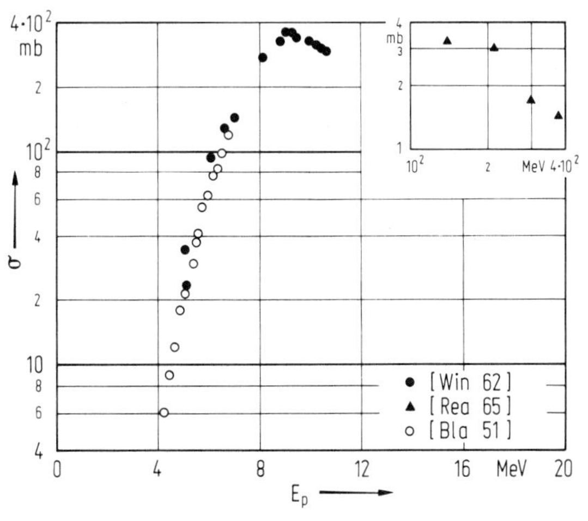

Fig. 527. Reaction $^{109}{}_{47}Ag \rightarrow {}^{109}{}_{48}Cd$

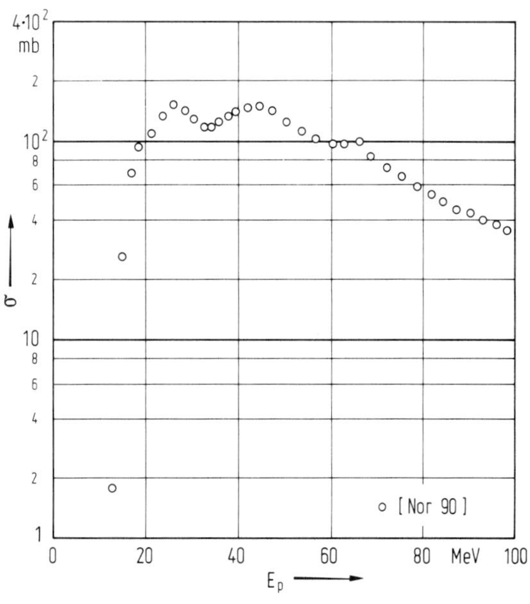

Fig. 528. Reaction $_{48}Cd \rightarrow {}^{(109m1+m2+g)}In_{49}$

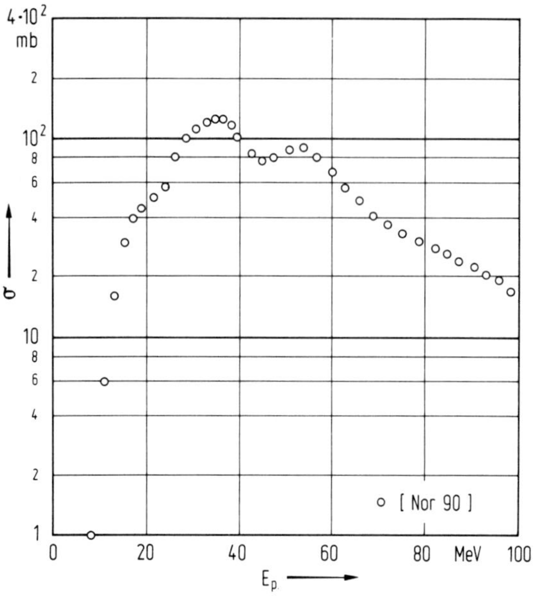

Fig. 529. Reaction $_{48}Cd \rightarrow {}^{110m}{}_{49}In$

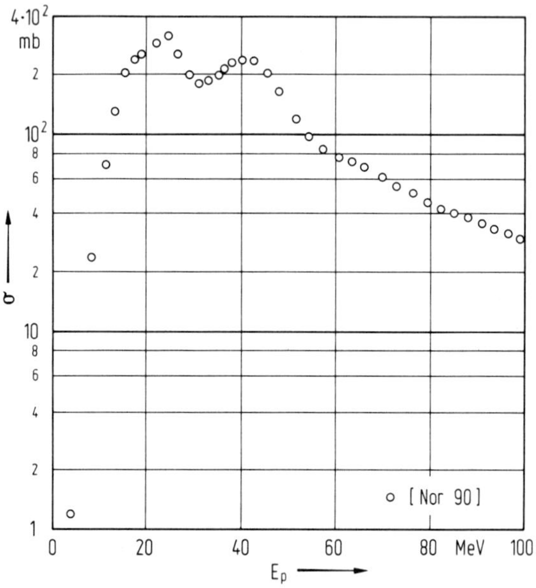

Fig. 530. Reaction $_{48}Cd \rightarrow {}^{111m+g}{}_{49}In$

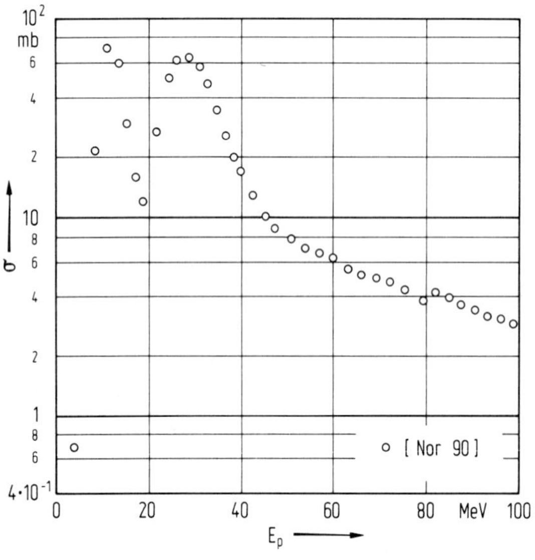

Fig. 531. Reaction $_{48}Cd \rightarrow {}^{114m}{}_{49}In$

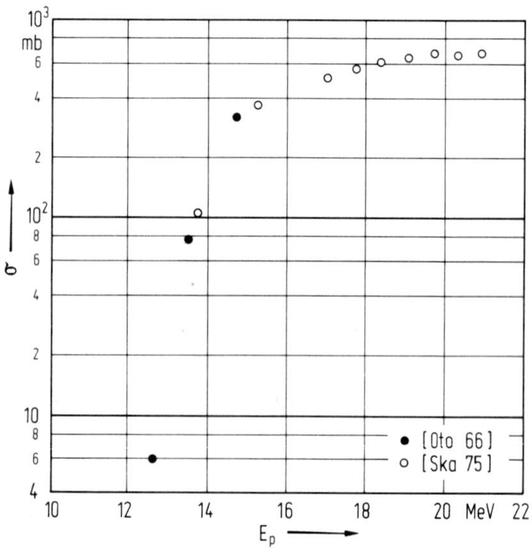

Fig. 532. Reaction $^{110}{}_{48}Cd \rightarrow {}^{109}{}_{49}In$

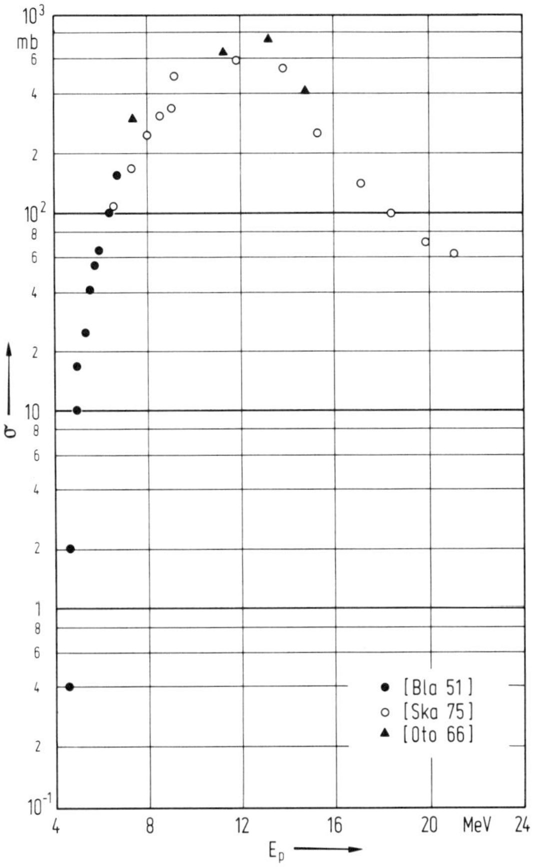

Fig. 533. Reaction $^{110}{}_{48}Cd \rightarrow {}^{110}{}_{49}In$

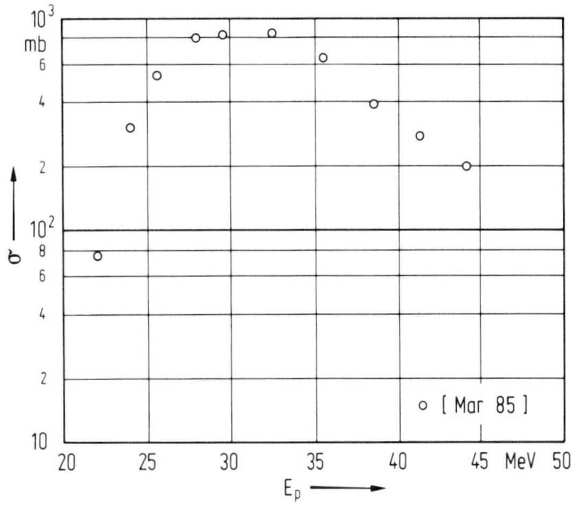

Fig. 534. Reaction $^{111}{}_{48}Cd \rightarrow {}^{109m+g}{}_{49}In$

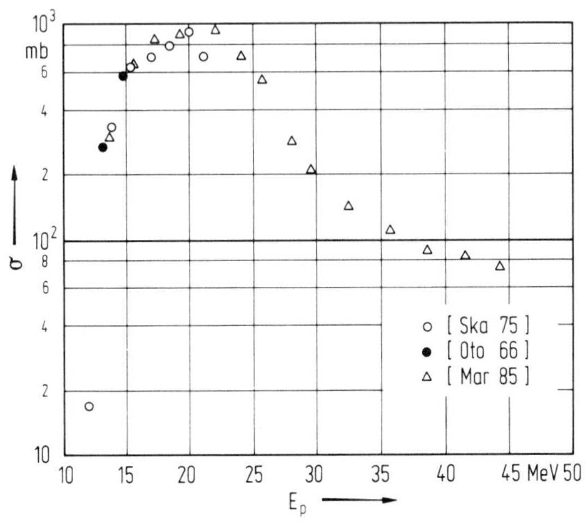

Fig. 535. Reaction $^{111}{}_{48}Cd \rightarrow {}^{110}{}_{49}In$

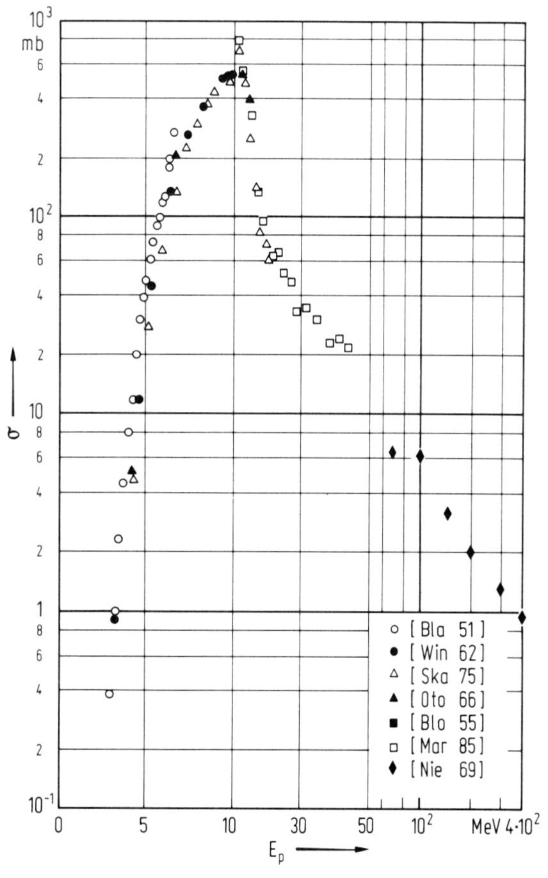

Fig. 536. Reaction $^{111}{}_{48}Cd \rightarrow {}^{111}{}_{49}In$

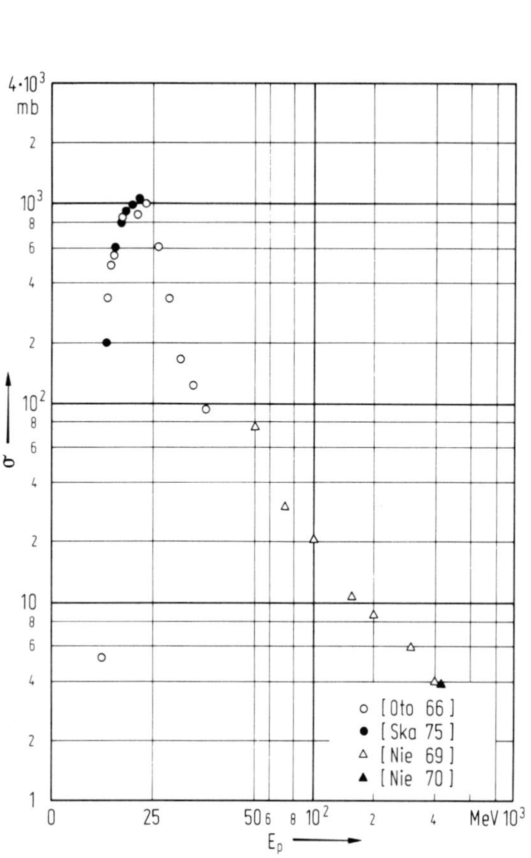

Fig. 537. Reaction $^{112}_{48}Cd \rightarrow {}^{111}_{49}In$

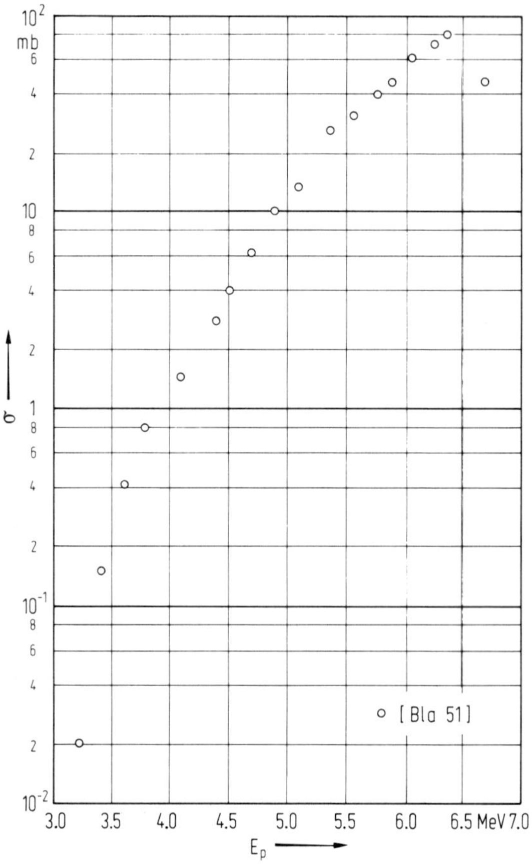

Fig. 538. Reaction $^{112}_{48}Cd \rightarrow {}^{112}_{49}In$

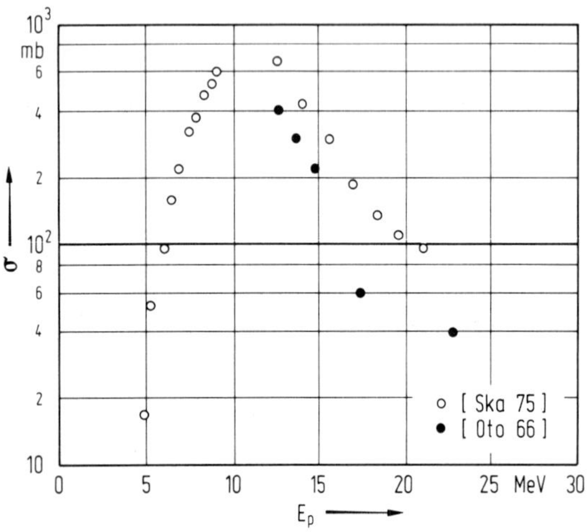

Fig. 539. Reaction $^{112}_{48}Cd \rightarrow {}^{112}_{49}In$

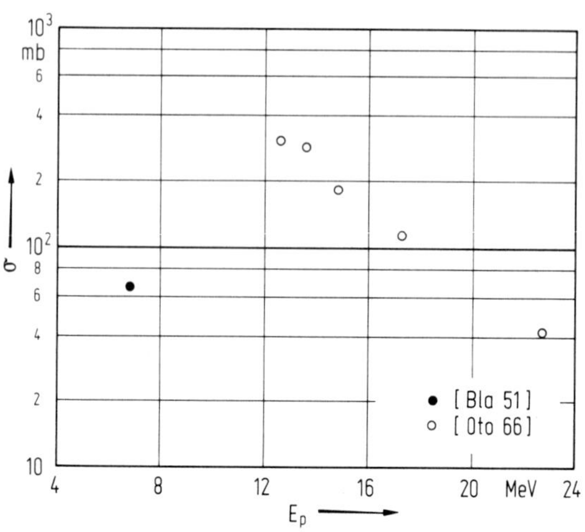

Fig. 540. Reaction $^{112}_{48}Cd \rightarrow {}^{112m}_{49}In$

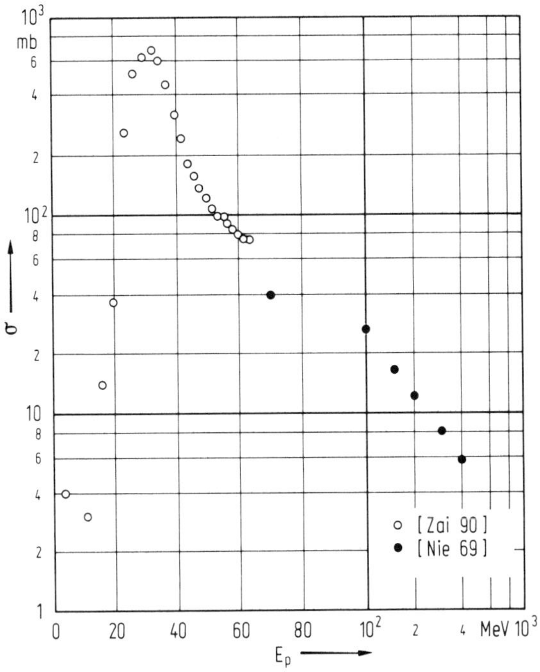

◀ Fig. 541. Reaction $^{113}_{48}Cd \rightarrow ^{111}_{49}In$

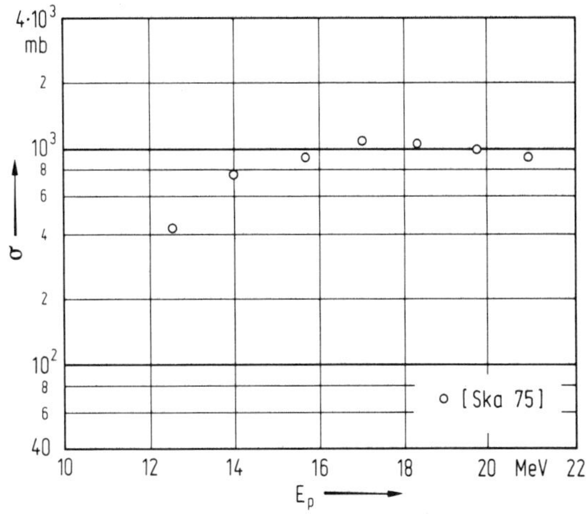

Fig. 542. Reaction $^{113}_{48}Cd \rightarrow ^{112}_{49}In$

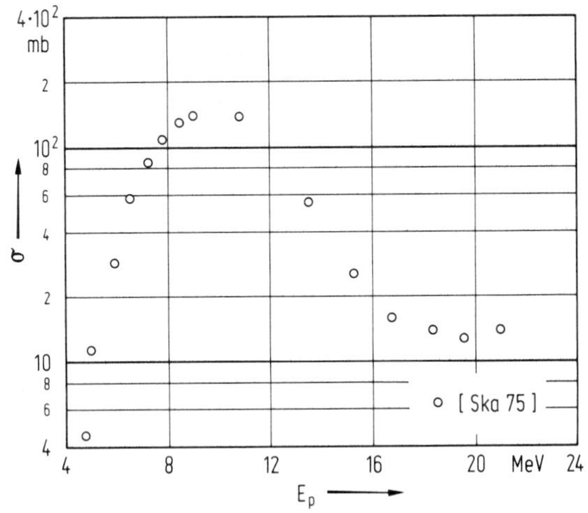

Fig. 543. Reaction $^{113}_{48}Cd \rightarrow ^{113m}_{49}In$

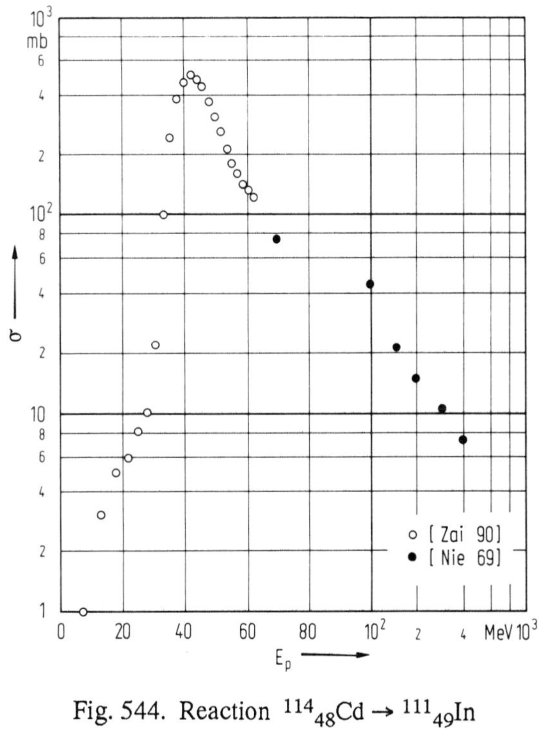

Fig. 544. Reaction $^{114}_{48}Cd \rightarrow ^{111}_{49}In$

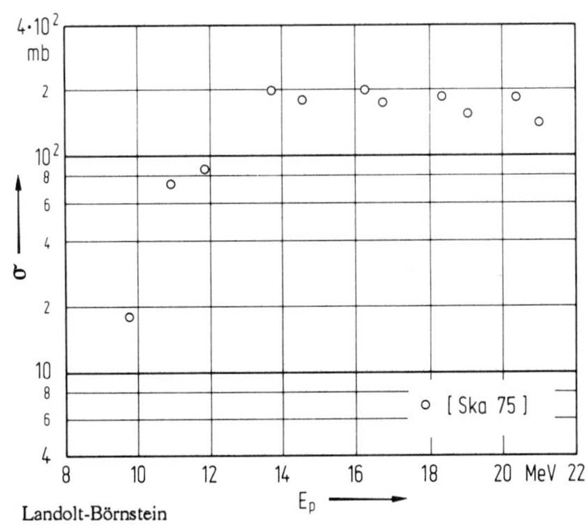

◀ Fig. 545. Reaction $^{114}_{48}Cd \rightarrow ^{113m}_{49}In$

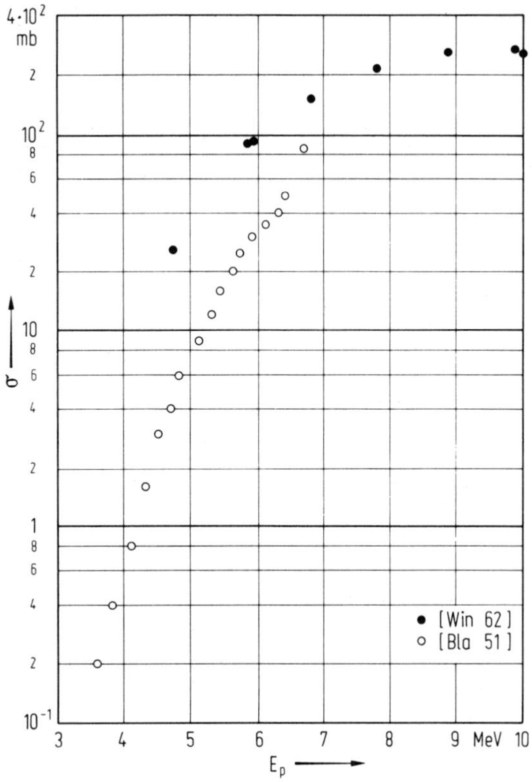

Fig. 546. Reaction $^{114}_{48}Cd \rightarrow {}^{114}_{49}In$

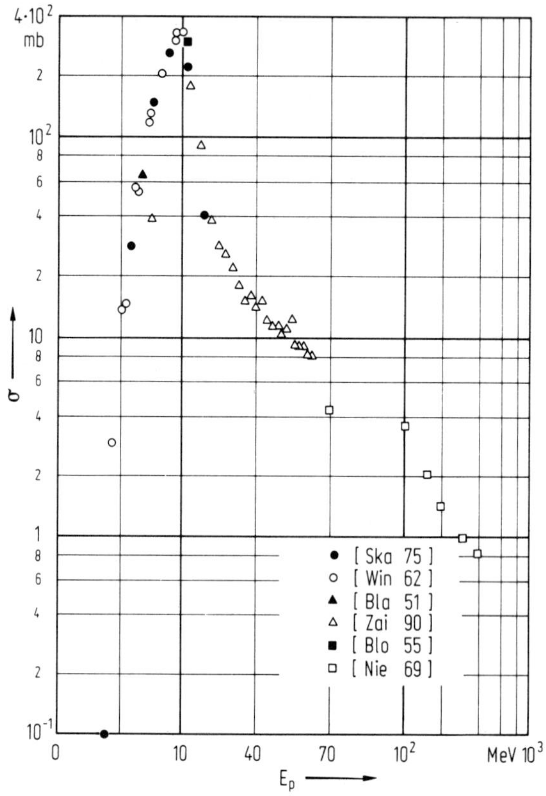

Fig. 547. Reaction $^{114}_{48}Cd \rightarrow {}^{114m}_{49}In$

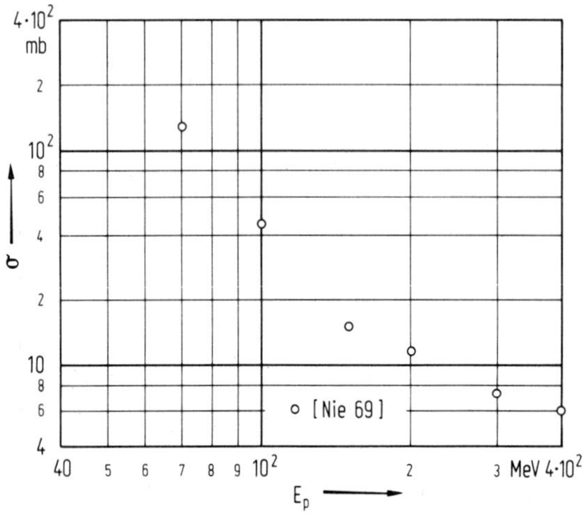

Fig. 548. Reaction $^{116}_{48}Cd \rightarrow {}^{111}_{49}In$

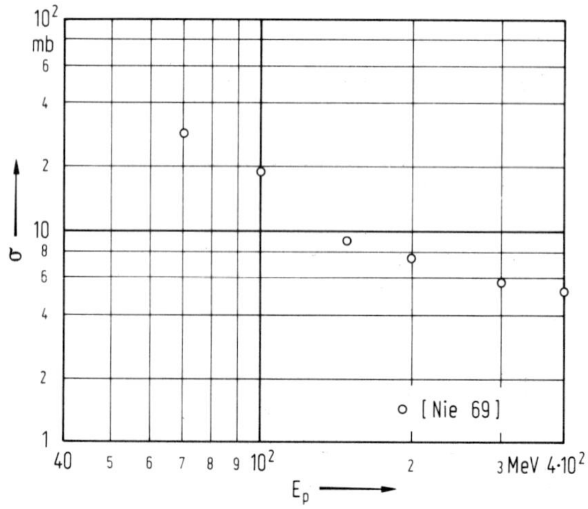

Fig. 549. Reaction $^{116}_{48}Cd \rightarrow {}^{114m}_{49}In$

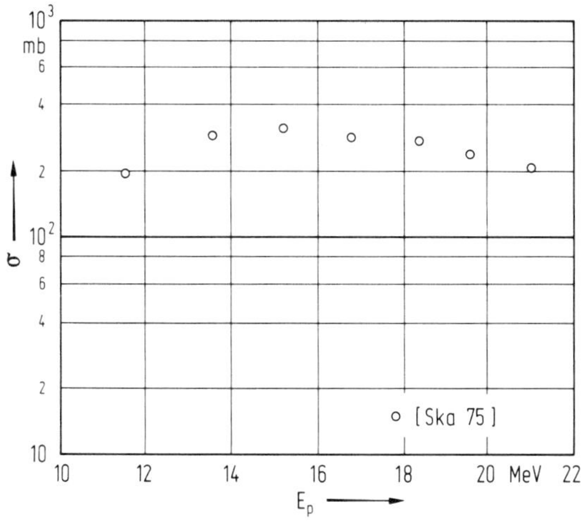

Fig. 550. Reaction $^{116}{}_{48}Cd \rightarrow {}^{115m}{}_{49}In$

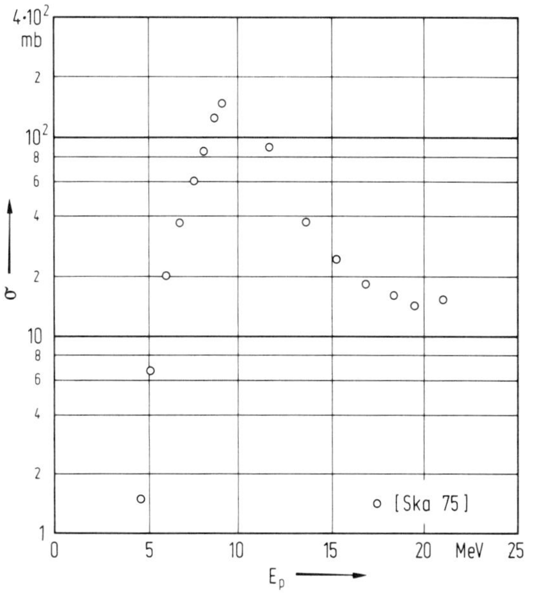

Fig. 551. Reaction $^{116}{}_{48}Cd \rightarrow {}^{116m}{}_{49}In$

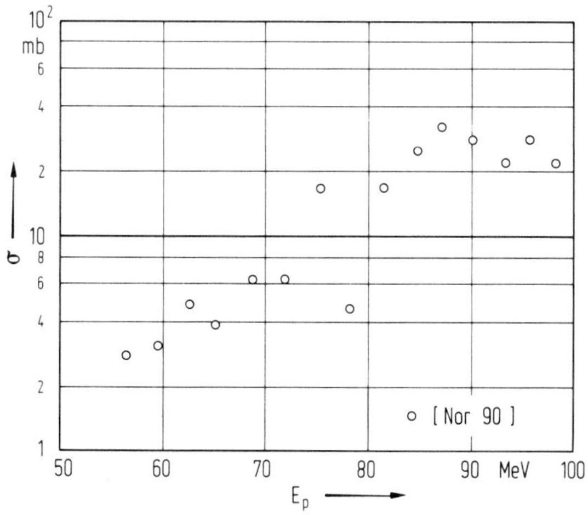

Fig. 552. Reaction $_{49}In \rightarrow {}^{109}{}_{50}Sn$

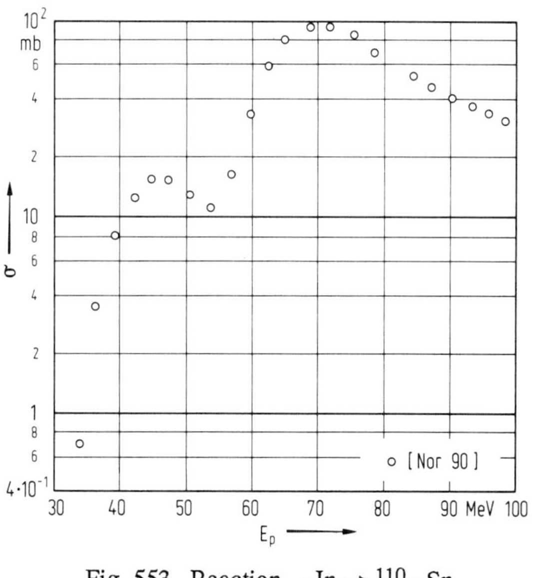

Fig. 553. Reaction $_{49}In \rightarrow {}^{110}{}_{50}Sn$

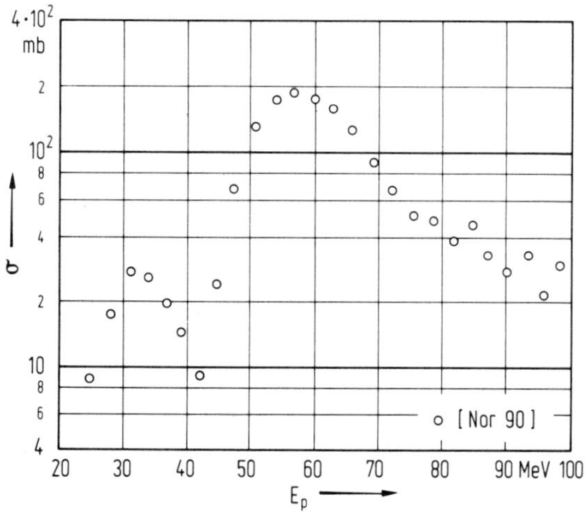

◀ Fig. 554. Reaction $_{49}In \rightarrow {}^{111}{}_{50}Sn$

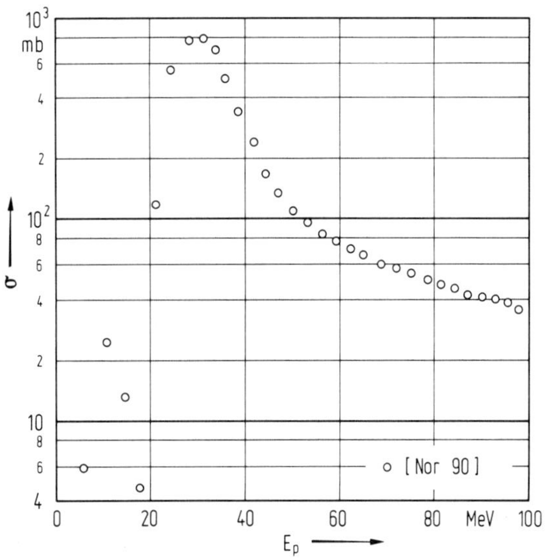

Fig. 555. Reaction $_{49}In \rightarrow {}^{113(0.91m+g)}{}_{50}Sn$

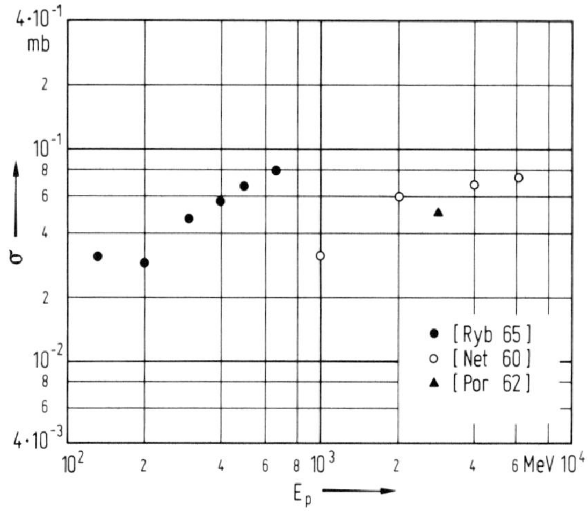

Fig. 556. Reaction $^{115}{}_{49}In \rightarrow {}^{115}{}_{48}Cd$

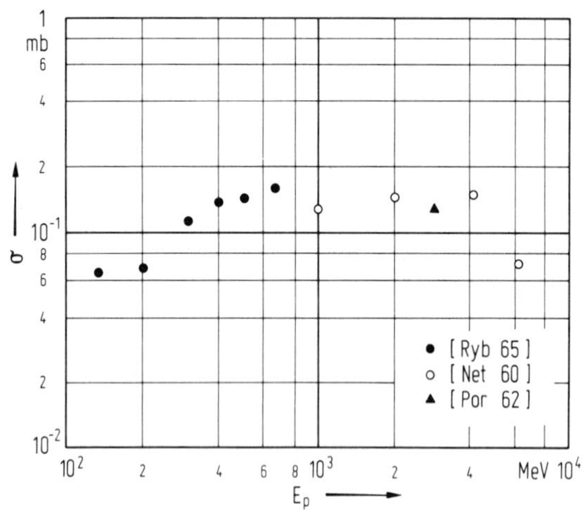

Fig. 557. Reaction $^{115}{}_{49}In \rightarrow {}^{115m}{}_{48}Cd$

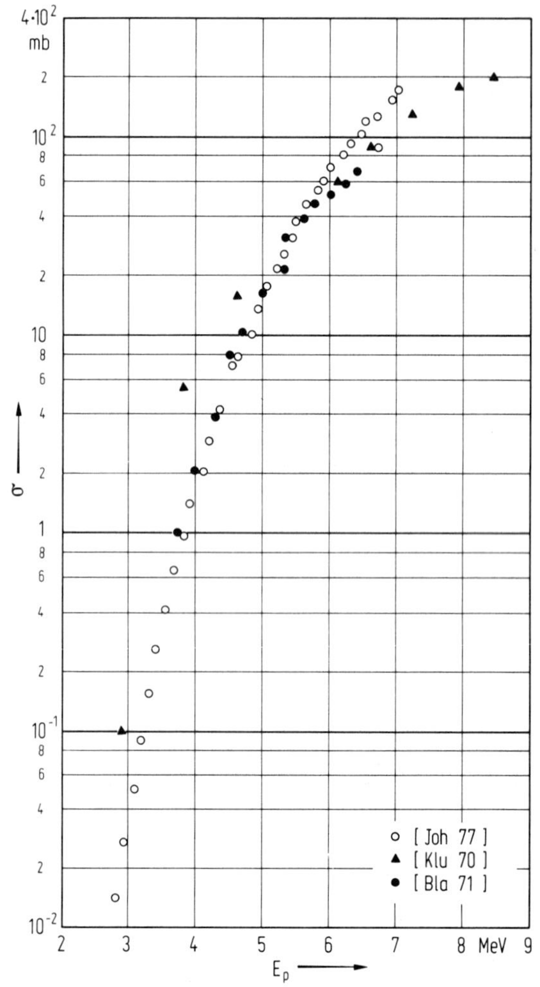

Fig. 559. Reaction $^{117}{}_{50}Sn \rightarrow {}^{117}{}_{51}Sb$

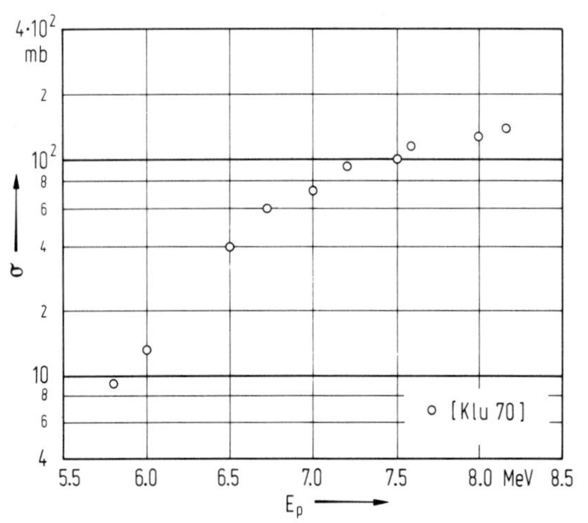

◀ Fig. 558. Reaction $^{116}{}_{50}Sn \rightarrow {}^{116}{}_{51}Sb$

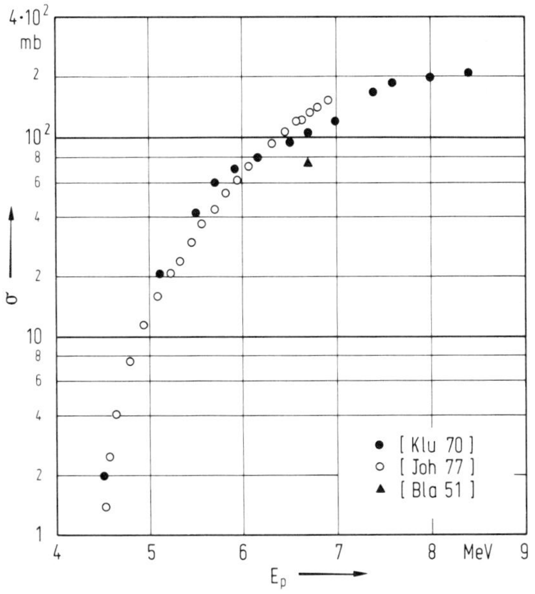

Fig. 560. Reaction $^{118}_{50}Sn \rightarrow ^{118}_{51}Sb$

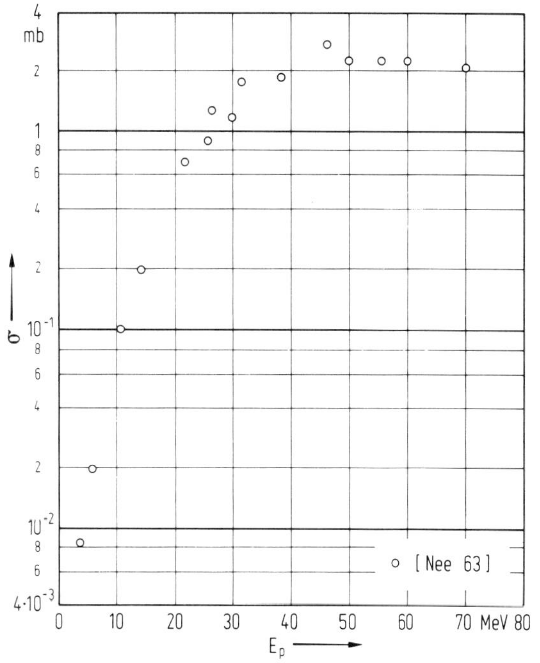

Fig. 562. Reaction $^{120}_{50}Sn \rightarrow ^{117m}_{49}In$

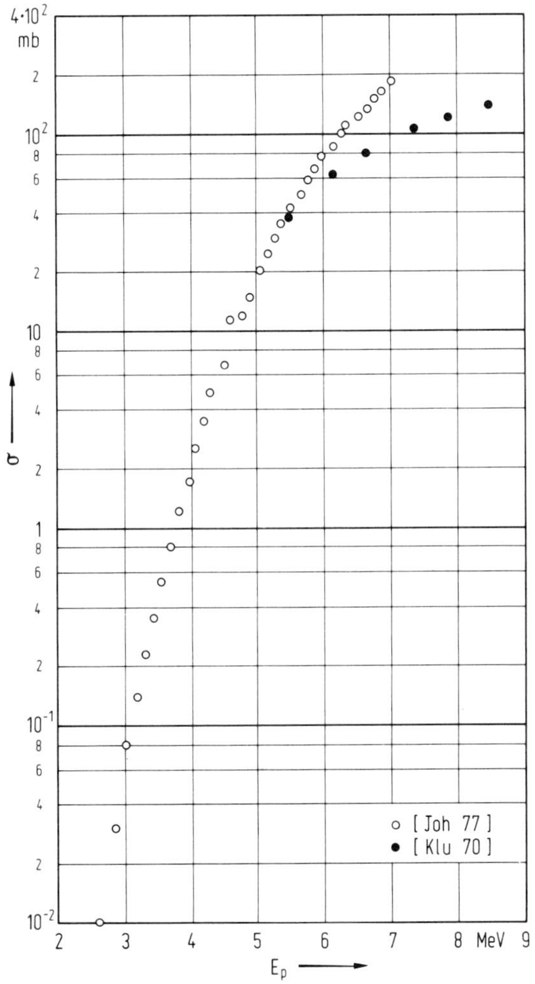

Fig. 561. Reaction $^{119}_{50}Sn \rightarrow ^{119}_{51}Sb$

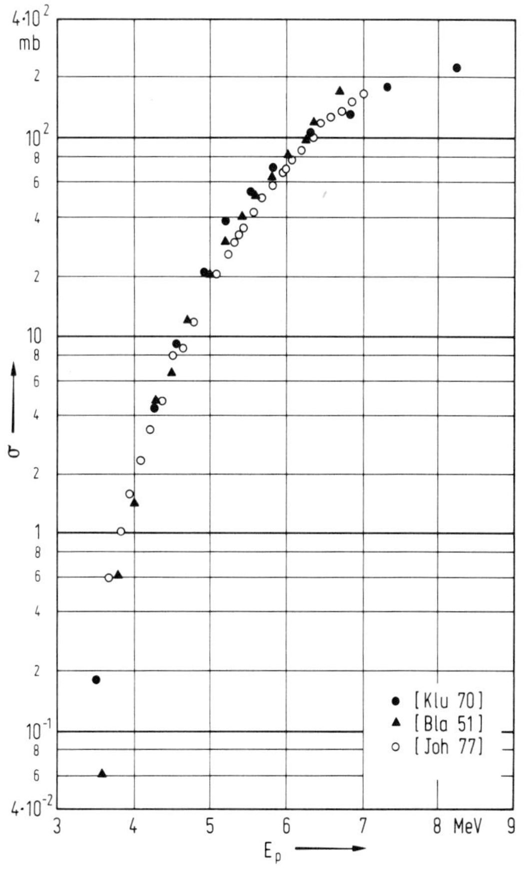

Fig. 563. Reaction $^{120}_{50}Sn \rightarrow ^{120}_{51}Sb$

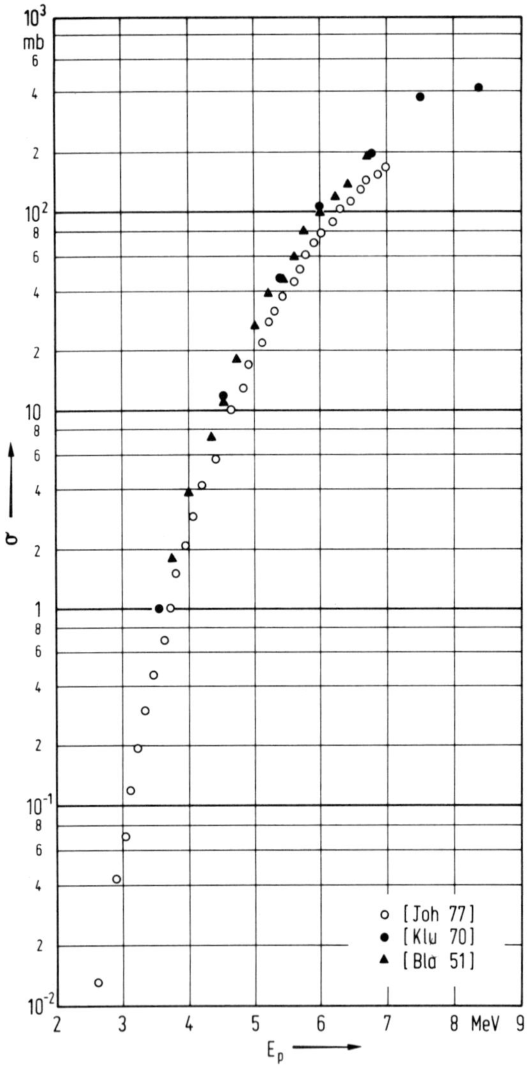

Fig. 564. Reaction $^{122}_{50}Sn \rightarrow ^{122}_{51}Sb$

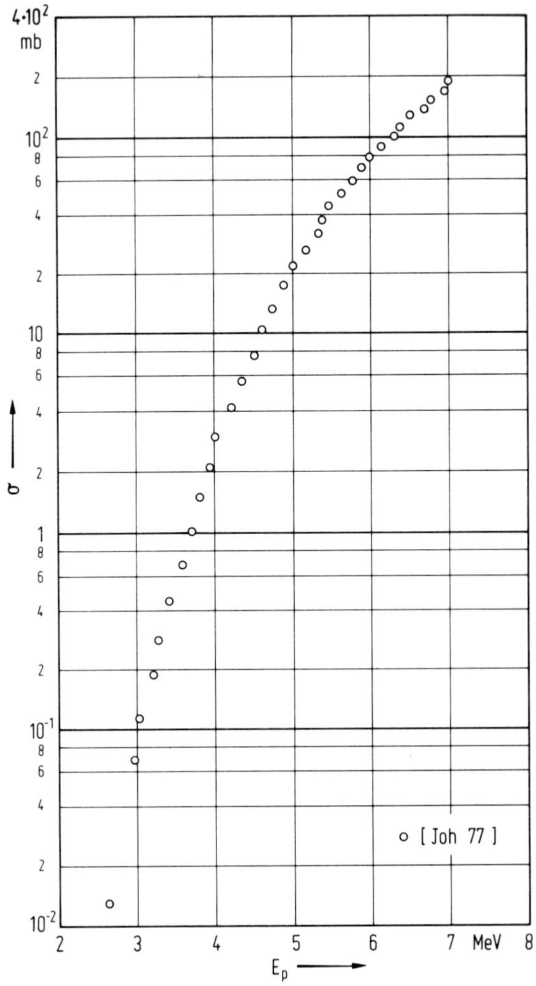

Fig. 565. Reaction $^{124}_{50}Sn \rightarrow ^{124}_{51}Sb$

Fig. 566. Reaction $_{51}Sb \rightarrow ^{117}_{52}Te$ ▶

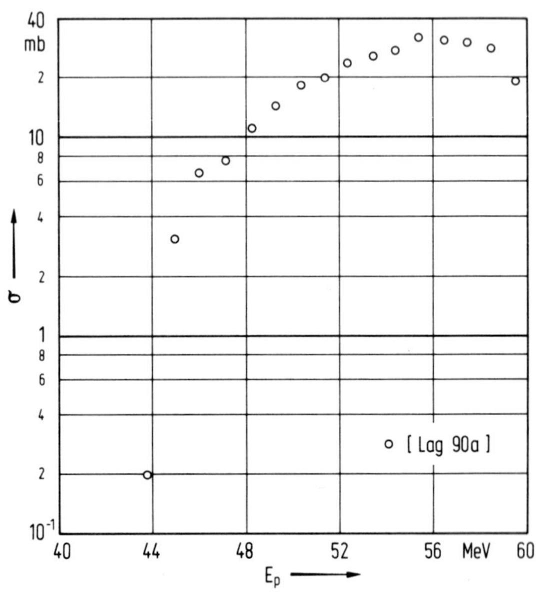

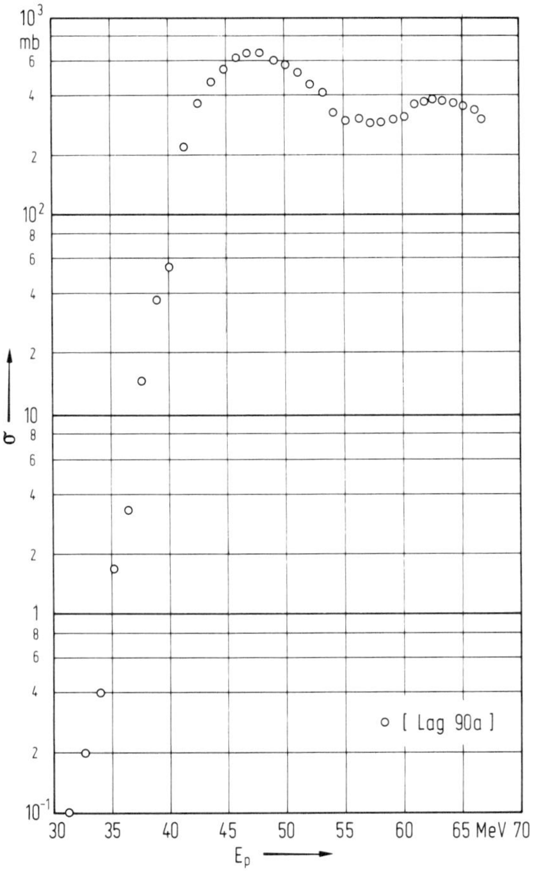

Fig. 567. Reaction $_{51}Sb \rightarrow {}^{118}_{52}Te$

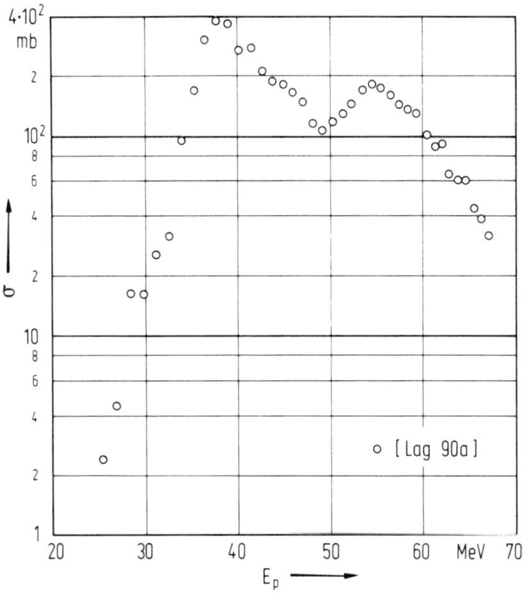

Fig. 568. Reaction $_{51}Sb \rightarrow {}^{119}_{52}Te$

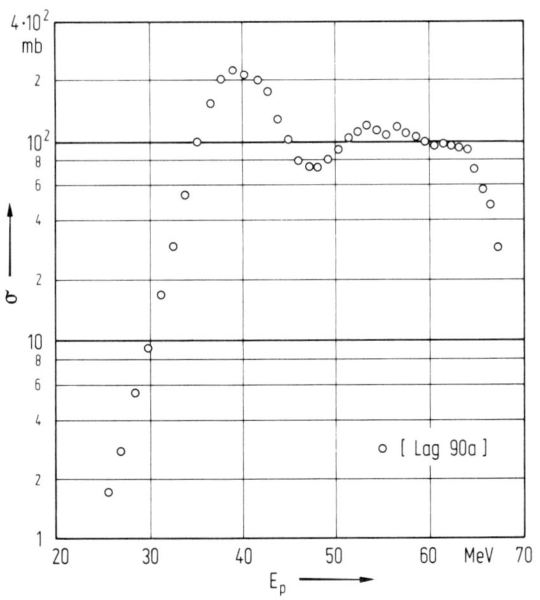

Fig. 569. Reaction $_{51}Sb \rightarrow {}^{119m}_{52}Te$

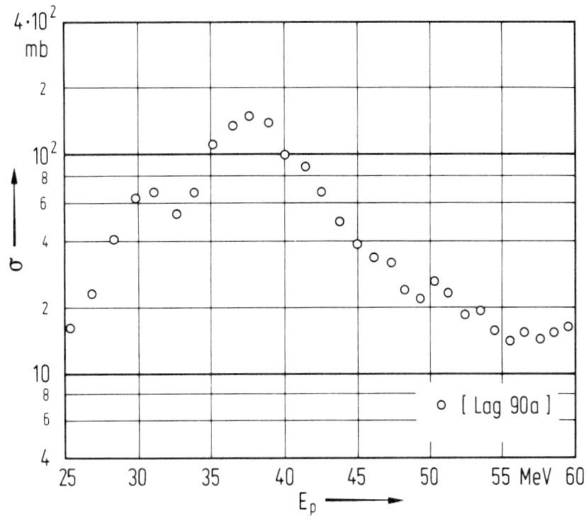

Fig. 570. Reaction $_{51}Sb \rightarrow {}^{121}_{52}Te$

Fig. 571. Reaction $_{51}Sb \rightarrow {}^{121m}_{52}Te$ ▶

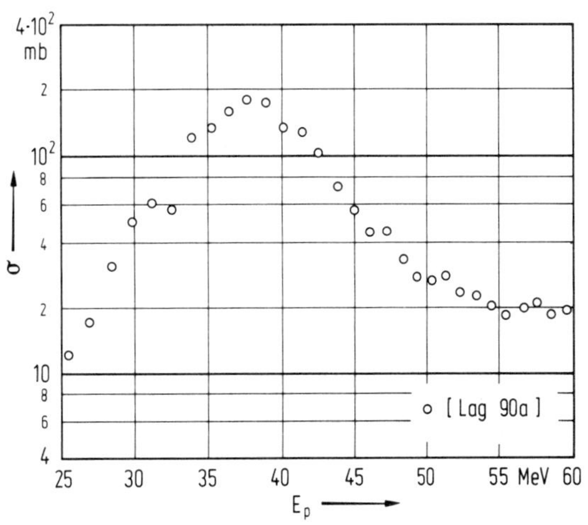

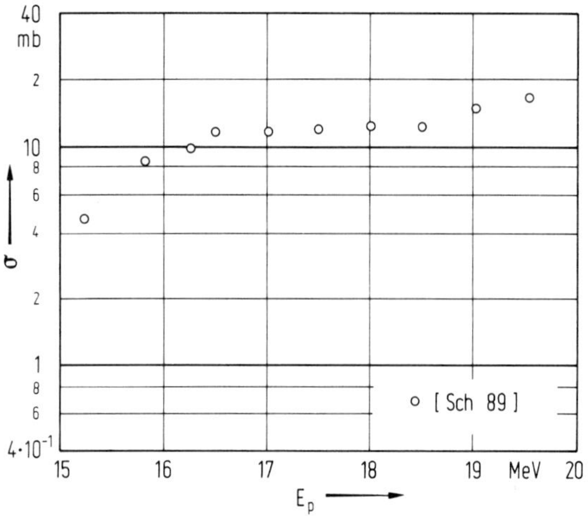

Fig. 572. Reaction $_{52}Te \rightarrow {}^{121}_{53}I$

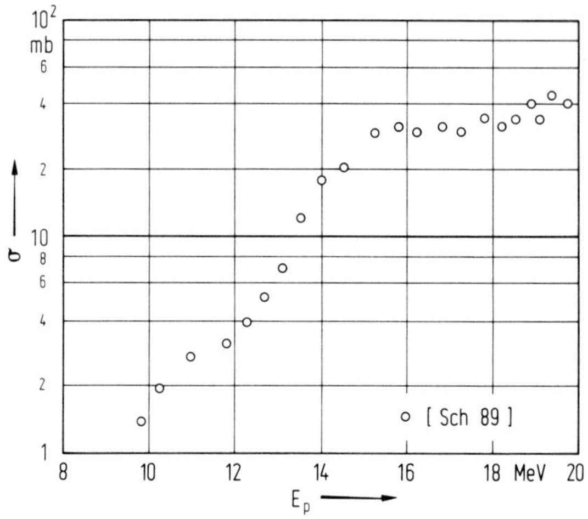

Fig. 573. Reaction $_{52}Te \rightarrow {}^{123}_{53}I$

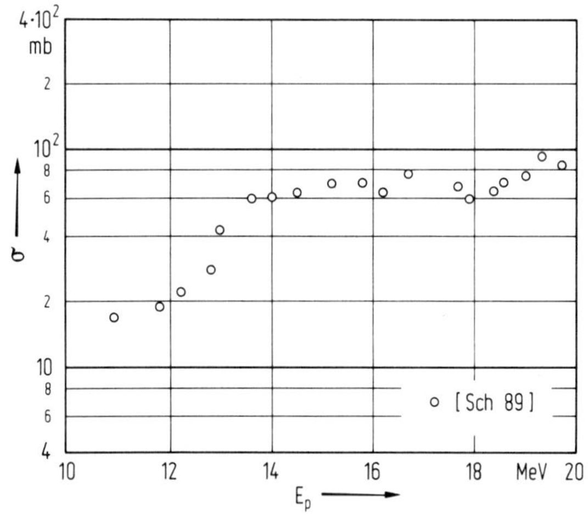

Fig. 574. Reaction $_{52}Te \rightarrow {}^{124}_{53}I$

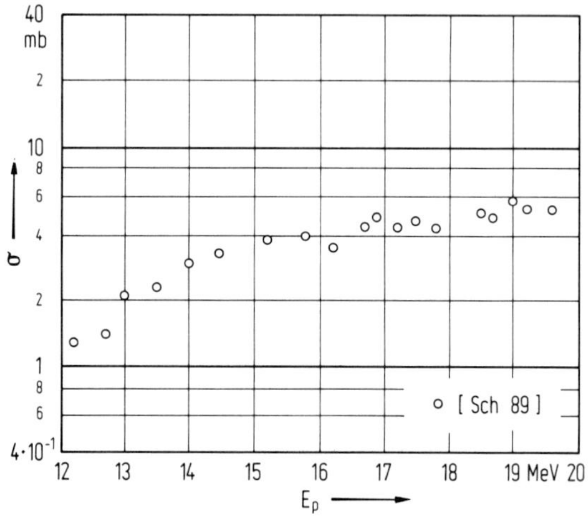

Fig. 575. Reaction $_{52}Te \rightarrow {}^{125}_{53}I$

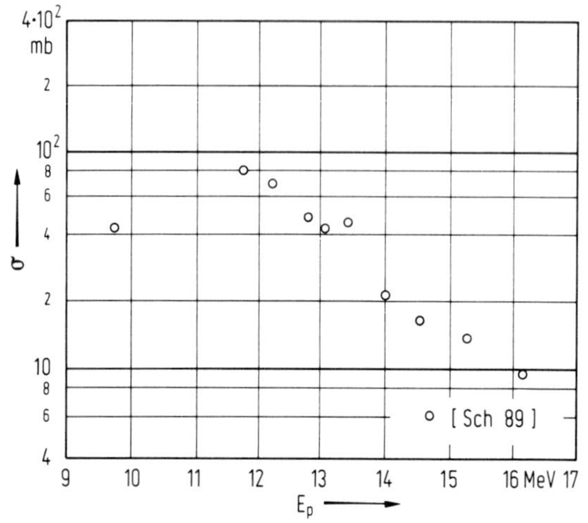

Fig. 576. Reaction $_{52}Te \rightarrow {}^{126}_{53}I$

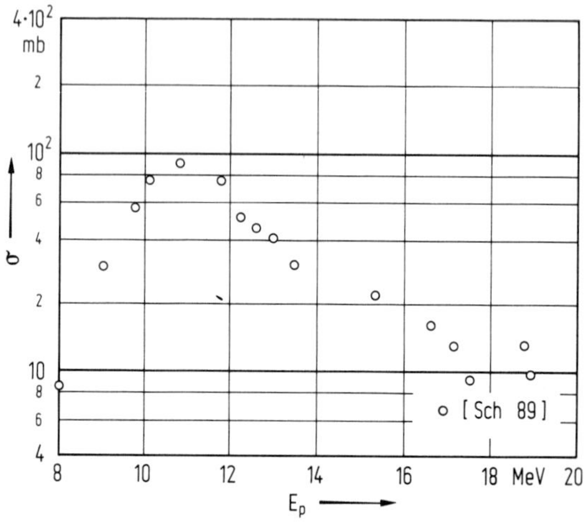

Fig. 577. Reaction $_{52}Te \rightarrow {}^{128}_{53}I$

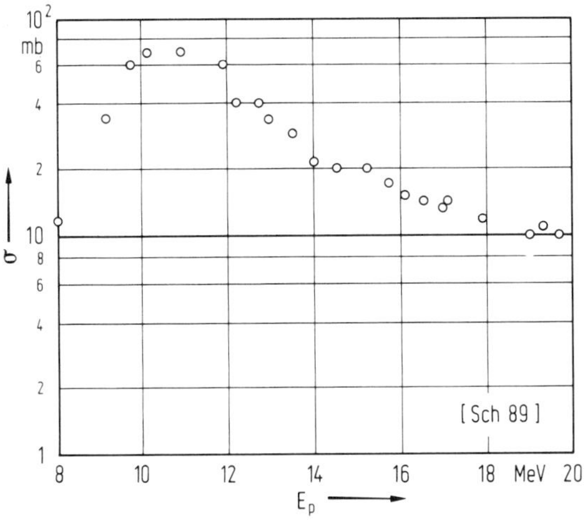

Fig. 578. Reaction $_{52}Te \rightarrow {}^{130}_{53}I$

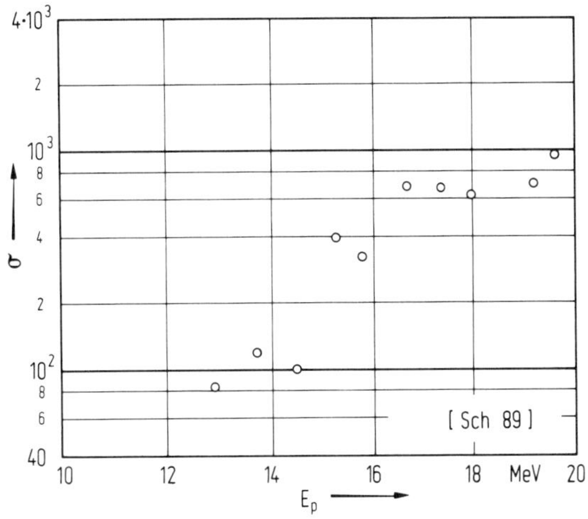

Fig. 579. Reaction $^{123}_{52}Te \rightarrow {}^{122}_{53}I$

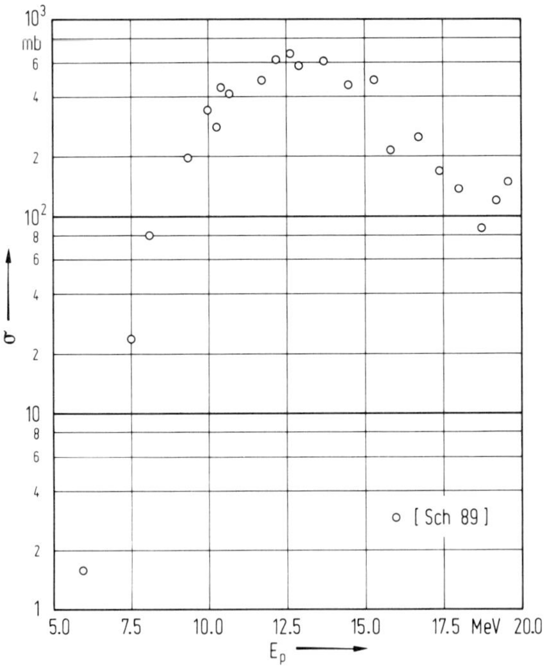

Fig. 580. Reaction $^{123}_{52}Te \rightarrow {}^{123}_{53}I$

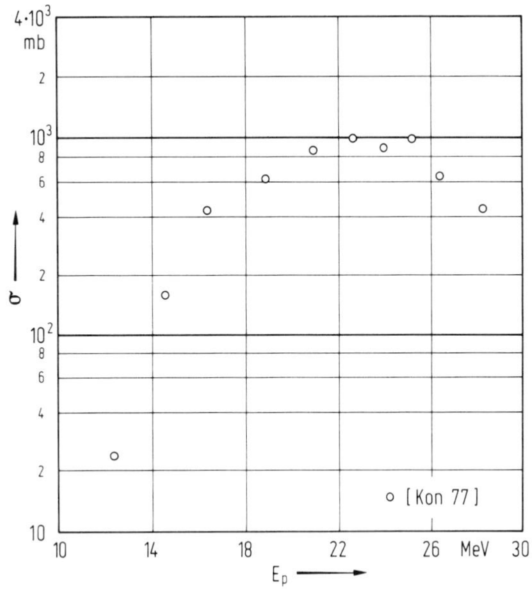

Fig. 581. Reaction $^{124}_{52}Te \rightarrow {}^{123}_{53}I$

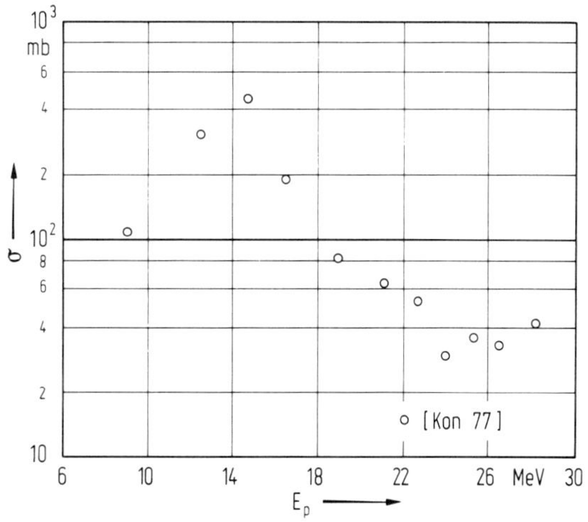

Fig. 582. Reaction $^{124}_{52}Te \rightarrow {}^{124}_{53}I$ ▶

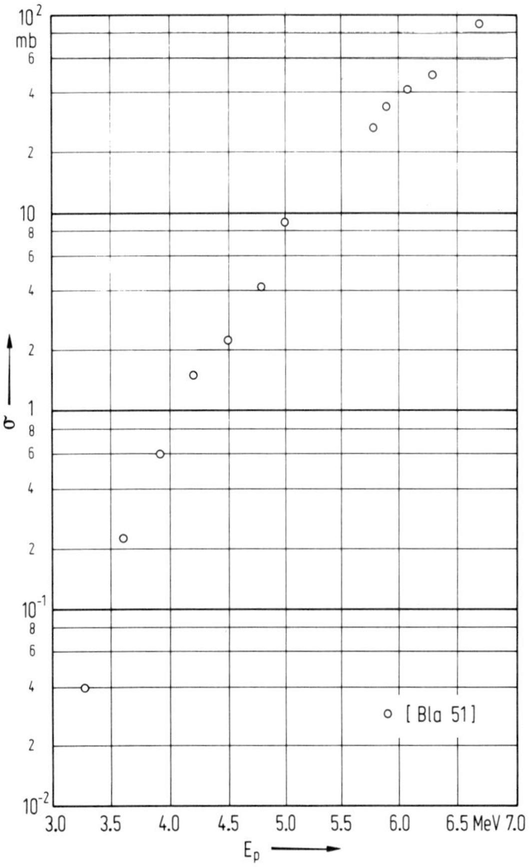

Fig. 583. Reaction $^{128}_{52}Te \rightarrow ^{128}_{53}I$

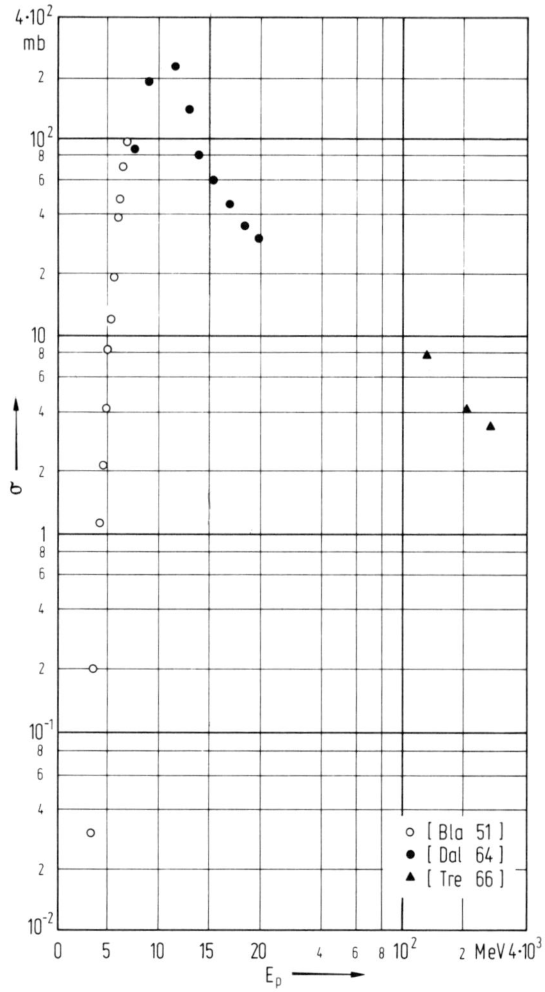

Fig. 584. Reaction $^{130}_{52}Te \rightarrow ^{130}_{53}I$

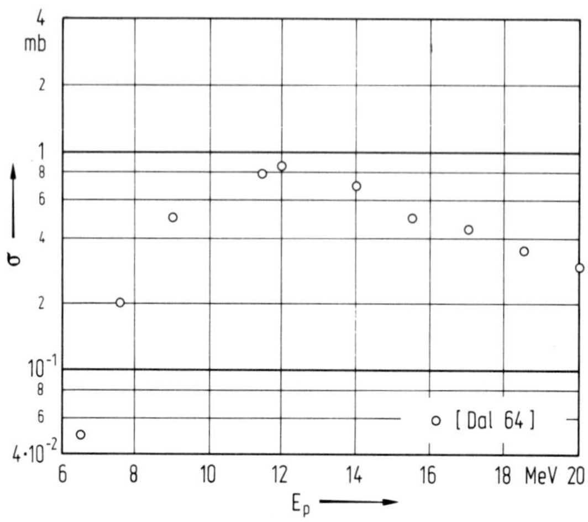

Fig. 585. Reaction $^{130}_{52}Te \rightarrow ^{131}_{53}I$

3.3 References for chapter 3

Adi 80 Adilbish, M., Fominykh, M.I., Khalkin, V.A., et al.: Cumulative Cross Sections of ^{200}Tl,^{201}Tl,^{202}Tl Production in 660 MeV Proton Induced Spallation of Lead and Bismuth. Radiochem. Radioanal. Lett. **45** (1980) 227.

Adi 80a Adilbish, M., Chumin, V.G., Khalkin, V.A., et al.: Medical ^{123}I Production from Radioxenon Formed in Spallation Reactions by 660 MeV Protons. Int. J. Appl. Radiat. Isot. **31** (1980) 163.

Alb 62 Albouy, M.G., Cohen, J.P., Gusakow, M., et al.: Spallation de l'Oxygene par des Protons de 20 à 150 MeV. Phys. Lett. **2** (1962) 306.

Alb 62a Albouy, M.G., Gusakow, M., Posse, N., et al.: Reaction (p,n) a Moyenne Energie. J. Phys. Radium **23** (1962) 1000.

Ale 63 Alexander, J.M., Baltzinger, C., Gazdik, M. F.: Cross Section and Recoil Studies of Reactions ^{238}U with Protons of 0.5 to 6.2 GeV. Phys. Rev. **129** (1963) 1826.

Ale 84 Alekseev, E.G., Guselnikov, V.S., Zaitsev, V.M., et al.: Formation of ^{76}Br and ^{77}Br Isotopes Using the Spallation Reactions at Proton Energy 1 GeV. Radiokhimiya **26** (1984) 667; Preprint LINP No. 839, Leningrad (1983).

Ale 91 Aleksandrov, V.N., Semenov, V.G., Sobolevsky, N.M.: Measurements and Calculations of Cross Sections for Spallation of Strontium and Cesium by 1 GeV Protons. Nuclear Transmutation of Nuclear Power Long-Lived Radioactive Waste. Obninsk, 1-5 July 1991. Workshop Abstracts, Obninsk (1991), p. 36.

And 56 Andre, C.G., Huizenga, J.R., Mech, J.F., et al.: Proton Cross Sections of ^{209}Bi. Phys. Rev. **101** (1956) 645.

Asa 85 Asano, Y., Mori, S., Noguchi, M., et al.: Spallation and Fission Yields in the Interactions of Tantalum, Tungsten, and Gold with 500-MeV Protons. J. Phys. Soc. Jpn. **54** (1985) 3734.

Asa 88 Asano, Y., Kariya, H., Mori, S., et al.: Spallation of Tantalum, Tungsten, and Gold by 12 GeV Protons. J. Phys. Soc. Jpn. **57** (1988) 2995.

Bak 58 Baker, E., Friedlander, G., Hudis, J.: Formation of ^{7}Be in Interactions of Various Nuclei with High-Energy Protons. Phys. Rev. **112** (1958) 1319.

Bat 84 Batiy, V.G., Skakun, E.A., Rastrepin, O.A., Rakivnenko Yu.N.: Excitation functions for Reactions ^{121}Sb (p,n) $^{121m+g}$Te and ^{123}Sb (p,n) ^{123m}Te. Izv. Akad. Nauk SSSR, Ser. Fiz. **48** (1984) 194.

Bat 88 Batiy, V.G., Skakun, E.A., Rakivnenko, Yu.N., Rastrepin, O.A.: Absolute and Relative Cross Sections of (p,n)-Reactions of Isomer Formation in Silver Isotopes. Yad. Fiz. **47** (1988) 609.

Bee 83 Beeley, P.A., Yaffe, L., Chatterjee, M., et al.: Isotopic and Neutron Yields of Products from the Assymmetric and Near Symmetric Proton Induced Fission of ^{233}U and ^{235}U. Phys. Rev. C **28** (1983) 1188.

Beg 71 Beg, K., Porile, N.T.: Energy Dependence of the Recoil Properties of Products from the Interaction of ^{238}U with 0.45 to 11.5 GeV Protons. Phys. Rev. C **3** (1971) 1631.

Bel 56 Bell, R.E., Skarsgard, H.M.: Cross Sections of (p,xn) Reactions in the Isotopes of Lead and Bismuth. Can. J. Phys. **34** (1956) 745.

Bel 63 Beljaev, B.I., Kaliamin, A.B., Murin, A.N.: Excitation Functions for Nuclear Reactions at ^{209}Bi Bombardment by Fast Protons.
Izv. Akad. Nauk SSSR, Ser. Fiz. **27** (1963) 923.

Bel 78 Beljaev, B.N., Domkin, V.D., Egorov, Yu.S., et al.: Nuclear Structure Influence on Nucleus Product Yields in Deep Spallation Reactions Induced by 1 GeV Protons.
Izv. Akad. Nauk SSSR, Ser. Fiz. **42** (1978) 2392.

Bel 88 Beljaev, B.N., Berezinsky, S.O., Domkin, V.D., et al.: Investigation of Production Cross Section in Deep Inelastic Interactions of Ir and U with Protons at 1 GeV and Reaction Mechanism Peculiarities.
Izv. Akad. Nauk SSSR Ser. Fiz. **52** (1988) 991.

Bey 66 Beyec, Y., Lefort, M., Peter, J.: Competition Fission-Evaporation de Neutrons Dans les Noyaux Issus du Bombardement de Bi par des Protons de 100 a 150 MeV.
Nucl. Phys. **88** (1966) 215.

Bey 67 Beyec, Y., Lefort, M.: Reactions Entre Protons de 30 a 155 MeV et Noyaux Lourds Complexes.
Nucl. Phys. A **99** (1967) 131.

Bim 66 Bimbot, R., Lefort, M.: Reactions Nucleaires (p,αxn) et (p,p αxn) sur le Bismuth aux Moyennes Energies, Presomptions en Faveur de la Projection de Sous-Structures Alpha.
J. Phys. Radium **27** (1966) 385.

Bir 71 Birattari, C., Gadioli, E., Grassistrini, A.M., et al.: (p,xn) Reactions Induced in ^{169}Tm, ^{181}Ta and ^{209}Bi with 20 to 45 MeV Protons.
Nucl. Phys. A **166** (1971) 605.

Bir 73 Birattari, C., Gadioli, E., Gadioli-Erba, E., et al.: Pre-Equilibrium Processes in (p,n) Reactions.
Nucl. Phys. A **201** (1973) 579.

Bir 86 Birgul, O., Yaffe, L.: The Fission of ^{197}Au by Protons of Energies of 52-85 MeV.
J. Radioanal. Nucl. Chem. Lett. **103** (1986) 153.

Bis 79 Biswas, S., Porile, N.T.: Formation of Isobaric Nuclides with A = 131 in the Interaction of ^{238}U with 0.8 to 11.5 GeV Protons.
Phys. Rev. C **20** (1979) 1467.

Bjo 67 Bjornholm, S., Borggrin, J., Westgaard, L., Karnaukhov, V.A.: Excitation Energy of Spontaneously Fissining Isomeric State in ^{240}Am.
Nucl. Phys. A **95** (1967) 513.

Bla 51 Blaser, J.P., Boehm, F., Marmier, P., Sherrer, P.: Fonctions d'Excitation de la Reaction (p,n). Part I, II, III.
Helv. Phys. Acta **24** (1951) 3, 441, 465.

Blo 55 Blosser, H.G., Handlley, T.H.: Survey of (p,n) Reactions at 12 MeV.
Phys. Rev. **100** (1955) 1340.

Bog 76 Bogatin, V.I., Litvin, V.F., Lozhkin, O.V., et al.: Isotopic Effects in High-Energy Nuclear Reactions and Isospin Correlations of Fragmentation Cross Sections.
Nucl. Phys. A **260** (1976) 446.

Bon 84 Bonardi, M., Birattapi, C.: Excitation Functions for the Au (p,3n) ^{195m,195}Hg Nuclear Reactions and ^{195m,195}Hg/^{195m}Au Generator Yields.
Int. J. Appl. Radiat. Isot. **35** (1984) 564.

Bou 60 Bouissieres, G., Hussonnois, M., Lefort, M.: Section Efficace de Formation du Protactinium 227 par Spallation du Thorium par des Protons de 155 MeV.
Compt. Rend. B **251** (1960) 2155.

Bru 62 Brun, C., Simonoff, G.N.: Competition Fission-Evaporation. Etude des Fonctions d'Excitation Dans Differents Noyaux de Protactinium.
J. Phys. Radium **23** (1962) 12.

Bru 62b Brun, C., Levort, M., Tarrago, X.: Contribution a l'Etude du Double Pick-up Indirect Mesure de la Production de Tritium par des Protons de 82 et 105 MeV Dans Diverses Cibles.
J. Phys. Radium **23** (1962) 167.

Bru 64 Brun, C., Lefort, M.: Production des Isotopes 196 a 202 du Polonium par Reactions Nucleaires (p,xn) sur le Bismuth.
J. Inorg. Nucl. Chem. **26** (1964) 1633.

Byc 82 Bychov, V.M., Manokhin, V.N., Paschenko, A.B., Pliaskin, V.I.: Cross sections of Neutron Induced Threshold Reactions.
Handbook, Energoatomisdat, Moscow, 1982.

Car 56 Caretto, A.A., Wiig, E.O.: Spallation of Yttrium by 240 MeV Protons.
Phys. Rev. **103** (1956) 236.

Car 58 Caretto, A.A., Hudis, J., Friedlander, G.: Production of ^{18}F and ^{24}Na in Irradiations of Various Targets with Protons Between 1 and 6 BeV.
Phys. Rev. **110** (1958) 1130.

Car 59 Caretto, A.A., Wiig, E.O.: Interaction of Yttrium with Protons of Energy Between 60 and 240 MeV.
Phys. Rev. **115** (1959) 1238.

Cha 73 Chang, S.K., Sugarman, N.: Cross Section and Recoil Properties of Copper Isotopes from Uranium Bombarded with 11.5 GeV Protons.
Phys. Rev. C **8** (1973) 775.

Cha 74 Chang, S.K., Sugarman, N.: Interaction of 300 GeV Protons with Uranium.
Phys. Rev. C **9** (1974) 1138.

Cha 77 Chan, K.C., Pathak, B.P., Nikkinen, L., et al.: Independent Yields of Indium and Gallium in the Proton Induced Fission of Natural Uranium.
J. Inorg. Nucl. Chem. **39** (1977) 1915.

Chu 69 Church, L.B., Caretto, A.A.: Study of (p,xn) Reactions at 400 MeV.
Phys. Rev. **178** (1969) 1732.

Chu 71 Church, L.B.: The ^{96}Zr (p,2n) ^{95}Nb Isomeric Ratio at 400 MeV.
Nucl. Phys. A **162** (1971) 643.

Chu 71a Chu, Y.Y., Franz, E.M., Friedlander, G., Karol, P.J.: Interaction of 28 GeV Protons with Uranium: Charge and Mass Distribution of Rare-Earth Products.
Phys. Rev. C **4** (1971) 2202.

Chu 72 Chu, Y.Y., Franz, E.M., Friedlander, G.: Secondary Effects Produced by Interaction of 28 GeV Protons with Heavy Element Targets.
Nucl. Phys. B **40** (1972) 428.

Chu 74 Chu, Y.Y., Franz, E.M., Friedlander, G.: Interaction of 28-GeV Protons with Tantalum: Charge Dispersion and Mass Yields of Light Rare-Earth Nuclides.
Phys. Rev. C **10** (1974) 156.

Chu 76 Chu, Y.Y., Franz, E.M., Friedlander, G., Karol, P.J.: Interaction of 28-GeV Protons with Thorium: Study of Charge Dispersions and Mass Yields.
Phys. Rev. C **14** (1976) 1068.

Chu 77 Chu, Y.Y., Friedlander, G., Husain, L.: Production of Nuclides with $43 < A < 51$ in the Interaction of 1 to 28.5 GeV Protons with V, Ag, In, Pb, and U Targets.
Phys. Rev. C **15** (1977) 352.

Chu 83 Chu, Y.Y., Zhou, M.L.: Comparison of the (p,xn) Cross Sections from ^{238}U, ^{235}U and ^{232}Th Targets Irradiated with 200 MeV Protons.
IEEE Trans. Nucl. Sci., NS-**30** (1983) 1153.

Coh 54 Cohen, B.L., Newman, E., Charpie, R.A., Handley, T.H.: (p,pn) and (p,αn) Excitation Functions.
Phys. Rev. **94** (1954) 620.

Coh 55a Cohen, B.L.: (p,γ) Cross Sections.
Phys. Rev. **100** (1955) 206.

Col 74 Colle, R., Kishore, R., Cumming, J.B.: Excitation Functions for (p,n) Reactions to 25 MeV on ^{63}Cu, ^{65}Cu and ^{107}Ag.
Phys. Rev. C **9** (1974) 1819.

Col 74a Colle, R., Kishore, R.: Excitation Functions for (p,n) Reactions on ^{79}Br and ^{127}I.
Phys. Rev. C **9** (1974) 2166.

Col 81 Cole, G.D., Porile, N.T.: Enhanced Fragment Emission in the Interaction of 18.5 GeV ^{12}C Ions with Complex Nuclei.
Phys. Rev. C **24** (1981) 2038.

Cre 63 Crespo, V.P., Alexander, J.M., Hyde, E.K.: Ejection of Large Fragments in High-Energy Nuclear Reactions.
Phys. Rev. **131** (1963) 1765.

Cur 59 Currie, L.A.: Tritium Production by 6 BeV Protons.
Phys. Rev. **114** (1959) 878.

Dal 64 Daly, P.J., Shaw, P.F.D.: Radiative Proton Capture Cross-Sections in Heavy Nuclei.
Nucl. Phys. **56** (1964) 322.

Dau 80 Dautet, H., Bischoff, G., Auria, J.M., Pate, B.D.: Yield of Deep Spallation Products of Medium to Heavy Mass Targets.
Can. J. Phys. **58** (1980) 891.

Del 63 Delaunau-Olkowsky, J., Strohal, P., Cindro, N.: Total Reaction Cross Section of Proton Induced Reactions.
Nucl. Phys. **43** (1963) 266.

Dik 77 Diksic, M., Yaffe, L.: A Study of ^{127}I (p,xn) and ^{127}I (p,pxn) Reactions with Special Emphasis on Production of ^{123}Xe.
J. Inorg. Nucl. Chem. **39** (1977) 1299.

Dik 79a Diksic, M., Galinier, J.L., Marshall, H., Yaffe, L.: Charge Dispersion Studies of Light-Mass Nuclides in the Fission of ^{232}Th by Protons of Energies 35-85 MeV.
J. Inorg. Nucl. Chem. **41** (1979) 795.

Dit 90 Dittrich, B., Herpers, U., Lupke, M., et al.: Determination of Cross-Sections for the Production of ^{7}Be, ^{10}Be and ^{22}Na by High-Energy Protons.
Radiochim. Acta **50** (1990) 11.

Dom 85 Dombsky, M., d'Auria, J.M., Kelson, I., et al.: Inclusive Measurement of ($p,\pi^{-}xn$) Double Charge Exchange Reactions on Bismuth from Threshold to 800 MeV.
Phys. Rev. C **32** (1985) 253.

Dos 61 Dostrovsky, I., Fraenkel, Z., Hudis, J.: Formation of ^{13}N in High-Energy Nuclear Reactions.
Phys. Rev. **123** (1961) 1452.

Eng 74 English, G., Porile, N.T., Steinberg, E.P.: Nuclear Reactions of Silver with 11.5 GeV Protons.
Phys. Rev. C **10** (1974) 2268.

Eng 74a English, G., Yu, Y.W., Porile, N.T.: Nuclear Reactions of Silver with 300 GeV Protons.
Phys. Rev. C **10** (1974) 2281.

Fel 83 Felice, P., Ocone, R., Rindi, A., et al.: High Energy Proton Interactions with Sr and Cu.
Nucl. Instrum. Methods Phys. Res. **212** (1983) 359.

Fin 54 Fink, R.W., Wiig, E.O.: Reactions of Cesium with 240 MeV Protons.
Phys. Rev. **94** (1954) 1357.

Fin 54a Fink, R.W., Wiig, E.O.: Reactions of Cesium with Protons at 60, 80, 100, 150 and 240 MeV.
Phys. Rev. **96** (1954) 185.

Fly 79 Flynn, D.S., Hershberger, R.L., Gabbard, F.: Sub-Coulomb Proton Absorption for Isotopes of Zirconium and Molybdenum.
Phys. Rev. C **20** (1979) 1700.

For 53 Forsthoff, C.W., Goeckermann, R.H., Naumann, R.A.: Formation and Decay of ^{93m}Mo.
Phys. Rev. **90** (1953) 1004.

Fow 62 Fowler, W.A., Greenstein, J.L., Hoyle, F.: Nucleosynthesis During the Early History of the Solar System.
Geophys. J. **6** (1962) 148.

Fra 66 Franz, E.M., Friedlander, G.: Cross Sections for Production of ^{149}Tb from Au by High Energy Protons.
Nucl. Phys. **76** (1966) 123.

Fra 71 Franz, E.M., Friedlander, G.: Gross Sections for the Production of ^{84}Rb, ^{86}Rb, ^{132}Cs, and ^{136}Cs from Thorium and Uranium by 28.5 GeV Protons.
Phys. Rev. C **4** (1971) 1671.

Fri 60 Friedlander, G., Yaffe, L.: Cross Sections for the Production of Cl, K, Ca, and Sc Isotopes in the Bombardment of Pb and U With 3.0-BeV Protons.
Phys. Rev. **117** (1960) 578.

Fur 67 Furukawa, M.: Excitation Functions for Proton-Induced Reactions of ^{140}Ce and ^{142}Ce up to E_p = 15 MeV.
Nucl. Phys. A **67** (1967) 253.

Gal 77 Galinier, J.L., Yaffe, L.: Nuclear Charge Distribution in the Region of Asymmetric Fission of ^{238}U by Protons of Encrgy 20 to 85 MeV.
J. Inorg. Nucl. Chem. **39** (1977) 1497.

Gar 76 Garber, D.I., Kinsley, R.R.: Neutron Cross Sections, Vol. 2, Curves.
BNL 325, Third Edition, Brookhaven, 1976

Gau 62 Gauvin, H., Lefort, M., Tarrago, X.: Emission d'Helions Dans les Reactions de Spallation.
Nucl. Phys. **39** (1962) 447.

Gau 63 Gauvin, H.: Reactions (p,2pxn) sur le Thorium 232 de 30 a 120 MeV.
J. Phys. Radium **24** (1963) 836.

Gra 63 Grant, G.R., Yaffe, L.: The Spallation of Iridium with Protons of Energies 9 à 87 MeV.
Can. J. Chem. **41** (1963) 2533.

Gra 83 Grant, P.M., Erdal, B.R., Brien, H.A.: Medium-Energy Spallation Cross Sections.
2. Mo Irradiation with 800-MeV Protons.
Int. J. Appl. Radiat. Isot. **34** (1983) 1631.

Gru 82 Grutter, A.: Cross Sections for Reactions with 593 and 540 MeV Protons in Al, As, Br, Rb, Y.
Int. J. Appl. Radiat. Isot. **33** (1982) 725.

Gus 62 Gusakow, M.: Des Reactions (p,pn) a Moyenne Energie.
Ann. Phys. (N.Y.) **7** (1962) 67.

Hal 81 Haldorsen, I., Engelsen, S., Eriksen, D., et al.: Production of ^{24}Na and ^{28}Mg by the Interaction of Uranium with 15 to 600 MeV Protons.
J. Inorg. Nucl. Chem. **43** (1981) 2197.

Han 62a Hansen, L.F., Jopson, R.C., Mark, H., Swift, C.D.: ^{181}Ta (p,n) ^{181}W and ^{197}Au (p,n) ^{197}Hg Excitation Functions between 4 and 13 MeV.
Nucl. Phys. **30** (1962) 389.

Hel 91 Helus, F., Zeisler, St., Gschwandtner, G., et al.: Production of ^{75}Br and ^{77}Br from Enriched Krypton: Excitation Functions.
J. Radiat. Nucl. Chem. Lett. **153** (1991) 417.

Hog 71 Hogan, J.J.: Study of the ^{141}Pr (p,xn) Reaction from 10 to 85 MeV.
J. Inorg. Nucl. Chem. **33** (1971) 3627.

Hog 72 Hogan, J.J.: ^{96}Mo (p,xn) Reaction from 10 to 80 MeV.
Phys. Rev. C **6** (1972) 810.

Hog 73 Hogan, J.J.: Isomer Ratios of Tc Isotopes Produced in 10 to 65 MeV Bombardments of ^{96}Mo.
J. Inorg. Nucl. Chem. **35** (1973) 705.

Hog 79 Hogan, J.J., Gadioli, E., Gadioli-Erba, E., Chung, C.: Fissionability of Nuclides in the Thorium Region at Excitation Energies to 100 MeV.
Phys. Rev. C **20** (1979) 1831.

Hor 80 Horiguchi, T., Noma, H., Yoshizawa, Y., et al.: Excitation Functions of Proton Induced Nuclear Reactions on ^{85}Rb.
Int. J. Appl. Radiat. Isot. **31** (1980) 141.

Hud 68 Hudis, J.: Production of Neon-24 and Sodium-24 in the 1 to 3 GeV Proton Bombardment of Copper, Silver, Gold, and Uranium Targets.
Phys. Rev. **171** (1968) 1301.

Hud 70 Hudis, J., Kirsten, T., Stoenner, R.W., Schaeffer, O.A.: Yields of Stable and Radioactive Rare Gas Isotopes Formed by 3 and 29 GeV Proton Bombardment of Cu, Ag, Au and U.
Phys. Rev. C **1** (1970) 2019.

Hun 59 Hunter, E.T., Miller, J.M.: Spallation of Bismuth by 380 MeV Protons.
Phys. Rev. **115** (1959) 1053.

Ilj 91 Iljinov, A.S., Semenov, V.G., Semenova, M.P., Sobolevsky, N.m., Udovenko, L.V.: Production of Radionuclides At Intermediate Energies.
Landolt-Börnstein, New Series, Group I, Vol. 13a. Berlin-Heidelberg-New York: Springer-Verlag, 1991.

Izu 91 Izumo, M., Matsuoka, H., Sorita, T., et al.: Production of ^{95m}Tc with Proton Bombardment of ^{95}Mo.
Appl. Radiat. Isot. **42** (1991) 297.

Jam 54 James, R.A.: Excitation Functions of Proton-Induced Reactions of ^{93}Nb.
Phys. Rev. **93** (1954) 288.

Jod 55 Jodra, L.G., Sugarman, N.: High-Energy Fission of Bismuth. Proton Energy Dependence.
Phys. Rev. **99** (1955) 1470.

Joh 68 Johnson, C.H., Kernell, R.L., Ramavataram, S.: The ^{89}Y (p,n) ^{89}Zr Cross Section Near the First Two Analogue Resonances.
Nucl. Phys. A **107** (1968) 21.

Joh 77 Johnson, C.H., Bair, J.K., Jones, C.M., et al.: P-Wave Size Resonances Observed by the (p,n) Reaction for 2.6 to 7 MeV Protons Incident on Isotopes of Sn.
Phys. Rev. C **15** (1977) 196.

Kan 76 Kantelo, M.V., Hogan, J.J.: Charged-Particle Emission in Reactions of ^{90}Zr with 10 to 86 MeV Protons.
Phys. Rev. C **14** (1976) 64.

Kan 76a Kantelo, M.V., Hogan, J.J.: Radiochemical Study of Alpha Particle Emission in Reactions ^{202}Hg with 10 to 86 MeV Protons.
Phys. Rev. C **13** (1976) 1095.

Kat 68 Katcoff, S., Fickel, H.R., Wyttenbach, A.: Distribution of Radionuclides from the Interaction of 3 and 29 GeV Protons with Silver.
Phys. Rev. **166** (1968) 1147.

Kau 63 Kaufman, S.B.: Nuclear Charge Distributions in the Interaction of 2.9 GeV Protons with Heavy Elements.
Phys. Rev. **129** (1963) 1866.

Kau 76 Kaufman, S.B., Weisfield, M.W., Steinberg, E.P., et al.: Nuclear Reactions of ^{197}Au with 11.5 and 300 GeV Protons.
Phys. Rev. C **14** (1976) 1121.

Kau 80 Kaufman, S.B., Steinberg, E.P.: Cross-Section Measurements of Nuclides Formed by the Reaction of 0.20 to 6.0 GeV Protons with ^{197}Au.
Phys. Rev. C **22** (1980) 167.

Kav 61 Kavanagh, T.M., Bell, R.E.: Cross Sections of (p,pxn) Reactions in ^{197}Au.
Can. J. Phys. **39** (1961) 1172.

Kaw 76 Kawakami, H., Koike, M., Komura, K., et al.: Cross Sections and Reaction Mechanisms of (p,pxn) Reactions on ^{208}Pb in the 24 to 52 MeV Range.
Nucl. Phys. A **262** (1976) 52.

Kel 50 Kelly, E.L.: Preprint UCRL 1044 (1950).

Kha 68 Khan, A.H., Saha, G.B., Yaffe, L.: Yields of Some Light-Mass Nuclides Produced in the Proton Induced Fission of ^{238}U.
Can. J. Chem. **46** (1968) 3565.

Kje 56 Kjelberg, A., Pappas, A.C.: Fission of Uranium with 170 MeV Protons.
Nucl. Phys. **1** (1956) 322.

Klu 70 Klusharev, A.P., Skakun, E.A., Rakivnenko, Ju.N., Romany, I.A.: Excitation functions of (p,n) Reactions on some Sn Isotopes and Isomeric Ratios.
Yad. Fiz. **11** (1970) 953.

Kne 75 Kneff, D.W., Mann, F.M., Switkowski, Z.E.: The ^{100}Mo (p,n) ^{100}Tc Reaction Cross Section and its Use as a Normalization for Shuttle-System Measurements.
Nucl. Phys. A **250** (1975) 285.

Kon 77 Kondo, K., Lambrecht, R.M., Wolf, A.P.: Iodine-123 Production for Radiopharmaceutical XX. Excitation Functions of the ^{124}Te (p,2n) ^{123}I and the Effect of Target Enrichment on Radionuclidic Purity.
Int. J. Appl. Radiat. Isot. **28** (1977) 395.

Kon 91 Kondratiev, C.H., Kuzmenko, B.A., Lobach, Yu.N., et al.: Radionuclides Production Cross Section at Interaction of Protons with Energy up to 70 MeV With Zirconium Nuclei.
At. Energ. **71** (1991) 325.

Kor 67 Korteling, R.G., Caretto, A.A.: Systematics of ^{24}Na and ^{22}Na Production with 400 MeV Protons.
J. Inorg. Nucl. Chem. **29** (1967) 2863.

Kor 70c Korteling, R., Kiefer, R.: Production of ^{32}P and ^{33}P from Various Targets with 550 MeV Protons.
Phys. Rev. C **2** (1970) 957.

Kov 91 Kovach, Z., Tarkanui, F., Qaim, S.M., Stoklin, G.: Excitation Functions for the Formation of Some Radio-Isotopes of Rubidium in Proton Induced Nuclear Reactions on Kr_{nat}, ^{82}Kr and ^{83}Kr with Special Reference to the Production of ^{81}Rb (^{81m}Kr) Generator Radionuclide.
Appl. Radiat. Isot. **42** (1991) 329.

Koz 88a Kozma, P., Kliman, J., Leonard, M.: Residues from the Reactions of 9 GeV/c Protons and Deuterons with ^{181}Ta.
Czech. J. Phys. B **38** (1988) 973.

Koz 89a Kozma, P., Damdinsuren, C., Chultem, D., Tumendemberel, B.: Nuclear Reactions of Tantalum with 3.65A GeV ^{12}C Ions and 3.65 GeV Protons.
Preprint JINR E1-89-259 (1989) Dubna.

Koz 90 Kozma, P., Damdinsuren, C.: Nuclear Reactions of Medium and Heavy Target Nuclei with High-Energy Projectiles. 2. Fragmentation of Ag_{nat} and ^{197}Au by 3.65A GeV ^{12}C Ions and 3.65 GeV Protons.
Czech. J. Phys. **40** (1990) 38.

Kra 66 Krasnov, N.N., Dmitriev, P.P.: Excitation Functions and Yields for ^{181}Ta (d,2n) ^{181}W and ^{181}Ta (p,n) ^{181}W Reactions.
At. Energ. **20** (1966) 154.

Kru 55 Kruger, P., Sugarman, N.: High-Energy Fission of Heavy Elements. Nuclear Charge Dependence.
Phys. Rev. **99** (1955) 1459.

Kud 82 Kudo, H., Muramatsu, H., Nakahara, H., et al.: Fission Fragment Yields in the Fission of ^{232}Th by Protons at Energies 8 to 22 MeV.
Phys. Rev. C **25** (1982) 3011.

Kur 89 Kurenkov, N.V., Malinin, A.B., Sebyakin, A.A., Venikov, N.I.: Excitation Functions of Proton Induced Nuclear Reactions on ^{124}Xe: Production of ^{123}I.
J. Radioanal. Nucl. Chem. Lett. **135** (1989) 39.

Lad 60 Ladenbauer, I.M., Winsberg, L.: Interaction of High-Energy Protons and Alpha-Particles with Iodine-127.
Phys. Rev. **119** (1960) 1368.

Laf 66 Lafleur, M.S., Porile, N.T., Yaffe, L.: Formation of ^{7}Be in Nuclear Reactions Induced by 85 MeV Protons.
Can. J. Chem. **44** (1966) 2749.

Lag 78 Lagunas-Solar, M.C., Jungerman, J.A., Peek, N.F., Theus, R.M.: Thallium-201 Yields and Excitation Functions for the Lead Radioactivities Produced by Irradiation of Natural Thallium with 15 to 60 MeV Protons.
Int. J. Appl. Radiat. Isot. **29** (1978) 159.

Lag 79a Lagarde-Simonoff, M., Simonoff, G.N.: Cross Sections and Recoil Properties of ^{83,84,86}Rb Formed by 0.6 to 21 GeV ^{1}H Reactions with Targets of Y, Ta, U.
Phys. Rev. C **20** (1979) 1498.

Lag 80 Lagunas-Solar, M.C., Jungerman, J.A., Paulson, D.W.: Thallium-201 Yields and Excitation Functions for the Lead Radioactivities Produced by Irradiation of Thallium-205 with 34 to 60 MeV Protons.
Int. J. Appl. Radiat. Isot. **31** (1980) 117.

Lag 83 Lagunas-Solar, M.C., Avila, M.J., Navarro, N.J., Johnson, P.C.: Cyclotron Production of Nocarrier-Added ^{97}Ru by Proton Bombardment of ^{103}Rh Targets.
Int. J. Appl. Radiat. Isot. **34** (1983) 915.

Lag 84 Lagunas-Solar, M.C., Avila, M.J., Johnson, P.C.: Cyclotron Production of ^{101m}Rh via Proton-Induced Reactions on ^{103}Rh Targets.
Int. J. Appl. Radiat. Isot. **35** (1984) 743.

Lag 86 Lagunas-Solar, M.C., Carvacho, O.F., Harris, L.J., Mathis, C.A.: Cyclotron Production of ^{122}Xe (20.1 h) $\rightarrow$ ^{122}I (β^{+} 77%, EC 23%, 3.6 min) for Positron Emission Tomography. Current Methods and Potential Developments.
Appl. Radiat. Isot. **37** (1986) 835.

Lag 86a Lagunas-Solar, M.C., Carvacho, O.F., Liu, B.L., et al.: Cyclotron Production of High-Purity ^{123}I. 1. A Revision of Excitation Functions, Thin-Target and Cumulative Yields for ^{127}I (p,xn) Reactions.
Appl. Radiat. Isot. **37** (1986) 823.

Lag 87 Lagunas-Solar, M.C., Carvacho, O.F., Nagahara, L., et al.: Cyclotron Production of no Carrier Added ^{206}Bi (6.24 d) and ^{205}Bi (15.31 d) as Tracers for Biological Studies and for the Development of Alpha-Emitting Radiotherapeutic Agents.
Appl. Radiat. Isot. **38** (1987) 129.

Lag 90a Lagunas-Solar, M.C., Carvacho, O.F., Yang, S.-J., Yano, Y.: Cyclotron Production of PET Radionuclides: ^{118}Sb (3.5 min; β^{+} 75%; EC 25%) from High-Energy Protons on Natural Sb Targets.
Appl. Radiat. Isot. **41** (1990) 521.

Lag 91 Lagunas-Solar, M.C., Kiefer, P.M., Carvacho, O.F., et al.: Cyclotron Production of NCA ^{99m}Tc and ^{99}Mo. An Alternative Non-Reactor Supply Source of Instant ^{99m}Tc and ^{99}Mo $\rightarrow$ ^{99m}Tc Generators.
Appl. Radiat. Isot. **42** (1991) 643.

Lav 62 Lavrukhina, A.K., Moskaleva, L.N., Malyshev, V.V., Satarova, L.M.: Production of the Light Nuclei in Bombardment of Heavy Elements by 660 MeV Protons.
JETP **43** (1962) 3.

Lav 84 Lavielle, B., Regnier, S.: Cross-Section Measurements of Krypton-78 to 86 Isotopes in the Spallation of Palladium and Silver. A New Semi-Empirical Spallation Formula in the Y to Ag Target Mass Region.
J. Phys. (Paris) **45** (1984) 981.

Leb 75 Lebowitz, E., Greene, M.W.: Thallium-201 for Medical Use.
J. Nucl. Med. **16** (1975) 151.

Lef 61 Lefort, M.M.: Fonction d'Excitation de la Reaction Nucleaire ^{238}U (p,3n) ^{236}Np Entre 30 et 150 MeV.
Compt. Rend. B **253** (1961) 2221.

Lef 61a Lefort, M.M., Simonoff, G.N., Tarrago, X.: Reactions Nucleaires de Spallation Induites sur le Thorium par des Protons de 150 et 82 MeV.
Nucl. Phys. **25** (1961) 216.

Lef 62 Lefort, M., Simonoff, G.N., Tarrago, X.: Comparaison des Sections Efficaces des Reactions (p,n) et (p,pn) Induites sur le Thorium par des Protons de 30 a 150 MeV.
J. Phys. Radium **23** (1962) 123.

Lev 66 Levenberg, I., Pokrovsky, V., Tarasova, L., et al.: Reactions (p,pn) , (p,2n) and (p,n) on ^{89}Y Induced by High Energy Protons.
Nucl. Phys. **81** (1966) 81.

Lev 84 Levkovskii, V.N., Reutov, V.F., Botvin, K.V.: Helium Accumulation in Mo Irradiated by Protons in the 15 to 30 MeV Energy Region.
Radiat. Eff. **80** (1984) 223.

Lev 91 Levkovskii, V.N.: Middle Mass Nuclides (A = 40 - 100) Activation Cross Sections by Medium Energy (E = 10 - 50 MeV) Protons and α-Particles (Experiment and Systematics).
Inter-Vesi, Moscow, 1991, 215 pages.

Lig 64 Ligonniere, M., Vassent, B., Bernas, R.: Sections Efficaces de Production du Beryllium 7, Dans l'Aluminium, le Vanadium, le Tantale et l'Or par des Protons de 155 et 550 MeV.
Compt. Rend. B **259** (1964) 1406.

Lin 54 Lindner, M., Osborne, R.N.: Some Studies of the Products of the High-Energy Fission Process.
Phys. Rev. **94** (1954) 1323.

Lin 56 Lindner, M., Osborne, R.N.: Nonfission Inelastic Events in Uranium and Thorium Induced by High-Energy Protons.
Phys. Rev. **103** (1956) 378.

Lin 59 Linder, B., James, R.: Cross Sections for Nuclear Reactions Involving Nuclear Isomers.
Phys. Rev. **114** (1959) 322.

Lun 79a Lundqvist, H., Malmborg, P., Langstrom, B., Chiengmai, S.N.: Simple Production of ^{77}Br$^-$ and ^{123}I$^-$ and their Use in the Labelling of [^{77}Br]Br UdR and [^{123}I]I UdR.
Int. J. Appl. Radiat. Isot. **30** (1979) 39.

Mar 51 Marquez, L., Perlman, I.: Observations on Lithium and Beryllium Nuclei Ejected from Heavy Nuclei by High Energy Particles.
Phys. Rev. **81** (1951) 953.

Mar 52 Marquez, L.: The Yield of ^{18}F from Medium and Heavy Elements with 420 MeV Protons.
Phys. Rev. **86** (1952) 405.

Mar 58 Markowitz, S.S., Rowland, F.S., Friedlander, G.: Reactions (p,pn) at Proton Energies from 0.3 to 3.0 BeV.
Phys. Rev. **112** (1958) 1295.

Mar 85 Marten, M., Schiiriug, A., Scobel, W., Probst, H.J.: Preequilibrium ^{111}Cd (p,xn) Reactions.
Z. Phys. A **322** (1985) 93.

May 64 May, M., Yaffe, L.: Spallation Products Produced by Bombardment of Dysprosium with Protons of Energies 9 - 87 MeV.
J. Inorg. Nucl. Chem. **26** (1964) 479.

Mcg 71 McGee, T., Rao, C.L., Yaffe, L.: Nuclear Charge Dispersion of Light Mass Elements in the Fission of ^{232}Th by Protons of Medium Energy.
Nucl. Phys. A **173** (1971) 595.

Met 59 Metzger, A.E., Miller, J.M.: Study of Thallium, Lead and Bismuth Nuclei Produced in the Bombardment of Gold with 380 MeV Protons.
Phys. Rev. **113** (1959) 1125.

Met 73 Metag, V., Lee, S.M., Liukkonen, E., et al.: Neutron-Fission Competition Near Threshold: the Influence of Shells and Pairing on the Decay of the ^{241}Cm Compound Nucleus.
Nucl. Phys. A **213** (1973) 397.

Mic 89 Michel, R., Dittrich, B., Herpers, U., et al.: Proton Induced Spallation at 600 MeV.
Analyst **114** (1989) 287.

Mil 74 Milazzo-Colli, L., Braga-Marcazzan, G.M., Milazzo, M., Signorini, C.: Preformation Probability of Alpha-Clusters in Rare Earth Nuclei Measured by Means of the (p,α) Reaction.
Nucl. Phys. A **218** (1974) 274.

Mil 75 Milazzo-Colli, L., Braga-Marcazzan, G.M., Milazzo, M.: Further Measurements of the Probability of Alpha-Cluster Preformation by Means of (p,α) Reactions in Heavy Elements.
Nuovo Cimento A **30** (1975) 632.

Min 85 Min, D.V., Shukolukov, A.Yu., Meshik, A.N., Gorin, B.D.: Mass-Spectrometrical Determination of Xenon Isotopes Yields in Reactions of Spallation of Er, Tm, Ta, W and Re by 1 GeV Protons.
Geokhimiya **2** (1985) 261.

Miy 73a Miyano, K., Nakahara, H.: The Radiative Proton Capture Cross Sections in ^{209}Bi with E_p from 10 to 50 MeV.
J. Phys. Soc. Jpn. **35** (1973) 1277.

Miy 74 Miyano, K., Sekikawa, M., Kaneko, T., Nomoto, M.: Reactions on ^{209}Bi Induced by Intermediate Energy Protons and the Effect of Direct Reactions.
Nucl. Phys. A **230** (1974) 98.

Miy 78 Miyano, K., Ando, T., Kudo, H., et al.: The (p,n) Reaction on ^{209}Bi and Pre-Compound Process.
J. Phys. Soc. Jpn. **45** (1978) 1071.

Mol 77 Molecke, M.A., Caretto, A.A.: (p,pxn) and (p,xn) Reactions and Mechanisms on ^{133}Cs at 550 MeV.
Phys. Rev. C **15** (1977) 719.

Mor 62 Morrison, D.L., Caretto, A.A.: Excitation Functions of (p,xp) Reactions.
Phys. Rev. **127** (1962) 1731.

Mus 88 Mustafa, M.G., West, H.I., O'Brien, H., et al.: Measurements and a Direct-Reaction-Plus-Hauser-Feshbach Analysis of ^{89}Y (p,n) ^{89}Zr, ^{89}Y (p,n) ^{88}Zr, and ^{89}Y (p,pn,) ^{88}Y Reactions up to 40 MeV.
Phys.Rev. C **38** (1988) 1624.

Nag 90 Nagame, Y., Sueki, K., Baba, S., Nakahara, H.: Isomeric Yield Ratios in Proton-, ^{3}He-, and Alpha-Particle-Induced Reactions on ^{197}Au.
Phys. Rev. C **41** (1990) 889.

Nar 68 Narang, V.P., Yaffe, L.: (p,n) and (p,pxn) Reactions in ^{127}I.
Can. J. Chem. **46** (1968) 3171.

Nee 63 Need, J.L., Linder, B.: Cross Section for Formation of the Isomeric Pair ^{117}In and ^{117m}In in the ^{120}Sn (p,α) Reactions.
Phys. Rev. **129** (1963) 1298.

Ner 55 Nervik, W.E., Seaborg, G.T.: Tantalum Spallation and Fission Induced by 340 MeV Protons.
Phys. Rev. **97** (1955) 1092.

Net 60 Nethaway, D.R., Winsberg, L.: Interaction of High-Energy Protons with Indium.
Phys. Rev. **119** (1960) 1375.

New 73a Newton, G.W.A., Robinson, V.J., Skarestad, M., et al.: Rare Earth Yields from Bombardment of Tungsten Targets with 24 GeV Protons.
J. Inorg. Nucl. Chem. **35** (1973) 1435.

Nie 69 Nieckarz, W.J., Caretto, A.A.: Production of ^{111}In and ^{114m}In from the Separated Isotopes of Cadmium Using 70 to 400 MeV Protons.
Phys. Rev. **178** (1969) 1887.

Nie 70 Nieckarz, W.J., Caretto, A.A.: Nuclear Structure Effects in (p,2n) Reactions on the Separated Isotopes of Cadmium at 400 MeV.
Phys. Rev. C **2** (1970) 1917.

Nik 80 Nikkinen, L., Pathak, B.P., Lessarad, L., Grant, I.S.: Independent Yields of Rb, Tn, and Cs Isotopes in the Proton-Induced Fission of ^{232}Th.
Phys. Rev. C **22** (1980) 617.

Nog 91 Noguchi, M., Miura, T., Kondo, K., et al.: Production Cross Sections of Tritium in High Energy Nuclear Reactions with 12 GeV Protons.
Appl. Radiat. Isot. **42** (1991) 577.

Nor 90 Nortier, F.M., Mills, S.J., Steyn, G.F.: Excitation Functions and Production Rates of Relevance to the Production of ^{111}In by Proton Bombardment of Cd_{nat} and In_{nat} up to 100 MeV.
Appl. Radiat. Isot. **41** (1990) 1201.

Ohk 83 Ohkubo, Y., Porile, N.T., Orth, C.J., Liu, L.C.: Excitation Functions of the ^{127}I (π,πxn) Reactions in the Region of the (3,3)-Resonance.
Phys. Rev. C **27** (1983) 1146.

Olk 60 Olkowsky, J., Gratot, I., Pape, M., Cohen, L.: Determination de Sections Efficaces de Reactions Procedant par Interaction Directe Dans la Region des Terres Rares.
J. Phys. Radium **21** (1960) 405.

Olk 61 Olkowsky, J., Lepape, M., Gratot, I.: Fonction d'Excitation de ^{141}Pr (p,n) ^{141}Nd Jusqu'a 11.1 MeV.
Nucl. Phys. **23** (1961) 164.

Olk 61a Olkowsky, J., Gratot, I., Pape, M.: Sections Efficaces des Reactions (p,n) et (p,2n) sur ^{150}Nd.
Nucl. Phys. **24** (1961) 84.

Oto 66 Otozai, K., Kume, S., Mito, A., et al.: Excitation Functions for the Reactions Induced by Protons on Cd up to 37 MeV.
Nucl. Phys. **80** (1966) 335.

Paa 76 Paans, A.M.J., Vaalburg, W., Herk, G.V., Woldring, M.G.: Excitation Function for the Production of ^{123}I via the ^{127}I (p,5n) ^{123}Xe Reaction.
Int. J. Appl. Radiat. Isot. **27** (1976) 465.

Pan 66 Panontin, J.A., Caretto, A.A.: Half-Life of ^{101}Ag and Formation Cross Section from Ag and In.
J. Inorg. Nucl. Chem. **28** (1966) 1795.

Pan 68 Panontin, J.A., Porile, N.T.: Charge Distribution in the Symmetric Fission of ^{238}U by 450 MeV Protons.
J. Inorg. Nucl. Chem. **30** (1968) 2017.

Pan 70 Panontin, J.A., Porile, N.T.: Charge Distribution and Recoil Properties in the Fission of ^{238}U by 11.5 GeV Protons.
J. Inorg. Nucl. Chem. **32** (1970) 1775.

Pat 58 Pate, B.D., Foster, J.S., Yaffe, L.: Distribution of Nuclear Charge in the Proton-Induced Fission of ^{232}Th.
Can. J. Chem. **36** (1958) 1691.

Pat 61 Pate, B.D., Poskanzer, A.M.: Spallation of Uranium and Thorium Nuclei with BeV-Energy Protons.
Phys. Rev. **123** (1961) 647.

Pee 79 Peek, N.F., Hegedus, F.: The Production of Xenon Isotopes with Protons of Energies from 320 to 590 MeV.
Int. J. Appl. Radiat. Isot. **30** (1979) 631.

Pol 80 Polak, P., Aten, A.H.W.: Fission of ^{237}Np by Medium Energy Protons.
J. Inorg. Nucl. Chem. **42** (1980) 641.

Por 62 Porile, N.T.: Simple Nuclear Reactions of Indium with 30 and 2.9 GeV Protons.
Phys. Rev. **128** (1962) 1916.

Por 64a Porile, N.T., Church, L.B.: Isobaric Yield Distributions in the Interaction of 1.8 GeV Protons with ^{96}Zr, ^{96}Mo, and ^{96}Ru.
Phys. Rev. **133** (1964) B310.

Por 66 Porile, N.T.: Yields of Products with A = 100 to 117 in the Interaction of 3 and 28 GeV Protons with Uranium and Lead.
Phys. Rev. **148** (1966) 1235.

Por 78 Porile, N.T., Dropesky, B.J., Williams, R.A.: Emission of ^{24}Na Fragments in the Interaction of ^{197}Au with Intermediate-Energy Pions and Protons.
Phys. Rev. C **18** (1978) 2231.

Por 79 Porile, N.T., Gole, G.D., Rudy, C.R.: Nuclear Reactions of Silver with 25.2 GeV ^{12}C Ions and 300 GeV Protons.
Phys. Rev. C **19** (1979) 2288.

Pos 71 Poskanzer, A.M., Butler, G.W., Hyde, E.K.: Fragment Production in the Interaction of 5.5 GeV Protons with Uranium.
Phys. Rev. C **3** (1971) 882.

Pre 91 Prescher, K., Peiffer, F., Stueck, R., et al.: Thin-Target Cross Sections of Proton-Induced Reactions on Barium and Solar Cosmic Ray Production Rates of Xenon-Isotopes in Lunar Surface Material.
Nucl. Instrum. Methods Phys. Res. B **53** (1991) 105.

Qai 79 Qaim, S.M., Weinreich, R., Ollig, H.: Production of ^{201}Tl and ^{203}Pb via Proton Induced Nuclear Reactions on Natural Thallium.
Int. J. Appl. Radiat. Isot. **30** (1979) 85.

Rao 63 Rao, C.L., Yaffe, L.: Nuclear Reactions Induced in Tantalum by Protons of Energy up to 84 MeV.
Can. J. Chem. **41** (1963) 2516.

Ray 63 Rayudu, G.V.S., Yaffe, L.: Reactions Produced on Erbium by Protons of Energies between 6 and 87 MeV.
Can. J. Chem. **41** (1963) 2544.

Rea 65 Read, J.B.J., Miller, J.M.: Nuclear-Structure Effects in High-Energy (p,n) Reactions.
Phys. Rev. **140** (1965) B623.

Reg 82 Regnier, S., Lavielle, B., Simonoff, F., Simonoff, G.N.: Nuclear Reactions in Rb, Sr, Y and Zr Targets.
Phys. Rev. C **26** (1982) 931.

Reu 69 Reuland, D.J., Caretto, A.A.: The Measurement of some (p,n) and (p,2n) Reaction Cross Sections at 400 MeV.
J. Inorg. Nucl. Chem. **31** (1969) 1915.

Rie 61 Riehl, C.: Reactions (p,pxn) sur le Cesium.
J. Phys. Radium **22** (1961) 770.

Ryb 65 Rybakov, B.N., Maltseva, N.S.: Radiochemical Investigation of ($p,p\pi^+$) Reaction for ^{115}In.
Radiokhimiya **7** (1965) 90.

Sac 67 Sachdev, D.R., Porile, N.T., Yaffe, L.: Reactions of ^{88}Sr with Protons of Energies 7 to 85 MeV.
Can. J. Chem. **45** (1967) 1149.

Sah 66 Saha, G.B., Porile, N.T., Yaffe, L.: Reactions (p,xn) and (p,pxn) of ^{89}Y with 5 to 85 MeV Protons.
Phys. Rev. **144** (1966) 962.

Sak 85 Sakamoto, K., Dohniwa, M., Okada, K.: Excitation Functions for (p,xn) and (p,pxn) Reactions on Natural $^{79+81}$Br, $^{85+87}$Rb, ^{127}I and ^{133}Cs up to E_p = 52 MeV.
Int. J. Appl. Radiat. Isot. **36** (1985) 481.

Sau 81 Sauvageon, H.: Production of ^{38}Ar and ^{39}Ar in the Interaction of Gold and Thorium with 1, 2.5 and 24 GeV Protons.
Phys. Rev. C **24** (1981) 2667.

Sau 82 Sauvageon, H., Regnier, S., Simonoff, G.N., et al.: Production of Kr and Xe Isotopes by Interaction of ^{232}Th with 0.15 to 24 GeV Protons.
Phys. Rev. C **25** (1982) 466.

Sau 83 Sauvageon, H., Regnier, S., Simonoff, G.N.: Production of Kr and Xe Isotopes by Interaction of ^{197}Au with 0.15 to 24 GeV Protons.
Z. Phys. A **314** (1983) 181.

Sch 70 Schall, R.F., Caretto, A.A.: Nuclear Structure Effects in (p,2n) Reactions on the Separated Isotopes of Tellurium at 300 MeV.
Phys. Rev. C **2** (1970) 1924.

Sch 76 Scheidemann, O., Porile, N.T.: Production of Sc Nuclides in the Interaction of ^{238}U with 1 to 300 GeV Protons.
Phys. Rev. C **14** (1976) 1534.

Sch 79 Schoen, N.C., Orlov, G., McDonald, R.J.: Excitation Functions for Radioactive Isotopes Produced by Proton Bombardment of Fe, Co and W in the Energy Range from 10 to 60 MeV.
Phys. Rev. C **20** (1979) 88.

Sch 89 Scholten, B., Qaim, S.M., Stocklin, G.: Excitation Functions of Proton Induced Nuclear Reactions on Natural Tellurium and Enriched ^{123}Te: Production of ^{123}I via the ^{123}Te (p,n) ^{123}I Process at a Low-Energy Cyclotron.
Appl. Radiat. Isot. **40** (1989) 127.

Sem 91 Semenov, V.G.: Radionuclide Production Cross Sections at Irraditation of Rh by Protons with Energies up to 100 MeV.
Private Communication (1991).

Ska 75 Skakun, E.A., Klucharev, A.P., Rakivnenko, Ju.N., Romany, I.A.: Excitation Functions for (p,n) and (p,2n) Reactions on Separated Cadmium Isotope Targets.
Izv. Akad. Nauk SSSR, Ser. Fiz. **39** (1975) 24.

Ska 87 Skakun, E.A., Batiy, V.G., Rakivnenko, Ju.N., Rastrepin, O.A.: Excitation Functions and Isomer Ratios for up to 9 MeV Proton Interactions with Zr and Mo Isotope Nuclei.
Yad. Fiz. **46** (1987) 28.

Ska 87a Skakun, E.A., Batiy, V.G., Rakivnenko, Ju.N., Rastrepin, O.A.: Cross Sections of (p,γ) Reactions on Isotopes ^{54}Fe, ^{112}Sn, and ^{114}Cd at Proton Energy up to 9 MeV.
Yad. Fiz. **45** (1987) 614.

Ste 58 Stevenson, P.C., Hicks, H.G., Nervik, W.E., Nethaway, D.R.: Further Radiochemical Studies of the High-Energy Fission Products.
Phys. Rev. **111** (1958) 886.

Ste 91 Steyn, G.F., Mills, S.J., Nortier, F.M., Haasbroek, F.J.: Integral Excitation Functions for Kr_{nat} + p up to 116 MeV and Optimization of the Production of ^{81}Rb for ^{81m}Kr Generators.
Appl. Radiat. Isot. **42** (1991) 361.

Str 61 Strohal, P.P., Caretto, A.A.: Excitation Functions of (p,2nucleon) Reactions.
Phys. Rev. **121** (1961) 1815.

Sug 54 Sugarman, N., Duffield, R.B., Friedlander, G., Miller, J.M.: Disintegration of Bismuth by 2.2 BeV Protons.
Phys. Rev. **95** (1954) 1704.

Suk 74 Suk, H.C., Crawford, J.E., Moore, R.B.: Absolute (p,xn) Cross Sections Measurements of ^{232}Th.
Nucl. Phys. A **218** (1974) 418.

Sum 90 Summerer, K., Bruchle, W., Morrissey, D.J., et al.: Target Fragmentation of Au and Th by 2.6 GeV Protons.
Phys. Rev. C **42** (1990) 2546.

Sym 78 Syme, D.B., Wood, E., Blair, I.M., et al.: Yield Curves for Cyclotron Production of ^{123}I, ^{125}I and ^{121}I by ^{127}I (p,xn) Xe → (β) → I Reactions.
Int. J. Appl. Radiat. Isot. **29** (1978) 29.

Tar 62 Tarrago, X.: Etude des Reactions (p,αxn) sur le Thorium Bombarde par des Protons de 155 MeV.
J. Phys. Radium **23** (1962) 1.

Tar 91 Tarkanui, F., Qaim, S.M., Stocklin, G., et al.: Excitation Functions of (p,2n) and (p,pn) Reactions and Differential and Integral Yields of ^{123}I in Proton Induced Nuclear Reactions on Highly Enriched ^{124}Xe.
Appl. Radiat. Isot. **42** (1991) 221.

Tew 55 Tewes, H.A.: Excitation Functions for some Proton Induced Reactions on Thorium.
Phys. Rev. **98** (1955) 25.

Til 63 Tilbury, R.S., Yaffe, L.: Nuclear Isomers ^{197m,g}Hg, ^{195m,g}Hg and ^{196m,g}Au Formed by Bombardment of Gold with Protons of Energies from 8 to 60 MeV.
Can. J. Chem. **41** (1963) 2634.

Til 63a Tilbury, R.S., Yaffe, L.: ^{131m}Ba - a New Isomer.
Phys. Rev. **129** (1963) 1709.

Tom 84 Tominaka, T., Ban, S., Ikeda, H., et al.: Spallation of Niobium by 12 GeV Protons.
Nucl. Phys. A **414** (1984) 385.

Tra 72 Tracy, B.L., Chanmont, J., Klapisch, R., et al.: Rb and Cs Isotopic Cross Sections from 40 to 60 MeV Proton Fission of ^{238}U, ^{232}Th, ^{235}U.
Phys. Rev. C **5** (1972) 222.

Tre 66 Treyte, W.J., Caretto, A.A.: Study of (p,n) Reactions between 100 and 400 MeV.
Phys. Rev. **146** (1966) 836.

Uns 61 Unseren, E., Wiig, E.O.: Excitation Functions of the (p,2p6n) and (p,3p5n) Reactions for 60, 100, 150, and 240 MeV Protons on Enriched Zirconium-90.
Phys. Rev. **122** (1961) 1875.

Van 60 Vandenbosch, R., Huizenga, J.R.: Isomeric Cross-Section Ratios for Reactions Producing the Isomeric Pair ^{197,197m}Hg.
Phys. Rev. **120** (1960) 1313.

Ver 67 Verdieck, E.V., Miller, T.M.: Radiative Capture and Neutron Emission in ^{139}La + α and ^{142}Ce + p.
Phys. Rev. **153** (1967) 1253.

Vin 81 Vincent, J.S., Dougan, A.H., Ruster, D.L., Blue, J.W.: A Facility for the Production of Iodine 123 by Spallation of Caesium.
J. Radioanal. Chem. **65** (1981) 17.

Wag 85 Wagner, B.J., Ward, T.E., Viola, V.E.: Radiochemical Measurements of 200 MeV Proton Induced Fission of ^{133}Cs.
Phys. Rev. C **31** (1985) 1934.

Wan 86 Wang, C.F., Porile, N.T.: Nuclear Reactions of Silver with 0.8 TeV Protons.
Phys. Rev. C **33** (1986) 2183.

War 61 Ware, W.R., Wiig, E.O.: Excitation Functions of the (p,pn) and (p,2p) Reactions on ^{142}Ce at 60 to 233 MeV.
Phys. Rev. **122** (1961) 1837.

War 81 Ward, T.E., Singh, P.P., Friesel, D.L., et al.: Radiochemical Study of the Combined (p,π_0) and (p,γ) Reactions on Bismuth with Protons from 62 to 480 MeV.
Phys. Rev. C **24** (1981) 588.

Wil 75 Wilkins, S.R., Shimose, S.T., Hines, H.H., et al.: Excitation Functions and Yields for ^{123}I Production Using the ^{127}I (p,5n) ^{123}Xe Reaction.
Int. J. Appl. Radiat. Isot. **26** (1975) 279.

Win 62 Wing, J., Huizenga, J.R.: (p,n) Cross Sections of ^{51}V, ^{52}Cr, ^{63}Cu, ^{65}Cu, ^{107}Ag, ^{109}Ag, ^{111}Cd, ^{114}Cd and ^{139}La from 5 to 10.5 MeV.
Phys. Rev. **128** (1962) 280.

Win 64 Winsberg, L.: Recoil Studies of Nuclear Reactions Induced by High-Energy Particles. 1. Production of ^{149}Tb.
Phys. Rev. **135** (1964) 1105.

Wol 56 Wolfgang, R., Baker, E.W., Caretto, A.A., et al.: Radiochemical Studies of the Interaction of Lead with Protons in the Energy Range 0.6 to 3.0 BeV.
Phys. Rev. **103** (1956) 394.

Yu 74 Yu, Y.W., Porile, N.T.: Interaction of ^{238}U with 300- and 11.5-GeV Protons: Cross Sections and Recoil Properties in the A $\approx$ 130 Mass Region.
Phys. Rev. C **10** (1974) 167.

Zai 62 Zaitseva, N.G., Kuznetsova, M.Ja., Min, N.B., Khalkin, B.A.: Investigation of the (p,xn) and (p,2pxn) Nuclear Reactions on Separated Tellurium Isotope Targets.
JETP **43** (1962) 1672.

Zai 87 Zaitseva, N.G., Knotek, O., Khan, K.C., et al.: Formation of ^{123}I in Irradiation of Sodium Iodine by 100 MeV Protons.
Radiokhimiya **3** (1987) 391.

Zai 90 Zaitseva, N.G., Knotek, O., Kovalev, A., et al.: Excitation Functions and Yields for ^{111}In Production Using ^{113,114}Cd,Cd$_{nat}$ (p,xn) ^{111}In Reactions with 65 MeV Protons.
Appl. Radiat. Isot. **41** (1990) 177.

Zai 91 Zaitseva, N.G., Deptula, C., Knotek, O., et al.: Cross Sections for the 100 MeV Proton-Induced Nuclear Reactions and Yields of some Radionuclides Used in Nuclear Medicine.
Radiochim. Acta **54** (1991) 57.